1004259220

VARIORUM COLLECTED STUDIES SERIES

Science, Technology and Learning
in the Ottoman Empire

Professor Ekmeleddin İhsanoğlu

Ekmeleddin İhsanoğlu

Science, Technology and Learning in the Ottoman Empire

Western Influence, Local Institutions,
and the Transfer of Knowledge

ASHGATE

VARIORUM

Published in the Variorum Collected Studies Series by

Ashgate Publishing Limited
Gower House, Croft Road,
Aldershot, Hampshire
GU11 3HR
Great Britain

Ashgate Publishing Company
Suite 420
101 Cherry Street
Burlington, VT 05401–4405
USA

Ashgate website: http://www.ashgate.com

ISBN 0–86078–924–1

British Library Cataloguing-in-Publication Data
Ihsanoglu, Ekmeleddin
    Science, technology and learning in the Ottoman empire :
    Western influence, local institutions, and the transfer of
    knowledge. – (Variorum collected studies series)
    1. Science – Turkey – History 2. Science and civilization
    3. Islam and science 4. Turkey – Civilization – 1288-1918
    5. Turkey – Civilization – 1288-1918 – European influences
    I. Title
    509.5'61'09

US Library of Congress Cataloging-in-Publication Data
Ihsanoglu, Ekmeleddin.
    Science, technology, and learning in the Ottoman Empire : Western influence,
local institutions, and the transfer of knowledge / Ihsanoglu, Ekmeleddin.
        p. cm – (Variorum collected studies series ; 773)
    Includes bibliographical references and index.
    ISBN 0-86078-924-1 (hardback)
        1. Science–Islamic countries–History. 2. Technology transfer–Islamic
countries–History. 3. Turkey–History–Ottoman Empire, 1288-1918. I. Title. II.
Collected studies ; 773.

Q127.I742147 2003
509'.17'671–dc21   2003050260

Printed and bound in Great Britain by TJ International Ltd, Padstow, Cornwall

VARIORUM COLLECTED STUDIES SERIES CS773

CONTENTS

This volume contains xiv + 338 pages

PREFACE

The papers and studies collected here relate to the cultural, intellectual and scientific aspects of Ottoman history. Their aim is to attract attention to areas of Ottoman history which have been neglected, as compared to studies on Ottoman politics, social and economic life. But even while stating this, it remains imperative to touch upon political, economic and social factors in order to shed light upon the socio-cultural background of the period. In other words, the research is not be limited to scientific activities. We shall try to examine and clarify the subjects from various aspects, considering factors which influenced the development of scholarly thought such as structures, institutions, wars and ideologies that swept through whole communities.

In considering the scientific activities that were carried out through the six-century history of the Ottoman Empire (1299-1923), it may be argued that the history of Ottoman science witnessed several distinctive trends. Though the historical evolution of Ottoman science shared many features common to the history of scientific endeavour in other parts of the Muslim world, beyond the boundaries of the Ottoman Empire, there were also some important differences, and the Ottomans may be considered "pioneers" in some areas.

The Ottoman state acquired additional dynamism in political and cultural fields because it provided peace and shelter to the cultures of rival Muslim and Christian ethnic groups within its lands, and because it shared common boundaries as well as an active relationship with European countries where modern cultural and scientific traditions developed. The state, though it was built upon the cultural heritage of previous Turkish and Moslem states, developed as a new model because of the socio-cultural structure of the region and its own dynamics. The state's structure, its social formation, educational and scholarly life, the understanding of economics that showed a distinct progress as compared to the Eastern and Western examples, was instrumental in its development from a small principality to a great empire that ruled over three continents for six hundred years.

The scientific activities of the Ottoman world are constituted from various scientific traditions: first, the Islamic tradition inherited by the Ottoman Turks that was carried on by the Arabs, who were part of the Ottoman Empire; then the European peoples such as Bosnians and Albanians who were newly converted to Islam, as well as the tradition of different Christian peoples living in Anatolia and the Balkans; and lastly the contributions made by native Jewish scholars and the Jewish emigrants from Andalusia. The Ottoman world had the necessary grounds for the interaction of all these different traditions. The Ottoman Empire also held vast lands in Europe and, as a result of its contact with European science from a very early time, this new scientific tradition spread in the Ottoman lands for the first time outside its own original cultural environment.

Inspired by the medieval Islamic scientific tradition at the beginning of its history, Ottoman science soon become influential in old scientific and cultural centres of the Islamic world. As early as the sixteenth century and more widely in the seventeenth century, Ottomans became acquainted with European science and started to make translations and adaptations from Western sources, becoming instrumental in spreading this new scientific tradition through the Islamic world. Thus, the Ottomans, who in their day represented the whole Islamic world, were able to combine Islamic scientific tradition with the newly emerging Western science. At the turn of the nineteenth century, however, the Ottomans opted for the exclusive practice of European science over the Islamic, and the Islamic scientific tradition gradually faded away as a new tradition emerged in accordance with modern Western scientific norms.

In the observations above, we have been trying to draw an outline of the history of Ottoman studies and survey its past, only to find out that the images reflected from these studies differ greatly from those that have been inherited from the previous generations. A. Adnan Adıvar and Aydın Sayılı were among those who first studied this subject. Though these two scholars came from different generations and traditions, they held similar opinions about Ottoman scholarly and cultural life. Adnan Adıvar (1882-1955) belonged to the last generation of the Ottoman cultural elite and he graduated from the medical school in İstanbul. He was highly accomplished and knowledgeable in European scholarship through his extensive knowledge of French. His work entitled *Science chez les Turcs Ottomanes* (Paris

1939; Turkish 1st edition Istanbul 1943) is the first scholarly book on Ottoman scientific activities.

Adıvar wrote this book in exile in Paris where he did not have any access to the rich collections in the libraries and archives of Istanbul. He was one of the patriots who saw the decline of the Ottoman State and lived through several wars and crushing defeats, and suffered the loss of Ottoman territories and misfortunes of World War I. He contributed to the foundation of the Republic of Turkey but had to live in exile because of internal political conflicts. He always carried a "negative" outlook towards Ottoman history under the effects of the past events and the psychological stress of an exile; his book was written in this state of mind. In it he expressed two opinions, the first one being that Ottoman science "was the continuation of inadequate and sometimes incorrect Arabic and Persian science". Secondly, he maintained that "from the viewpoint of Western science, there was a strong barrier as if the Ottomans never had any contact with the West".

Aydın Sayılı (1913-1993) belonged to the first generation of intellectuals of the young Republic of Turkey and readily acknowledged the Republic's renouncing of its Ottoman heritage. As well as not being very appreciative of this heritage, he was also in agreement with the viewpoints expressed by his mentor George Sarton regarding Islamic science. In his outstanding work, Sarton maintained that science in the Islamic world was worthy of merit only in its first centuries and this "golden age" lasted until *ca* 1000 A.D. Sayılı did not consider the Ottoman scientific activities that began in the fourteenth century as a subject worthy of extensive research. Adıvar's influence, too, can be seen in the studies that he undertook.

In the papers presented here the validity of these viewpoints on Ottoman science, which were dominant until the end of 1970s, is discussed. Based on methodical examination of numerous manuscripts and archival material, they attempt to present new viewpoints in the field of history of science, and maintain that scientific traditions which developed in sub-cultural milieus are also worth examining, alongside the focused research on scientific activities in the major civilizations conducted by Sarton and his followers.

The first serious studies on the history of Ottoman science started on an institutional basis in Istanbul at the beginning of 1980. Institutions such as the Research Centre for Islamic History, Art and Culture (IRCICA) founded in 1980, University of Istanbul, Faculty of Letters, Department of History of Science (1984-2001), and the Turkish Society for History of Science (TBTK) founded in 1989, jointly organized national and international symposia dedicated mostly to different aspects of science and technology in the Ottoman world. These meetings had a significant role in presenting the findings of different research projects and papers to historians of science all over the world. Following the publication of their proceedings, there was an increasing interest in the subject and new studies were started in various countries.

The series *History of Ottoman Scientific Literature*, published by IRCICA on the subjects of astronomy, mathematics and geography in 1997, 1999 and 2000 respectively, provides a wide range of information about the scholarly works produced in the Ottoman world and their authors; so far 6 volumes have been published with two volumes on each subject. It is planned that the series will continue, and separate volumes on music, military science, medicine and natural sciences will follow. These studies have also attracted the attention of readers outside academic circles. The scientific journal *Nature* in its 13 August 1998 issue No, 394, announced these books to its readers under the heading "Eclipsed no more" and wrote: "It is a massive enterprise, inspired to some extent by Joseph Needham's works on Chinese science and running into several volumes. Its nature and scope, as well as the problems involved in such an undertaking are well illustrated by the first two volumes on Ottoman astronomy.....The result is a monumental achievement. It not only provides us with a true picture of the extent of Ottoman scientific activity but also turns the standard view on its head."

In addition to the activities that have been carried out in Istanbul, studies on Ottoman history of science were manifested on an international level and for the first time an independent symposium was held within the framework of the XX International Congress of History of Science in 1997 in Liège, Brussels. This was the symposium on "Science, Technology and Industry in the Ottoman World" and it added a new dimension to the above-mentioned studies. Its proceedings were published under the title: *Science, Technology and Industry in the Ottoman World, Proceedings of the XX International Congress of History of Science*, vol. VI, ed. E. İhsanoğlu, A. Djebbar and F. Günergun, (Brepols, 2000). Another symposium within the framework of the XXI International Congress of History of Science was organized by E. İhsanoğlu, E. Nicolaïdis and K. Chatzis in Mexico City on July 8-14, 2001, on the theme of "Cultural Diversity: from the Ottoman Empire to the National States". In the meantime, another project is planned jointly by three institutions in France, Turkey and Greece (LATTS/ENPC, IRCICA, and INR/NHRF) on the transfer of technology and science from Western Europe to the East Mediterranean during the nineteenth century.

There are two aspects of the history of Ottoman science that await attention. Firstly, there is the need for the publication of critical editions of scientific texts in order to expand our knowledge of the level of science in the Ottoman period. The second important aspect is to research scientific activities of different ethnic or religious communities within their own cultural environment, which will help us to find out the reciprocal effects of these distinct traditions. These subjects and many others related to the history of Ottoman culture, science and technology that have not been studied before, are now available to researchers. To an extent, this present book aims to draw a general picture of the history of science and show the role it played in bringing two different civilizations closer. It also attempts to survey the scientific activities conducted mainly by the Muslims that formed the majority of the population in the Ottoman Empire. Readers that wish to have further information on history of Ottoman science may consult the two detailed chapters in *History of Ottoman State, Society and Civilization* published by IRCICA (2002).

Here I would like to renew my thanks to my dear colleagues, friends and students who contributed to my research projects throughout the years and whose names I have repeatedly acknowledged in my works. I am indebted to my publisher Dr John Smedley, who undertook to publish this volume and pursued the project with enthusiasm; I thank him for his support and encouragement.

Ekmeleddin İhsanoğlu

İstanbul, December 2002

ACKNOWLEDGEMENTS

Grateful acknowledgement is made to the following institutions and publishers for their kind permission to reproduce the papers included in this volume: The Japan-Netherlands Institute (I, XI), Brepols Publishers (III), The Thomas Jefferson University Press (IV), The ISIS Press (V), The Harrassowitz Publishing Press (VI, IX), Peeters (VII), Turkish Historical Society (VIII), Centre for Neohellenic Research (X), IRCICA (XII).

PUBLISHER'S NOTE

The articles in this volume, as in all others in the Variorum Collected Studies Series, have not been given a new, continuous pagination. In order to avoid confusion, and to facilitate their use where these same studies have been referred to elsewhere, the original pagination has been maintained wherever possible.

Each article has been given a Roman number in order of appearance, as listed in the Contents. This number is repeated on each page and is quoted in the index entries.

I

Some Remarks on Ottoman Science and its Relation with European Science & Technology up to the End of the Eighteenth Century

The Turkish author Adnan Adıvar has tried to explain the Ottoman relation to Western science and technology in his book *Science of the Ottoman Turks* expressing as his opinion that sixteenth century Turkey was firmly barred to sciences and that it seemed as if there was no contact with the West. Other historians have also discussed this problem and generally concur with this point of view. The blame for creating this supposed barrier between the Ottomans and Western sciences, was put upon Ottoman religous institutions and the Ulema.

However, subsequent studies and, specifically, the research which we ourselves conducted over the past ten years on the relationship between the Ottomans and Western science and technology, resulted in a very different and more convincing picture. In order to acquire a good understanding of this relationship within a rational context, it is necessary to have an overall view of the state of Ottoman science in this period.

The Ottoman state emerged in history as a small principality at the turn of the fourteenth century. It expanded into the lands of the Byzantine Empire in Anatolia and the Balkans, ruled over the Arab world after 1517 and eventually conquered a large area extending from central Europe to the Indian Ocean. Hence, the Ottoman Empire became the most powerful state in the Islamic world. The Ottomans preserved and enriched the cultural and scientific heritage of the Islamic world by giving it a new dynamism and vigor. It seems safe to assume that Islamic science progressed and kept pace with advances in the economic, social and cultural areas until the end of the sixteenth century when the magnificence and power of the Ottoman Empire was at its peak — a period known as the classical age. In the seventeenth century, the Ottoman Empire declined vis à vis the European states, which caused changes in state policy and social life in line with the emerging superiority of the West. Concurrently with these changes, some decline can be observed in Ottoman science as well, and as a result the Ottomans were forced to follow the direction of the West in the field of science as well as in many other fields.

I

The Ottomans, heirs to the illustrious Islamic civilisation and science as well as keepers of the rich Seljuk scientific tradition, were actively involved in other Arabic and Islamic cultural centres such as Iran, Egypt, Syria and Turkestan. Hundreds of students were educated in the *medreses* which were established in a short period and became centers of a new scholarly environment in the Muslim world, due to the teaching activities of great men of learning. Until the end of the sixteenth century, the Ottomans had 324 *medreses* spread all over the empire. The sciences which were taught in the *medreses* were divided into two parts: *ulūm-ı āliye* (ancillary sciences), also known as *cüz'ıyyat* (special sciences), and *ulūm-ı ᶜāliyye* (high sciences). *Cüz'ıyyat* or *ulūm-ı āliye* consisted of grammar and syntax (*sarf ve nahiv*), rhetoric (*belagat*), logic (*mantık*), Islamic theology (*kelam*), arithmetic (*hesab*), geometry (*hendese*), astronomy (*heyet*), and philosophy (*hikmet*). *Ulūm-ı ᶜāliyye* comprised commentary on the Kur'an (*tefsir*), traditions of the Prophet (*hadis*), and Muslim canonical jurisprudence (*fıkh*).

European travelers who observed Ottoman scientific and educational life, tended to compare it with that of contemporary Europe. This enables us to reach a better understanding of Ottoman science. The Italian nobleman Comte de Marsigli stayed in İstanbul for eleven months between the years 1679-1680 and later on lived among the Ottomans for many years. During the second siege of Vienna, he was kept in the Turkish headquarters as a prisoner of the ex-financier Temeşvarlı Ahmed Paşa. He came to İstanbul again in 1692 and was nominated to the Sublime Porte (*Bāb-ı Ālī*). He states:

> One of their major occupations is study, and there is no reason for most of the Christians to blame them for being illiterate and hardly listening to the Kur'an. One should correct this idea; it results from the limited knowledge that we have about the oriental languages which are used by them. These languages were once taught in our universities, when sciences seemed to reappear among us; but this practice was not continued, though that was the purpose of the Ancients; from there come our unjust prejudices which discredit our knowledge and scholarship.

In order to protect the readers from wrong ideas, he feels the need to make an explanation:

> In Constantinople, and even in the major cities of the Ottoman Empire — and in order to better convince the readers, let me say that this also applies to many people other than Turks — among Persians and Arabs

I

of Arabia, there is not one scholar who does not know the three languages (Turkish, Arabic and Persian), Turkish being a combination of the two others.

This view of the Comte de Marsigli, who did not refrain from criticizing the Ottomans in certain matters, clearly indicates the difference from the West in some points of scientific and educational life and confirms the above-mentioned case. Comte de Marsigli further says:

In their schools, the principles of their false religion are taught first; one learns about matters of faith and develops a capacity to judge. Those who wish to make further progress in literature continuously exercise writing prose and poetry, and later write their history with insight and great accuracy, which is even boring, because they care too much for describing every detail and circumstance. They take great interest in logic and other fields of ancient philosophy, and especially in medicine.

Alchemy is very pleasant to them. They mix their medicines according to the old prescriptions of Avicenna and Dioscorides, and have a certain knowledge of botanics. They very seriously study geometry, astronomy, geography and ethics; I can give, as evidence, a catalogue of more than eighty-six thousand authors of the last century, which I have in my library in Bologna, compiled for the use of scholars.

On the great interest of the Turks in geography, he says:

As a matter of fact, we will never have perfect maps of the Ottoman, Persian, Tartar and Arab Empires if we do not make use of the translation of these authors' works on geography.

The following statements of Comte de Marsigli illuminate the reasons why printing started late in Ottoman society. Moreover, they refute the unfounded convictions which are still repeated today.

Turks do not always get their works printed; but this is not, as commonly believed, because printing is forbidden or because their works are not worth printing. They do not want to prevent all the copyists whose number reached ninety thousand when I was in Constantinople, from earning their living; and this is what Turks themselves told the Christians and Jews who wanted to introduce printing techniques to the empire in order to make profits.

Comte de Marsigli, in expressing his amazement at, and appreciation of the Turks, was clearly aware of 'the completely false and unfounded ideas and remarks about the Turks':

> But, to conclude this subject, how to explain that the government of an empire as vast as that of the Turks succeeded to maintain it at this size, though it did not have steady revenues, the sole means which could contribute to its establishment and development. I can say there is no other government in the world which has more accurate registries, in anything related to treaties with foreign states, in any field regarding the estates, the protocol and rules observed, sending of orders, decrees, officers in service and, as I said, all that is related to finance. What I reported seems sufficient to eliminate the prejudices that may exist, and I leave it to those who read my manuscripts to extract what is in surplus, ...

Western observations of the Ottoman *medreses* continued to appear in later periods as well. Some Ottoman authors mention a gradual deterioration in the *medreses* around the end of the sixteenth century and in the seventeenth century, which occurred in the other institutions of the state as well. They agree that toward the end of the sixteenth century, the *medreses* did no longer maintain their former high quality as the result of acts contrary to the law concerning the *müderrises*. This deterioration had direct effects upon the *müderrises*, instruction and the students. One of these authors, Gelibolulu Mustafa Ālī Efendi, cites as reasons for the corruption of the *medrese* system the decreasing demand for knowledge, the appearance of the *mevālīzādes* and their promotion in a short period of time, access to the *ʿilmiye* career through patronage, the conferring of *kadılık* and the *müderrislik* positions in return for bribery, the difficulty to distinguish the learned from the ignorant as well as the decline in the number of written works. According to Kātib Çelebi (1609-1657), the decline resulted from the exclusion of rational sciences and mathematics from the curriculum of *medrese* education. Furthermore, many Ottoman intellectuals had critical remarks about the corruption of the *medrese* system, the excessive number of students, the irregularities of the *müderrises* and their revolts. Nevertheless, the Italian priest Toderini, who had the opportunity to become closely acquainted with the *medreses* during his stay in İstanbul between the years 1781-1786, shares the views of Comte de Marsigli about the Ottoman educational institutions. Toderini's words concerning the *ʿulemā* are noteworthy: 'The Ottoman scholars are knowledgeable and reliable since they do not have any uncultivated intellectual activities and they all know Arabic and Persian.' Toderini expanded his studies on Ottoman literature so as to include all kinds of scientific activities of the Ottomans. His numerous remarks about the *medreses*, which he calls 'academies,' and their curriculum, are worthy of attention. Toderini who examined the administration of the *medreses* and their *vakfs* was of the opinion that: 'From the viewpoint of scientific autonomy and other aspects they are more advanced than their counterparts in Europe.' The Ottoman intellectuals who noticed the decay in the institutions of their society criticized the system and at the same time looked for solutions. However, as it appears from the above-mentioned examples, foreign observers continued to appreciate and admire Ottoman institutions. These two approaches constitute an important factor in evaluating the decay in the Ottoman institutions.

From the seventeenth century onwards, the situation which until then had been favorable to the development of science gradually reversed. Weakening central authority, the disruption of political stability, lack of new conquests, perpetual territorial losses as well as the abundant flow of American silver to Europe led to a decrease in the revenues of the empire. Emerging economic and social difficulties resulted in economic and social disintegration which had negative effects on scientific activities. The factors which urged the scholars to pursue scientific studies in the previous periods gradually disappeared and the struggle to make a living assumed more importance instead.

At the same time that corruption in the organization and institutions of the Ottoman state appeared, irregularities occurred in the *ʿilmiye* career and the *medreses* and these were interpreted as the causes of the disintegration of the empire. As we have stated above, however, the decay of the *ʿilmiye* career and that of the other institutions of the state resulted from political, economic and social reasons.

Early Contacts with the West

The *medreses*, which were the most important institution of education and learning in the Ottoman state, were organized to meet the needs of the state and society. It was in their own science that the Ottomans found the solutions to intellectual and technical problems related to state and society and the answers to their principal intellectual concerns. For this reason, the Ottoman educational system was self-sufficient in every respect. The Ottomans, therefore, did not consider the transfer of science from contemporary European states necessary. Yet, they did not hesitate to follow and adopt new techniques which had evolved in the West. They almost immediately adopted particularly those techniques which concerned military technology, fire-arms and mining, and applied them successfully.

Since its establishment, the relations of the Ottoman state with Europe consisted mainly of political and military encounters. One of the manifest characteristics of these relations is the fact that the Ottomans considered themselves superior to the Europeans in every respect. The Ottomans were more powerful in military and economic terms than the European and Islamic states of the same period. The fact that the Ottomans possessed rich mineral resources, controlled the trade routes and came out victorious from all the battles which they fought, caused them to grow economically powerful and to feel superior.

The Ottomans also regarded themselves as morally superior to Europe, being the followers of Islam which they believed to be the last and the truest of faiths, and by virtue of being the heirs of the brilliant medieval Islamic civilisation. For these reasons, they were unwilling to recognize the importance of the new intellectual and scientific understanding that emerged in Western Europe during and after the Renaissance period, or to comprehend the effects of related developments on the progress of European societies, or, finally, the fact that all of these could be a potential danger to themselves. Yet, this did not prevent the Ottomans from adopting from the Western world any new elements that they needed.

It appears that the Ottomans were capable of following the developments in the West. With its lands extending from the Balkans to central Europe and its sovereignty in the Mediterranean and North Africa, the Ottoman Empire was close to European countries. Evidently, this geographical proximity facilitated the transfer of knowledge. Moreover, diplomats, European converts to Islam, travelers, merchants, seamen, prisoners, refugees and particularly the Jews and Moriscos who fled from religious oppression by the Inquisition in Spain and Portugal and found safety under the protection of the Ottomans, brought in a great deal of new scientific and technical information. The examples which we shall cite below indicate that the Ottomans were aware of the developments in the Western world. At the same time, they will show us their selective attitude toward these developments.

In the fifteenth and the sixteenth centuries, the Ottomans followed the developments in the West, particularly in the subjects of war technology, mining, geography and medicine, with various means and ways. Their direct geographical contact with Europe had an important influence upon their pursuit of these developments, as it enabled them to obtain technical knowledge from the West easily.

Sultan Mehmed the Conqueror invited Gentile Bellini, a painter of the Renaissance period, to İstanbul and had him paint his portrait. He also had the walls of his new palace decorated with frescos in the Renaissance style. Moreover, he invited master bronze casters and scabbard makers. These incidents show that immediately after the war with Venice, which had lasted sixteen years, the Ottomans established contacts with the West. Owing to these contacts, they were able to obtain whatever they needed from the West.

Apparently, no large difference existed in the sixteenth century between the Ottomans and Europe in regard to mining technology. In the 1580s the Ottomans used almost the same mining techniques as Europe. The Ottoman silver mines in Serbia were from the beginning regulated according to the Saxon mining laws and naturally, the Ottomans also used the legal and technical terminology related to mining. These mines were often operated under the surveillance of Saxon mining experts, who were employed by the state and received a certain share from the profit of the mine. This condition was meant to be an incentive in order to keep the production at a high level.

In extracting and operating the mines in their vast lands, the Ottomans employed their own mining experts besides the foreign specialists. They operated mines efficiently both in Rumelia and in Anatolia. Since those who were occupied in mining were exempt from all örfī (customary laws), mining became an attractive occupation.

Fire-arms

The use of fire-arms by the Ottomans is an important example which explains their attitude towards Western technology and its newest developments; it also clearly indicates the nature of the relationship between the Ottomans and the West.

The exact date when the Ottomans first used fire-arms is not definitely known. However, there is a great deal of evidence indicating that they adopted this technique very early from Europe. It is known that the Serbians, who had close relations with the states in central Europe towards the end of the fourteenth century and who learned about the progress made in the filed of

I

52

fire-arms from them, had an important role in the introduction of fire-arms to the Ottoman world. Moreover, there are documents indicating that during the same years, the Ottomans began to buy weapons from the West.

In the first half of the fifteenth century, the Ottomans made clear progress in the field of fire-arms and particularly artillery. At first, Ottoman artillery developed parallel with that in the Balkans and the other countries of Europe. In the fifteenth century, great progress in artillery techniques was made which, besides the presence of skilled metal smiths, seems to have been due to Sultan Mehmed the Conqueror's interest in ballistics and the fact that he and his commanders realized the value of these new weapons during battle more than contemporary Europeans.

Around the middle of the fifteenth century, the 'hand-gun,' which together with other fire-arms was manufactured largely in the regions of Serbia and Bosnia, aroused Ottoman interest. This was similar to their interest in cannon and the model of learning and adoption of this new technique took place in the same manner. A large number of guns was obtained from the armies which they fought in the Balkans. After the conquest of the Balkans, the Ottoman Turks simply continued their production in the same places. In 1480, different types of the arquebus and hand-guns, with the sign of *Kayı* inscribed on them, were manufactured in Samothrake. At first, most of the technical staff who worked in these factories were Christian Serbians.

Western techniques such as that of casting and using cannon were transmitted to the Ottoman world initially by Christian Bosnians and Serbians and subsequently by Italian and German experts. In the following years, French, English and Dutch (Flemish) technicians entered the service of the Ottomans as well.

In addition to these, the Ottomans employed a technical staff in the palace called *tāife-i efrenciyān* in order to transfer technology from the West. The identities and activities of these foreign technicians, who served the Ottomans in applying modern technology in various military and civilian projects, are quite interesting; it is, however, difficult to find extensive information about them. A Hungarian by the name of Urban who was a famous individual in this group was the first to cast cannon for the Byzantine emperor; later he changed sides and cast the cannon which Mehmed the Conqueror used during the siege of İstanbul. In the sixteenth century the Ottomans paid attention to training their own technical staff in the manufacture and use of fire-arms. Moreover, independently from foreigners, the state also endeavored to recruit its own musketeers.

Although it was not easy for the Ottomans to follow all the new developments in Europe initially, i.e., between the fifteenth and the seventeenth centuries, they were successful in following the developments in war techniques and fire-arms. The cannons which they used during the conquest of Crete and the siege of Candia had the same design as Italian cannons; in fact, it was an Italian master who established an arsenal in Candia. At the same time, the Ottomans in some respects used heavy artillery in a more advantageous manner and applied techniques which were unknown to the Europeans in transporting this artillery.

Comte de Marsigli states that when he was in İstanbul he saw the Turkish translation of a work on artillery by Pierre Sardi. Apparently, the Ottomans had the means to follow the military techniques of the West through the experts which they brought from the West and by translating the literature which appeared there. It was also observed that the guns which the Ottomans used during the Austrian wars in 1680 were of the same quality as the ones in the Austrian army; they were even superior in some respects, such as the range of the guns.

This situation gradually changed from the eighteenth century onwards when in addition to obtaining military equipment from the West, the Ottomans adopted the English method of preparing gunpowder. This is reflected by the expression of 'producing gunpowder according to the English method' which was often used in Ottoman documents. In the eighteenth century, however, simply copying guns as the Ottomans had done in the past was no longer sufficient to bridge the widening gap with Europe. In this period, the assimilation of the system of war to technical and tactical fields, proved difficult. The nature of warfare had changed. This change in the system of warfare was treated as part of the attempts of reform which began in the last quarter of the eighteenth century and completely changed the structure of the army and even of the state. It appears that the nature of this reorganization in the Ottoman army led to severe and incisive changes, and considerable agitations in the relation between state and society.

R. Murphey, who added a new dimension to the field of the relations of the Ottomans with Western technology, studied the reasons why the regression of the Ottomans appeared with a delay of a century after the beginning of the Technological Revolution in Europe. In his view, these reasons must be sought among the changes related to obtaining provisions rather than in the backward tactics of the Turks or in psychological obstacles.

Towards the end of the seventeenth century, starting from 1695, the Ottomans began to buy gunpowder from Europe, particularly from English merchants, at a high price. It appears that until that date the Ottomans were capable of meeting their own military needs. The Ottomans bought gunpowder from Europe not because their powder was of low quality, but due to their inability to meet the increasing demand. This is evident from the statement of Mehmed Emin Efendi, director of the government powder mills (*barutçu başı*): 'I cast better powder than the very first-rate powder of the Europeans.' Nevertheless, the Ottomans were constantly in need of European techniques. Approximately from the beginning of the eighteenth century onwards, an increasing number of European experts were employed in the army.

Political and economic factors need to be further explained in order to acquire a better understanding of the success of the transfers from the West to the Ottoman world in the field of fire-arms and weaponry. According to Mahmud Raif Efendi, a graduate and instructor of algebra and geometry in the *Mühendishāne* in the last quarter of the eighteenth century, the ministry of the powder mill (*baruthāne nāzırlığı*) had become an ordinary office since the Ottoman Empire had been in a state of peace for many years. As the situation constantly worsened, only half of the 3000 centals (300 tons) of gunpowder which was expected to arrive from the powder mills of İstanbul, Gallipoli and Salonica could be supplied. It was, moreover, of very poor quality. After the war with Russia in 1182 A.H. (1768 A.D.), the gunpowder which was produced in the Ottoman lands was only good for ceremonial cannon shots. In order to meet military needs, the Ottomans were obliged to buy gunpowder from foreign countries at high prices. A very small amount of this gunpowder was of good quality. It was both difficult to supply and impossible to obtain timely.

During the reign of Selim III (1789-1807), new regulations were made regarding military equipments. The Ottomans took precautions to prevent the sale abroad of saltpeter, one of the basic components of gunpowder. They also brought experts from Europe. During the rearrangement of the İstanbul powder mill, all the necessary implements and materials were concentrated there and it started to operate efficiently. In addition, the new Azadlı powder mill was established in Küçük Çekmece, where water mills, invented by an Armenian by the name of Arakel, were used. The two powder mills came to produce a total of ten thousand centals (1000 tons) of gunpowder annually, of the same quality as the powder produced in Holland and Great Britain. While the price of saltpeter doubled, the

other materials were bought at current prices; the payments of the workers in powder mills were tripled and the payments of the foreign masters
were raised to between 500-1000 *kuruş*.

When M. Raif Efendi evaluated this development in the technical field,
he gave the good news of the success achieved during the last years of the
eighteenth century with great joy to his readers, saying:

> We have reached such an excellent state that one ounce (30,5 g) of new gun-
> powder is eight times more powerful than the old one. We no longer
> need the foreigners, our warehouses are filled with sufficient supplies
> for use on the battlefield. We have even started to export them.

According to Mahmud Raif, the units which were established during the
reign of Selim III and trained with European methods equalled those of
the European nations as regards tactics. This new type of Ottoman bu-
reaucrat-technocrat intellectual who was a product and an enthusiastic
supporter of the reform attempts in the Ottoman army, unfortunately lost
his life in an unlucky incident as a result of the reactions which arose
against these reforms.

Clockmaking

Mechanical clocks, which had been constructed in Europe since the beginning of the fourteenth century, were not of much interest to the Ottomans
because of their high margin of error. It is known, however, that during
the first half of the sixteenth century, mechanical clocks were in fact being
used by the Ottomans. And in 1547, there were numerous clocks among
the gifts offered by the Austrians to the Ottomans. Consequently, a market for European clocks developed and demand for these clocks rose.

The European-made mechanical clocks were not practical to meet the
needs of the Ottomans in the calculation of the times of prayer and worship since they had a margin of error which was as high as thirty minutes.
In line with the classical Islamic tradition, the Ottomans used the hourglass, the water clock, and the sundial for timekeeping. The call to prayer,
informing the people of prayer times, limited the need for clocks. Another
factor limiting their use was their high price.

From the second half of the sixteenth century on, mechanical clocks
brought from European countries such as Holland, Germany, Hungary,
and France could be seen in the Ottoman world. These clocks were designed according to the tastes and needs of the Ottomans, showed the

phases of the Moon and had Ottoman numerals on the face. It was therefore easy for these European clocks to find a market. Another reason for the success of these clocks was probably the fact that the new European system of clockmaking and clocks was not foreign to Ottoman culture. The only difference was the division of the Ottoman system of timekeeping into two twelve-hour periods, one from sunrise to sunset and the other from sunset to sunrise. The clocks were therefore set in a daily practical way.

In the sixteenth century, some of the clockmakers of the *hās oda* in the palace made clocks for the sultan. Among them were masters Hasan and Pervane, both Muslim clockmakers related to the *hās oda*. Apparently, between the years 1574 and 1595, a few clockmakers served in the palace, including Muslims and European clockmakers. Since twenty-four clocks made between these dates have survived to our day, there must have been a great number of clocks in the palace.

During the last decade of the sixteenth century, European clockmakers made clocks in Galata district in İstanbul. Therefore, the Ottomans could closely follow clockmaking technology in Europe. The Ottoman astronomer Takiyüddīn al-Rāsid (1525-1585) wrote the first work on the construction of mechanical clocks in Europe entitled *al-Kawākib al-Durriya fī Bengāmāt al-Dawriyya* (İstanbul copy; 973 A.H./1563 A.D.). When he was asked, in 1561, to make a clock indicating the times of the call to prayer, he examined the numerous mechanical clocks in the treasury of Semiz Ali Paşa (d. 1565), the Grand Vezir of the period, and had discussions with European clockmakers about technical details.

Takiyüddīn examined three types of mechanical clocks: weight-driven, spring-driven, and watches, or clocks with lever-escapement. He also provided information about pocket watches and astronomical clocks which he developed based on the division of every hour into sixty minutes and every minute into five seconds. He is known to have constructed an astronomical clock which he used in order to be able to take more precise measurements in his observations. During the same years, Tycho Brahe, on the other hand, avoided using clocks in his observations since they were not as precise as they should have been.

We do not have any information as to whether further attempts at invention and development were made after Takiyüddīn's time. From the first half of the seventeenth century, however, Europeans who had settled in İstanbul, undertook the making of mechanical clocks. Genoese clockmakers who established a guild in Galata after 1630 supplied the necessary clocks. From the eighteenth century onwards, European clockmakers, who noticed that the Turks had a taste for clocks, began to construct clocks for this market. As a result of mass production by the Europeans, production in Galata folded when imported clocks became cheaper than domestic-made ones.

An analysis of the characteristics of the Ottomans' attitude towards Western technology shows first of all that they were capable of following closely the new techniques which developed in the West, and adopted those which they found necessary. Secondly, it appears that the Ottomans did not have any difficulty in adopting Western techniques in the necessary fields with a selective attitude. Thirdly, in learning and applying Western techniques, the Ottomans always brought and employed the experts and technicians which they needed from Europe. Finally, they resorted to European experts in order to train their personnel, who would then use the newly imported technology all around the empire. Thus, they attracted a large number of technicians and thanks to them tried to keep up with Europe. On the one hand, the Europeans attempted to prevent the entrance of European military technology into the Ottoman state, but, on the other hand, they took the needs of Ottoman society into consideration when they made clocks for which, in their view, the Ottoman Empire was a great market. The Ottomans did not deem it necessary to take measures which would enable them to produce and develop the new technology and act independently from the developments in Europe during their rivalry with the Europeans. This attitude of the Ottomans which lasted until the twentieth century clearly indicates their basic attitude towards Western technology.

Geography

The Ottomans needed knowledge of geography in order to determine the borders of their perpetually expanding territory and to establish control over military and commercial activities in the Mediterranean and the Black Sea. They benefitted both from the geographical works of classical Islam and from works of Western origin in obtaining this information. On the other hand, by supplementing their own observations, the Ottoman geographers produced original works as well. The importance of the Mediterranean trade routes decreased particularly due to the circumnavigation of the Cape of Good Hope by the Europeans. This situation negatively affected the Ottoman economy and caused the Ottoman policy of

conquest to turn to the Mediterranean. Consequently, the Ottomans started to give priority to navigation and geography.

The first source for Ottoman knowledge about geography is the
Samarkand tradition of astronomy and geography. The works on astrono-
my and geography of classical Islam which appeared in this tradition
greatly influenced Ottoman geography. During the reign of Mehmed the
Conqueror, the work of Ptolemy on geography was translated from the
Greek.

An examination of fifteenth century maps drawn by Muslim cartographers such as that of the Mediterranean drawn by İbrāhīm Kātibī in 1416
and Southwestern Europe by Tabib Mürsiyeli İbrāhīm in 1461, which are
among the first examples of maps of Western origin, shows that they benefitted from Western knowledge of cartography. The fact that these maps
were used by the Ottomans around the same date clearly indicates that
new geographical information was indirectly transmitted to the Ottomans
as well.

From the sixteenth century onwards, the science of Ottoman geography
began to produce its greatest works with the studies of Pīrī Reis. Pīrī Reis
was brought up by his uncle Kemal Reis, one of the famous Turkish cap-
tains, and accompanied him on numerous naval campaigns. When Kemal
Reis died in 1511, Pīrī Reis returned to his birthplace Gallipoli and prepared his first world map.

The map of Pīrī Reis which was presented to Sultan Selim in Egypt in
1517, was based on maps prepared by Europeans including Columbus'
map of America. The first map in the Ottoman world including this preliminary information about the New World represents Southwest Europe,
Northwest Africa, Southeast and Central America. This map is part of a
large scale world map. It is a portolano without the lines of latitudes and
longitudes, including the coasts and the islands with the purpose of de-
lineating the lines of division. In general, it is believed that maps of this
kind do not have a mathematical basis. Research, however, shows that
there were five centers of projection placed on the Atlantic Ocean in the
map of Pīrī Reis. Moreover, the lines of latitudes and longitudes can be
drawn very easily on this map.

In the brief explanations which he wrote on his first map, Pīrī Reis
states that he made use of thirty-four maps. Twenty of these maps were of
Alexander the Great's period where he explains: 'one fourth of the world
covered with continents is represented in it.' Eight of them were Islamic
maps called *caferiye* by Muslim geographers. Finally, there were an

Arabo-Indian map, four new maps drawn by the Portuguese and the map of Columbus.

Columbus made four voyages to America between the years 1489-1504 and drew maps during these journeys. According to the statement of Pīrī Reis, a sailor who had joined Columbus on three of his voyages was taken prisoner by Kemal Reis. Pīrī Reis took the map from the prisoner and also listened to his accounts about the voyages to the New World. This attitude clearly indicates that Pīrī Reis had a scientific interest. On the one hand, he searched the old literature, on the other hand, he followed the new developments, in addition noting down his own observations. The original map of Columbus has not been preserved, therefore the map of Pīrī Reis which was based on this map has both scientific and historical value. Pīrī Reis also compiled a book entitled *Kitāb-ı Bahriye* (1521) which comprises observations he noted down over many years, information from the geographical books of the time and his own maps. In this work, Pīrī Reis presents pictures and maps of the cities on the Mediterranean and the Aegean coasts and gives extensive information about navigation. His work also includes his observations and ideas about nautical astronomy. In this period, Pīrī Reis drew his second world map which was presented to Süleyman the Magnificent in 1528.

Only that part of the second map of Pīrī Reis is extant which contains the northern part of the Atlantic Ocean and the recently discovered areas of Northern and Central America. In comparison with the first map, the coastal lines have been delineated more neatly, the empty spaces have been filled in and unknown places were again left blank. Pīrī Reis drew one of the best maps of the time from the viewpoint of cartographic techniques. The fact that he followed new scientific developments and discoveries and left the unknown places blank are additional indications of his scientific approach.

Another work of the sixteenth century which contains information about the geographical discoveries and the New World is the book entitled *Tārih-i Hind-i Garbī*. This work, whose author is still unknown, was presented to Sultan Murad III in 1583. Based on Spanish and Italian geographical sources, it is important in showing that the Ottomans followed the geographical discoveries in the West. The work is in three parts; the third part, which is the most important one and which comprises two thirds of the whole book, relates the adventures of Columbus, Balboa, Magellan, Cretes and Pizarro during the sixty years from the discovery of America in 1492 until 1552.

I

60

The first and the second parts of the work are about the Old World and the Indian Ocean. The sources of the book are the works and maps of contemporary Muslim geographers. The author frequently quotes from al-Mas⁽ūdī and refers to Ibn al-Wardī, al-Ṭūsī, al-Ḵazwīnī, al-Suyūṭī and Imam Rāzī. In the third part about the New World, however, the author relates information from European books of geography without mentioning the name of any book or author.

In the 1580s, a considerable number of sources mentioned the discovery of America in detail. A few general titles which formed the basis of his translation could be cited:

1. Francisco López de Gómara, *Historia General de las Indias* which was first printed in Spanish in 1552;

2. Gonzalo Fernández de Oviedo y Valdés, *De la Natural Hystoria de las Indias,* first printed in Spanish in 1526;

3. *De Orbe Novo* by Peter Martyr d'Anghera (Pedro Martir d'Anghiera) first printed in 1516 in Latin; and

4. Agustín de Zárate, *Historia del Descubrimiento y Conquesta del Peru* which was first printed in Spanish in 1555.

There is clear proof that the author of *Tārih-i Hind-i Garbī* used these four Western sources, since the contents of the book completely agrees with these geographical works printed in the sixteenth century.

The name of the author does not appear in any part of the work; there is also no information in contemporary or later Ottoman manuscripts about this work and its author. T. Goodrich, who did the most extensive research on this subject, states that the author may have been a geographer or an astronomer. He may even have worked in Takiyüddīn's observatory in İstanbul, which functioned between 1575 and 1580, and he may have completed his work there. But, at present, there is no evidence to support this theory.

Goodrich believes that the author definitely received assistance in translating the sources. In his view, the person who assisted the author was probably one of the Jews or Muslims who settled in İstanbul after being exiled from Spain.

The fact that the work relates geographical discoveries which took place in 1552 gives us an idea of the extent to which the Ottomans were capable of following developments in the West.

A copy of the work which appears to be written during the first half of the seventeenth century, contains a writing in the margin which provides information about the sources of the book. On the basis of this information, as quoted from a book in a European language, the possibility exists that the source of *Tārih-i Hind-i Garbī* could be the work entitled *De Orbe Novo* by Pedro Martir, who wrote about the voyage of Columbus for the first time. Since the *Tārih-i Hind-i Garbī* relates events that took place up until 1552, however, it seems that apart from this book, the author used books which dealt with discoveries in later periods.

Ottoman geographers obtained information both on the West and on the East. During the period when maps of the Eastern world were not yet available in the West, the Ottomans seem to have been on a more advanced level than might have been expected.

The maps drawn by Ottoman geographers can be divided into three groups. First, maps which combine old and new information; second, copies of maps that are completely of European origin, but of which the originals no longer exist today; and third, original works by Ottoman geographers. Pīrī Reis' maps of the Mediterranean, for example, and some maps in the atlas of Ali Macar Reis based on his observations, are original works. Until now, three *Atlas-ı Hümāyuns* which appear to have been prepared for presentation to the sultan have been discovered. Because of the similarities among them, they were mentioned under the same name. The first *Atlas-ı Hümāyun* (Walter's Sea Atlas) which appears to have been prepared at its earliest between 1550 and 1560, consists of eight maps. These three atlases contain maps of the Black Sea, the Mediterranean, North Africa, Southern Europe, the coast of England, the Atlantic Ocean, and a world map, a total of seven to nine maps. These three atlases have almost the same characteristics; one can even claim that some of them were prepared by the same person. Although some maps are original, others are copies of works by European cartographers. Two among them are complete copies of European maps which no longer exist today, with only the place names having been changed. The old world map must have been taken from the work of Seydi Ali Reis entitled *al-Muḥīṭ*, which is about the Indian Ocean. These portolanos were prepared as a combination of old and new information. The atlases, made approximately between 1550 and 1567, are, to a large extent, based on Italian maps; they also contain original maps representing sections of İstanbul, Salonica and Gallipoli.

Apparently, cartography was organized as a profession in the Ottoman Empire. In the seventeenth century, fifteen individuals were occupied with the art of surveying in eight workshops in İstanbul and nearby areas. Evliyā Çelebi relates that these cartographers knew several languages — Latin in particular — and that they prepared nautical maps by making use of European geographical works and sold them to seamen. He states that geographical works from previous scholars such as the *Atlas Minor* and the *Papa Monte* were used in particular.

When Comte de Marsigli mentions the Ottoman maps of the seventeenth century, he states that maps of Turkey, Arabia, Iran and Turkestan did not exist then in Europe and translations of the Turkish geographical works were necessary in order to obtain them.

The Compass

It is known that the compass was widely used in the Eastern Mediterranean before the Ottomans became powerful navigators. We have not come across any research, however, pertaining to the spread of the compass among the Ottomans. Probably, Ottoman sailors used this device for determining directions from the very beginning. During the sixteenth century, Ottoman navigation was fully developed and there are signs indicating that the compass, together with the science of cartography, was transferred from Europe.

From Pīrī Reis' above-mentioned *Kitāb-ı Bahriye*, it appears that the compasses used by the Ottomans were of Western origin. When Pīrī Reis stresses the importance and necessity of maps and compasses for sailors, he states that those who do not know about them should not put out to sea because they will be in danger.

In his poem of ten couplets entitled *Der Beyān-ı Pusula*, Pīrī Reis defines and describes the compass and discusses the developed form of the compass rose. This device is known as *beyt el-ibre* or *daire* in contemporary Arab literature. In the *Kitāb-ı Bahriye* and later Ottoman geographical sources, it is called *pusula*, a term taken from the Italian word *bussola* seemingly indicating that the compass was brought to the Ottomans from the West.

Seydi Ali Reis, a famous Ottoman navigator, provides a more technical explanation of the compass in his work *al-Muḥīṭ*; he mentions the compasses used in Portugal and other parts of Europe, such as the *kible-nümās* brought from Germany (*Diyar-i Alman*), the *kible-nümā*, and compasses in Anatolia during that period. He does not give any explanation, however, as to whether or not they were manufactured domestically. Providing a detailed explanation of how compasses are produced, he also refers to problems that may prevent the device from working and repairs that may be necessary.

I

Subsequently, information about the compass exists in the works of
Kātib Çelebi and Evliyā Çelebi in the seventeenth century and İbrāhīm
Hakkı of Erzurum in the eighteenth century. Among them, Evliyā Çelebi
mentions the tradesmen of 'Pusulacıyan' which leads one to believe that
the Ottomans made the compass. In the eighteenth century, İbrāhīm
Müteferrika clearly states that compasses were produced in İstanbul by
the Ottomans. On this subject Müteferrika prepared a work entitled
Füyūzāt-ı Miknātısiyye in which he provides extensive information about
the compass and about magnetism.

In addition to the compass, which the Ottomans used during navigation, sources state that the Ottomans used a magnetic needle known as
lağım tapası in order to determine the right direction while digging the
gallery of a mine. This device is composed of a magnetic needle in a semi-
circle.

These first observations on the introduction and developments of the
compass in the Ottoman world indicate that the compass was adopted
from the Europeans by the Ottomans and used at an early date. İbrāhīm
Müteferrika states that the first magnetic declination was determined in
İstanbul in 1727. In his note entitled Tezyil-i Tābi (Publisher's supplement), he says:

> in the year 1140 A.H. (1727 A.D.), disputes arose on determining the
> niche of the mosque built by the late Kapudan Mustafa Paşa in the
> place called Bebek Bahçesi in the vicinity of Rumeli Hisarı. Although
> knowledgeable men prepared many compasses, rubuʿ, astrolabes and
> other astronomical instruments, the results yielded by the compass did
> not conform to those of the astronomical instruments and in fact, the
> difference between them was large. Therefore people did not consider
> the results of the compass but acted on the indication of other instru-
> ments to determine the niche of the mosque. However, a kıble-nümā
> was constructed containing an immense magnetic needle. Upon exami-
> nation it was observed that, contrary to what was expected, the needle
> had an inclination of eleven and a half degrees to the West.

This information should be considered important as it emphasizes the
studies of Ottoman astronomers based on observation and experiment. In
his Seyahatnāme, Evliyā Çelebi (d. 1681) states that the tradesmen of
pusulacıyan in the Ottoman Empire checked their compasses in the niche
of the Bayezid Mosque. This seems to indicate that the Ottomans knew
how to determine magnetic declination as well.

Medicine

Ottoman physicians were educated within the classical Islamic tradition in a master-apprentice type relationship, and they received their practical education in hospitals. During the reign of Süleyman the Magnificent (1520-1566), an independent *medrese* called Süleymaniye Tıb Medresesi (Dār al-Ṭibb) was established within the complex of Süleymaniye *medreses* for educating physicians to meet the military and civilian needs. Thus, medicine gained an educational institution of its own with the establishment of this separate *medrese*. Professionals and craftsmen related to medicine had an organization of their own.

In the Ottoman Empire of the sixteenth century many educated physicians began working in hospitals which were founded in the imperial capital and numerous other cities. At the same time, the on-going process of rendering Ottoman Turkish into a language of science, which had started in the previous century, was successful and Turkish became the language of medicine along with Arabic and Persian. As a result works on medicine written in Turkish began to appear.

The first contacts of the Ottomans with European medicine began rather early during the Renaissance. Giacomo di Gaeta, a Jewish Italian physician who had entered into the service of the sultan during the reign of Murad II (1421-1451), became private physician to Sultan Mehmed the Conqueror, converted to Islam and took the name of Yakup. According to Aşıkpaşazāde's statement, Mehmed II had a great liking for him, and appointed him chief physician (*hekimbaşı*) and conferred upon him the rank of *vezir*. Thanks to this Italian physician Yakup Paşa, from the end of the fifteenth century through the first half of the sixteenth century, Ottomans obtained the means to establish closer contacts with medicine during the Renaissance.

The Jews who were exiled by King Charles VI from France in 1394, and by King Ludwig X of Bavaria from Germany took shelter in Turkey. Later, Jews who were expelled from Spain in 1492 were guaranteed liberty of religion under the protection of the Ottoman sultan. Among those who took refuge in the Ottoman Empire were also physicians of Spanish, Portuguese and Italian origin. Probably, the achievements of Western medicine reached the Ottoman Empire through the physicians who came in with this second wave of migration.

The immigrant Jewish physicians settled in İstanbul and Salonica and brought with them new elements of European medicine which were dif-

ferent from classical Islamic medicine. This allowed them to make rapid
progress in their careers. These physicians, who managed to bring their
books with them, had been educated at the universities of Lisbon,
Coimbra and Alcala among others. Some of them succeeded in entering
the palace and rising to the rank of sultan's physician. Around the beginning of the seventeenth century there was a religious community of
'Jewish physicians' composed of forty-one individuals.

The Jewish physicians were exempt from some taxes and honored
with such privileges as horseback riding. Nicolas de Nicolay states that
there were physicians among them who were well versed in theoretical
and practical medicine, and that their superiority stemmed from their access to works of medicine in Hebrew, Arabic and Greek. In another work,
De Nicolay states that Jewish physicians were skillful because, in addition
to the above-mentioned languages, they were fluent in Latin, Italian and
Spanish.

Mūsā b. Hāmūn (d. 1554) who was the most famous of these physicians, was appointed physician to Süleyman the Magnificent, and wrote
one of the first works on dentistry in Turkish. A. Terzioğlu published a
facsimile of this work. According to him, it was prepared on the basis of
ancient Greek, Islamic and Uighur Turkish medical sources and with the
benefit of the works of the Ottoman physician Şerafettin Sabuncuoğlu (d.
1468) and other Ottoman and Western medical sources.

Besides this well-known work, Mūsā b. Hāmūn wrote another work in
Turkish which has not yet been examined in detail. This work, which contributed to transferring the medical knowledge of Europe, is entitled
Risāla fī Ṭabayiᶜ al-Adwiya wa İstiᶜmālihā. This short *risāla*, consisting of
four chapters, describes the characteristics of medicine. In his introduction, Mūsā b. Hāmūn explains that he prepared his work with the help of
'Islamic, European, Greek and Jewish' sources.

Shaban b. İshak al-İsraīlī (d. ca. 1600) also known as Ibn Jānī, who was
one of the lesser known Jewish authors of the time about whose life we
have little information, translated a *risāla* on medical treatment using tobacco from Spanish into Arabic. Ibn Jānī mentions methods of curing applied by the Spanish physician Mortarus, who lived at the end of the
sixteenth century. These methods are based on the use of tobacco leaves
and the juice obtained from them. Ibn Jānī states that he decided to write
this *risāla* when he saw that everyone, including women, smoked tobacco
and even saw a *risāla* in verse, which praised tobacco.

Thus throughout the sixteenth century, Jewish physicians, of whom we have only mentioned a few, contributed to Ottoman medicine. The Ottoman medical works written during this period were essentially based on classical Islamic medical literature. In order to determine the exact nature and degree of the contributions of Jewish physicians, one needs to compare in an analytical and detailed way the Ottoman medical works written during this period with contemporary Western medical sources or with earlier ones which they supposedly followed. This will enable a more exact evaluation of Ottoman medicine and its relations with Western medicine.

The best known channel through which elements of European medicine were introduced into the Ottoman Empire in the sixteenth century was formed by Jewish physicians at the Ottoman court. These Ottoman Jewish doctors subsequently lost their contact with Europe. From the seventeenth century onwards, however, Greek physicians who were Ottoman subjects maintained the Ottoman contact with the West in the area of medicine. Most of these physicians were educated in Italian universities. Western medicine was disseminated in the Ottoman world through the less obvious influence of the missionaries, merchants, travellers and consular doctors from the sixteenth century onwards. Michael Dools maintains that before the period of translations, the effectiveness of the transmission of Western medicine by Europeans in the Ottoman world can be judged, to some extent, by the response to the new diseases from the West which appeared in the sixteenth century. There is information on whooping cough and syphilis and the treatments that were recommended by Westerners by the early sixteenth century. Since local physicians were unable to cure these diseases, treatment began with medicines that were used in Europe. Thus, Ottoman physicians began including descriptions of these epidemics and Western medications in their works. Al-Antākī's (d. 1599) work *Tadhkira* (976 A.H./1568 A.D.) is one of the first examples of books presenting classical Islamic medicine together with European knowledge. From the sixteenth century onwards, new physicians and diseases of Western origin led to the emergence of new medical ideas and methods in order to prevent and cure diseases in the Ottoman world.

G. Russell states that some of the Jewish physicians who came from Europe to the Ottoman Empire were educated at the universities of Padua and Salamanca. Owing to their knowledge of modern anatomy they were on a different level from Ottoman physicians; these physicians brought their books of medicine and anatomy along with them. Thus, besides traditional anatomical drawings, we find some features of sixteenth century

European anatomy in Ottoman medical literature of the seventeenth century.

Şemseddin İtakī's book on anatomy entitled *Risāle-i Teşrīh-i Ebdān* (1632) portrays these features. İtakī's drawings show the principle of selective copying and the degeneration of the newly copied information. The medical texts of the seventeenth century reveal that European medicine was transmitted but that it was information belonging to a century earlier. Another aspect of the transmission of Western medicine to the Ottoman world then was that the sources of this information were relatively old works and that they were not completely understood. The works of İtakī and his successors brought about a gradual change, but the gap between European and Ottoman knowledge of anatomy was only bridged in the nineteenth century.

Nil Sarı states that from the seventeenth century onwards the new medical doctrines which were put forward by Paracelsus and his followers in the sixteenth century began to be observed in the Ottoman literature of medicine. Among the most prominent followers of this new trend in medicine which developed under the names of *Ṭibb-i Cedīd* and *Ṭibb-i Kimyaī*, she enumerates Salih b. Nasrullah (d. 1669), Ömer b. Sinan al-İznikī (eighteenth century) and Ömer Şifā'ī (d. 1742). Salih b. Nasrullah, in his work *Nuzhat al-Abdān*, quotes from European physicians who were the representatives of new medicine and gives compositions of medicines. Al-İznikī, in his work *Kitāb-i Künüz-i Hayāt al-İnsān Kavānin-i Etibbā-i Feylesofān* quotes from Arab, Persian, Greek and European physicians and here, too, presents old and new medicine together. Likewise, Ömer Şifā'ī in his work *al-Cevher al-Ferīd* states that the compositions of medicines were taken from the books of Latin doctors, and that he translated them from European languages into Turkish. In this way, the new medicine of European origin existed side by side with traditional medicine until the beginning of the nineteenth century.

With Şānizāde Atāullah Mehmed Efendi's work entitled *Mir'āt al-Abdān fī Tashrihi Aʿḍā al-İnsan* (1820) this tradition disappeared. Both the text and the surgical pictures reflect eighteenth century European medicine which treated the human body as a machine. In his work, which was copied from the works of the best contemporary anatomists, Şānizāde does not mention traditional medicine.

Finally, within the time period of this conference, following this information about geography and medicine, I wish to present briefly the results of my study on the introduction of modern astronomy from the West to the Ottoman world between mid-seventeenth and mid-nineteenth centuries.

We have observed that in this period of two centuries, the contacts of the Ottomans with Western science developed in four stages. Two of them can be observed until the end of the eighteenth century:

1. Awareness and familiarity
2. Utility and application

The Ottomans were capable of following the developments in Europe closely. Since the Ottoman astronomers had their own rich experience and were aware of the fact that Muslim astronomers had made great contributions to astronomy during the Middle Ages, they did not accept this new European science right away. They accepted it only after observing its compatibility with their own science.

While Europe was shaken by the heliocentric theory of Copernicus, the Ottoman astronomer Tezkireci Köse İbrāhīm Efendi treated the basic concepts of this theory only on the level of a technical detail since the transition from the geocentric to the heliocentric system, i.e. the change of coordinates did not bring about any changes from the viewpoint of practical calculations. Another reason for this approach may have been the fact that Ottoman society in this period was not inclined to debates on systems of the universe, probably due to the absence of religious dogmatism on this issue.

Subsequent works translated by the Ottomans were al-Dimashḳī's translation of Janszoon Blaeu's work in Latin entitled briefly as *Atlas Major* which al-Dimashḳī completed in 1685, and Osman b. Abdülmennan's translation of Bernhard Varenius' work from Latin, which he entitled *Tercüme-i Kitab-ı Coğrafya*, in 1751. These translations show that Turkish scientists did not have any prejudice against the concepts of new astronomy, or, particularly, against heliocentricity. Since the European writers whose works these two translators translated were in favor of the geocentric system, they, too, transmitted the information in this manner. Osman b. Abdülmennan, although he essentially followed the original text, preferred the heliocentric system on rational grounds.

Thanks to the establishment of a printing house (1728), works which addressed a broad mass of readers were published. This made possible the introduction of concepts of new astronomy to broader masses of people. İbrāhīm Müteferrika, known as the founder of the state-financed printing house in Ottoman society and who was a European convert to Islam, presented these new concepts in a very cautious manner in his supplement to Kātib Çelebi's

Cihannümā which he printed in 1732. The reason for his prudence was his awareness of the strong reactions to these concepts on the part of the church and some religious circles in Christendom.

Müteferrika also translated Andreas Cellarius' book entitled *Atlas Coelestis* from Latin one year after he published the *Cihannümā*. In this translation, he is much less prudent than in the *Cihannümā*, since there was no reaction similar to the one in Europe against such views in the Ottoman Empire.

Until this stage, the concepts of new astronomy had been dealt with in a professional way and with understanding by scholarly circles and by astronomers in İstanbul. In the middle of the eighteenth century, these concepts were introduced to a new audience and presented to the people of Anatolia in an interesting and contradictory style. İbrāhīm Hakkı of Erzurum, a famous mystic of the eighteenth century, explained the new concepts very clearly, and presented them based on his reading of the printed book of Müteferrika. He gave preference to the heliocentric system in a very manifest and definite way. In another part of his book, however, he presented an understanding of astronomy based entirely upon unscientific and legendary explanations as well as fictitious traditions under the heading of 'Islamic astronomy.'

As it appears, İbrāhīm Hakkı addressed 'the common people' and 'the intellectuals' in his work *Mārifetnāme* which was compiled in 1757. He had to be cautious of some official ʿulemā who envied his popularity and renown. In the above pages, we have mentioned the rumor according to which *müftī* Kadızāde Mehmed opposed İbrāhīm Hakkı and blamed him with violating the *şerīʿat*. Probably in order to protect himself from such unjust accusations, İbrāhīm Hakkı included contradictory views in his work. For the same reason, he wrote his books ʿUrwat al-İslām and Hay'at al-İslām in 1777 both based on religious sources as well as the section on 'Islamic astronomy' in his *Mārifetnāme*. This regression from his scientific views in the *Mārifetnāme,* observed in his later works, also appears to be a defense against some narrow-minded ʿulemā.

In this first stage of awareness and familiarity, the Ottomans received some information and preliminary observations about the scientific developments in Europe through their ambassadors who visited European capitals. Although in most cases they did not penetrate deeply into the matter, some of these ambassadors were closely interested in these developments on a technical level. For instance, the Ottoman ambassador Yirmisekiz Mehmed Çelebi visited the Observatory of Paris in 1721; ambassador Hattī Mustafa Efendi visited the Observatory of Vienna in 1748. While Hattī Mustafa Efendi's visit was an ordinary diplomatic one, Yirmisekiz Mehmed Çelebi's visit was different. Due to his interest in astronomy, Yirmisekiz Mehmed Çelebi, who was not content with his first visit to the Observatory of Paris, paid a second visit to this observatory. He had a detailed discussion with Cassini, the director of the observatory, and examined the modern astronomical instruments which were new to him. He also consulted Cassini about the observations which were contrary to Uluğ Bey's tables and received a written report from Cassini on these matters. Upon his return to İstanbul, Yirmisekiz Mehmed Çelebi presented this report to Ottoman astronomers.

Later, when the second stage of transfer began in the second half of the eighteenth century, Sultan Mustafa III ordered that 'the most recent and the most perfect' books on European astronomy be brought from the Academy of Sciences in Paris. Hence, in this same period, Halifezāde İsmāil Efendi translated the work by the French astronomer Clairaut (d. 1765) under the title of *Tercüme-i Zīc-i Kılaro* in 1768 and another work by Cassini (d. 1756) under the title of *Tuhfe-i Behīc-i Rasīnī Tercüme-i Zīc-i Kasini* in 1772. Since the translation of these astronomical tables vitally effected Ottoman calendar making, Sultan Selim III ordered that calendars be organized in conformity with Cassini's astronomical tables. From then on Uluğ Bey's tables were no longer in use.

Cassini's tables were widely used, but the big margin of error in the calculations of these tables caused important mistakes in the preparation of calendars. Therefore, the French astronomer Lalande's astronomical tables entitled *Tables Astronomiques* were translated first into Arabic in 1814, and later into Turkish before the year 1826. Upon Sultan Mahmud II's order, calendars were organized according to these tables. These translations of astronomical tables by the astronomers Durret, Clairaut, Cassini and Lalande between the seventeenth and the nineteenth centuries show that the Ottomans were informed about the literature on astronomical tables in the West.

The present stage of our research indicates that in these two stages, first, contrary to Adnan Adıvar's statement, the Ottoman Empire was not 'barred to' the sciences of the West. Secondly, it has been clearly established that the Ottoman religious institutions and the ʿulemā had no such role as preventing the contact of the Ottomans with Western science. Moreover, we can observe that in these two stages, in general, the scholars who were educated in the *medrese* were influential. Thirdly, the prin-

I

Ottoman Science   71

cipal sources and theoretical works in the West which brought about fundamental changes in astronomy escaped the attention of the Ottomans. They preferred to translate works necessary for timekeeping and calendar making. This marks the practical aspect of Ottoman science which has been its primary characteristic.

Finally, understanding the context and nature of the interest of the Ottomans in the developments of Western science and technology would add a new dimension to the theories put forward about the spread of Western science. Evidently, the theories of Basalla and Pyenson about the spread of Western science are not applicable to the Ottoman case, since there is no center-periphery and colonial power-recipient pattern. The matter discussed here is the attitude of a powerful empire towards the developments occurring outside its domain. The matter is not a question of the influence of the West on the Ottoman state, but rather of the interest of Ottomans in the West. For these reasons, construction of a theoretical model for the relationship of Ottomans with Western science would mean to construct a *sui generis* model or paradigm.

Selected Bibliography

Adnan Adıvar, *Osmanlı Türklerinde İlim* (4th ed.; İstanbul 1982).

A. Afetinan, *Pīrī Reis'in Hayatı ve Eserleri* (Ankara 1974).

Cevdet Memduh Alpar, *Onbeşinci Yüzyıldan Bu Yana Türk ve Batı Kültürlerinin Karşılıklı Etkileme Güçleri Üstüne Bir İnceleme* (Ankara 1981).

Michael Dools, 'Medicine in Sixteenth-Century Egypt,' Ekmeleddin İhsanoğlu ed.,*Transfer of Modern Science & Technology to the Muslim World* (IRCICA; İstanbul 1992).

Evliyā Çelebi, *Evliyā Çelebi Seyahatnamesi* ed. Ahmed Cevdet, I (Dersaadet: İkdam Matbaası 1314).

Avram Galante, *Türkler ve Yahudiler* (2nd ed.; İstanbul 1947).

Idem, 'Medicins Juifs au service de la Turquie.' *Histoire des Juifs de Turquie* IX (İstanbul: ISIS, n.d.).

Thomas C. Goodrich, *The Ottoman Turks and the New World, A Study of Tarih-i Hind-i Garbi and Sixteenth Century Ottoman Americana* (Wiesbaden 1990).

Idem, 'The Earliest Ottoman Maritime Atlas-Walters Deniz Atlası,' *Archivum Ottomanicum* XI (1986 [1988]).

Fatma M. Göçek, *East Encounters West, France and Ottoman Empire in the Eighteenth Century* (New York & Oxford 1987).

R. Hillenbrand, 'Madrasa,' *The Encyclopaedia of Islam* [*EI*] V (new edi.; Leiden 1985) 1144-1145.

İbrāhīm Müteferrika, *Füyuzat-ı Mıknatısiye* (Dar el-Tıbaat el-Amire 1145 (1732)).

I

72

İhsanoğlu, Ekmeleddin. ed., *Catalogue of Islamic Medical Manuscripts in the Libraries of Turkey* (IRCICA; İstanbul 1984).

Idem, 'Development of the Ottoman Science,' (18th International Congress of History of Science, 1-9 August 1989; Hamburg).

Idem, 'Introduction of Western Science to the Ottoman World: A Case Study of Modern Astronomy,' *Transfer of Modern Science and Technology to the Muslim World* (İstanbul 1992).

Idem, 'Ottoman Science in the Classical Period and Early Contacts with European Science and Technology,' *Transfer of Modern Science and Technology to the Muslim World* (İstanbul 1992).

Idem, 'Ottoman and European Science, Science and Empire: European Expansion and Scientific Development,' Patrick Petitjean, Catherine Jami and Anne-Marie Moulin coord.,*Boston Studies in the Philosophy of Science* (Kluwer Academic Publishers B.V.; Dordrecht 1991).

Halil İnalcık, *The Ottoman Empire: The Classical Age 1300-1600* (New York 1973).

Idem, 'Mehmed II,' *İslâm Ansiklopedisi* [İA] VII (1972) 535.

Bernard Lewis, *The Muslim Discovery of Europe* (London 1982).

Kātip Çelebi, *Mizanu'l-Hakk fi İhtiyari'l-Ahakk (İstanbul 1280)* ed. O. Şaik Gökyay *(İstanbul 1972)*.

Luigi Ferdinando Marsigly, *Stato Militare dell'Imperio Ottomano = L'Etat Militaire de l'Empire Ottoman* (2 parts Amsterdam 1732; reprint; Graz 1972).

Wolfgang Mayer, *Topkapı Sarayı Müzesindeki Saatlerin Kataloğu* (n.p n.d.).

Rhoads Murphey, 'The Ottoman Attitude Towards the Adoption of Western Technology: the Role of the Efrenci Technicians in Civil and Military Applications,' *Contributions à l'histoire économique et sociale de l'Empire Ottoman (Collection TURCICA III)* (Louvain 1983).

V.J. Parry, 'Barud,' *EI* I (Leiden 1979).

Djurdjica Petrovic, 'Firearms in the Balkans on the Eve of and After the Ottoman Conquests of the Fourteenth and Fifteenth Centuries,' V.J. Parry & M.E. Yapp ed., *War, Technology and Society in the Middle East* (London 1975).

Pīrī Reis, *Kitâb-ı Bahriye* (Türk Tarihi Araştırma Kurumu; İstanbul 1935).

Ahsan Jan Qaisar, 'Response of Turkey and Other Asian Countries to European Clocks and Watches During 16th and 17th Centuries: A Comparative Study,' *First International Congress on the History of Turkish-Islamic Science and Technology, Proceedings* IV (İstanbul Teknik Üniversitesi 14-18 September 1981).

Gül Russell, 'The Owl and the Pussycat: The Process of Cultural Transmission in Anatomical Illustration,' *Transfer of Modern Science and Technology to the Muslim World* (İstanbul 1992).

Nil-Bedizel Zülfikar Sarı, 'The Paraselsusian Influence on Ottoman Medicine in the Seventeenth and Eighteenth Centuries,' *Transfer of Modern Science and Technology to the Muslim World* (İstanbul 1992).

Aydın Sayılı, *The Observatory in Islam* (Ankara 1960).

Seydī Ali Reis, *Muhit* (Nuruosmaniye Library, Ms. no. 2948, fols. 15b-17a).

Osman Necip Sipahioğlu, *Türkiye'de Jeomagnetizm Çalışmaları* (2nd ed.; n.p. 1985).

Sevim Tekeli, *The Clocks in Ottoman Empire in the 16th Century and Taqi al Din's 'The Brightest Stars For the Construction of the Mechanical Clocks'* (Ankara 1966).

Abbe Toderini, *De la Litterature des Turcs* II (Paris 1789).

D. Uçar, 'Mürsiyeli İbrahim'in 1461 Tarihli Haritası Hakkında Bir Araştırma,' *First International Congress on the History of Turkish-Islamic Science and Technology, Proceedings* III (İstanbul Teknik Üniversitesi 14-18 September 1981).

İsmail H. Uzunçarşılı, *Osmanlı Devletinin İlmiye Teşkilatı* (Türk Tarih Kurumu Yayınları, 2nd ed.; Ankara 1984).

II

THE INTRODUCTION OF WESTERN SCIENCE TO THE OTTOMAN WORLD: A CASE STUDY OF MODERN ASTRONOMY (1660-1860)

Historians of science in general acknowledge the role of Muslim astronomers of the Middle Ages in the birth of modern astronomy as a science. This new astronomy gradually developed when a small but significant part of the astronomical tradition of the Muslims was transmitted to Europe and flourished there as a result of the studies of such great scientists as Copernicus, Tycho Brahe, Galileo, Kepler, Descartes and Newton. The astronomy of Ptolemy-Aristotle was replaced by this new science. The contacts of Muslims with this new astronomy have not been examined until now; moreover, the way in which these contacts began has not been investigated sufficiently. How did this contact take place? What was the attitude of Muslim astronomers to the new astronomy? How was the heliocentric system of Copernicus, which caused such great disputes in Europe, received in the Islamic world? What was the manner and the level of the transmission of this new science of astronomy? Besides, how did Muslims accommodate this new Western astronomy during the two centuries of their first contacts with European science? What were the differences between Muslim attitudes towards Greek and Indian astronomy during the 7th and 8th centuries and their attitudes towards this new encounter which began in the 17th century?

Most of these questions still await answers and need to be researched. There is no modern study dealing with the introduction of this new astronomy to the Islamic world. Within the framework of the history of Ottoman science, however, very short references have been made to the subject in Adnan Adıvar's work *Osmanlı Türklerinde İlim,*[1] and the present author has treated this matter with respect to the relationship of religion and science in a paper entitled "Some Critical Notes on the Introduction of Modern Sciences to the Ottoman State and the Relation between Science and Religion up to the End of the Nineteenth Century".[2] This new study does not claim to answer all of the above questions.

---

* For the 'Note on Transliteration' see the last page of this paper.

[1] Adnan Adıvar, *Osmanlı Türklerinde İlim,* 4th edn. Edited by Aykut Kazancıgil and Sevim Tekeli. Istanbul, 1982. The first French edition of this work is *La Science chez les Turcs Ottomans,* Paris, 1939.

[2] For the English text of this paper first presented in the Vth CIEPO meeting in 1984 in Cambridge, see E. İhsanoğlu, "Some Critical Notes on the Introduction of Modern Sciences to the Ottoman State and the Relation between Science and Religion up to the End of the Nineteenth Century", *Turcica, variation IV,* Istanbul, 1987, pp. 235-251. For the expanded Turkish text see *Toplum ve Bilim,* 29/30, Summer-Spring, 1985, pp. 79-102.

Its purpose is to seek answers to some of them by presenting the preliminary results of investigations over several years on the basic original sources, to pose new questions and to draw the attention of colleagues to this field.

In this work, we examine the known astronomical works of the Ottoman period in more detail than previous researchers and make a critical evaluation thereof. On the other hand, in the course of the research on Ottoman scientific literature, previously-unknown works on astronomy and cosmography have been found and examined. Moreover, the originals of some of these works in European languages which were available to us have been examined. On the whole, in addition to its new findings, this research also contributed to a better understanding of the scientific activities in the Ottoman period and to a more accurate assessment of the first contacts between the Ottomans and Western science.

The first contact of the Ottomans with the concept and the theories of modern astronomy seems to have begun with the translation of the astronomical tables of the French astronomer Noel Durret in the 1660s. These contacts, which continued with the translation of Western literature on geography into Ottoman Turkish in the 17th and 18th centuries, lasted in the second half of the 18th century, again with the translation of French astronomical tables. Setting apart these technical works for the interest of narrow circles, Müteferrika with his supplement to Kātib Çelebi's *Cihannümā* (printed in 1732), and Ibrāhīm Hakkı of Erzurum in his work *Mārifetnāme* (written in 1757) began to popularize the new concepts of modern astronomy for the broad masses of readers. In the 1830s, new astronomical concepts and knowledge were introduced into the Ottoman educational system in a relatively detailed manner as a result of the efforts to modernize the educational program of the *Mühendishāne* thanks to the efforts of *Başhoca* Ishak Efendi. Later in mid-19th century when the harmony was recognized between Islam and the cosmological perception based on the new astronomy, which replaced the old astronomy, some hesitations advanced in the 18th century disappeared and agreement ensued.

In the following pages we shall present the materials in chronological order, starting with the evidence of the first contact and continuing until mid-19th century, relying on our analysis of the literature of astronomy and geography in manuscript and printed books in three centres of the Islamic lands, namely, Istanbul and Cairo, both in the Ottoman world, and Baku, where Ottoman influence was also felt.[**]

[**] Following the first publication of this article, books in the history of Ottoman scientific literature series began to be published. These are: 1. *Osmanlı Astronomi Literatürü Tarihi (History of Astronomy Literature During the Ottoman Period) (OALT)*, Prepared by E.İhsanoğlu, R. Şeşen, C. İzgi, C. Akpınar, İ. Fazlıoğlu, edited by E. İhsanoğlu, 2 vols, IRCICA: Istanbul 1997; 2. *Osmanlı Matematik Literatürü Tarihi (History of Mathematical Literature During the Ottoman Period) (OMLT)*, prepared by E. İhsanoğlu, R. Şeşen, C. İzgi, edited by E. İhsanoğlu, 2 vols., IRCICA: Istanbul 1999; 3. *Osmanlı Coğrafya Literatürü Tarihi (History of Geographical Literature During the Ottoman Period (OCLT)*, prepared by E. İhsanoğlu, R. Şeşen, M. S. Bekâr, G. Gündüz, A. Hamdi Furat, edited by E. İhsanoğlu, 2 vols., Istanbul 2000. For an overview of the development of Ottoman science, its institutions and literature see *History of the Ottoman State, Society & Civilisation*, ed. E. İhsanoğlu, vol. 1, Istanbul 2001; vol. 2, Istanbul 2002. See the chapters titled "Ottoman Educational and Scholarly-Scientific Institutions" and "The Ottoman Scientific-Scholarly Literature" by E. İhsanoğlu in volume 2 of this work.

The First Encounter:
Tezkireci Köse İbrāhīm of Szigetvar and *Sajanjal al-aflāk*

Among the works which previously escaped attention in the search for the sources of history of Ottoman science, and about which there is scarce information, is Tezkireci Köse İbrāhīm Efendi's[3] translation of the book titled *Nouvelle Théorie des Planetes* by the French astronomer Noel Durret (d. ca. 1650).[4]

Illus. 1.) Cover page of Noel Durret's work entitled
Nouvelle Théorie des Planetes which was printed in Paris in 1635.

[3] Köse İbrāhīm Efendi of Szigetvar, whose original profession consisted of writing official memoranda, at the same time occupied himself with astronomy. We learn from the introduction of the *Sajanjal,* that he translated from French, and that he came to Istanbul where he settled and established friendship with Müneccimbaşı (chief astronomer) Müneccimek Efendi. Again we learn from the introduction that he compiled another work himself apart from the *Sajanjal.* Tezkireci mentions this work, saying: "For the proofs I compiled a different and new *risāla* with all its operations which is easier than *Almagest,* as well as compiling a work for ephemerides that are used internationally which came out to be more graceful and succinct than all" (fol. 2a). Bursalı Tāhir Bey cites an astronomer by the name of İbrāhīm b. Mehmed in *Osmanlı Müellifleri (OM),* III, 253. Tāhir Bey saw the astrolabe which was designed and manufactured in the Köprülü Mehmed Paşa *medrese* in Istanbul in 1098/1686 by this scholar in the Melāmī Tekke in Skopje. The fact that this İbrāhīm Efendi and Tezkireci Köse İbrāhīm of Szigetvar were contemporaries and colleagues calls to mind this question: Whether these two İbrāhīms could possibly be the same person.

[4] The death date of Noel Durret is not stated definitely in the sources. *Biographie Universelle* gives it as 'ca. 1650'. We know that Durret was alive in 1647 from a work which he published in this year. On the basis of this fact, it is highly possible that he might have died in the year 1648 or 1649 (*Biographie Universelle,* Paris 1814, XII, pp. 371-372; Jérôme de Lalande, *Bibliographie Astronomique avec l'Histoire de l'Astronomie depuis 1781-jusqu'à 1802,* Paris 1803, pp. 205-206, 208, 210, 212, 224; *Dictionary of Scientific Biography,* VII, . 308a). For a list of Durret's works see *Bibliographie Astronomique* by Lalande and *Biographie Universelle* (the pages cited above).

Tezkireci, a translator originally from Szigetvar (today in Hungary) who settled in Istanbul, translated this work in the period 1660-1664 under the name of *Sajanjal al-aflāk fī ghāyat al-idrāk* (The Mirror of the Heavens and the Limit of Perception). This is the first book which treats the Copernican system in the literature of Ottoman science; it also contains the first diagram illustrating this system. (See illustration 2).

According to what Tezkireci Köse Ibrāhīm relates in the introduction of the work, he had the original of this work brought to him, translated it into Arabic and showed this translation to Müneccimek Şekībī Mehmed Çelebi (d. 1078/1667), the chief astronomer *(başmüneccim)* in Istanbul at the time. At first Müneccimek did not approve of the work, saying: "Europeans have many vanities similar to this one." But when Tezkireci Köse prepared an ephemeris using the French tables and saw that it was in conformity with Ulugh Beg's Tables, Müneccimek came to appreciate the work, copied it for himself and bestowed upon the translator a benefaction. Afterwards, in 1663, when Köprülü Fāzıl Ahmed Paşa went on the Uyvar campaign - Uyvar is the city of Nové Zamky in present day Slovakia - Tezkireci Köse joined

Illus.2) Illustrations of the systems of Copernicus, Ptolemy and Tycho Brahe, starting from the bottom, in the *Sajanjal al-aflāk fī ghāyat al-idrāk*, (The Library of Kandilli Observatory, nr. 403, fols. 1b-2a).

him and during the stay of the army at the winter quarters in Belgrade, he worked on this translation again upon the encouragement of *Kādīasker* (chief judge) Ünsī Efendi (d. 1075/1664). Tezkireci thus re-examined all the solar, lunar and planetary mean motions of the *zīj* which was compiled according to the meridian of Paris in conformity with the sexagesimal system; he abbreviated the tables and arranged them according to the signs of the zodiac *(abrāj)*. He entitled the work *Sajanjal al-aflāk fī ghāyat al-idrāk* and gave a copy of it to *Kādīasker* Ünsī Efendi. Later, upon the wish of some of his friends, he translated the introduction of the work from Arabic into Turkish, with a few explanations left in Arabic. He thus gave the work its final form. In the introduction, after a brief account of the history of astronomy, the author presents explanations

4

related to astronomy and the book. These explanations, arranged in twenty-four subchapters *(ta'līm)*, are followed by tables.[5]

The Introduction of the Sajanjal:[6] "The ancients would follow the motions of the stars not by means of *zījes* but through observations. For the first time Hipparchus compiled a *zīj* in 140 B.C. which shows the ephemerides of the Sun and the Moon. This *zīj*, however, which did not follow the method of the previous ones, did not give the ephemerides of the other stars, and at the same time had numerous mistakes of observation, was used for 285 years; after this period, Ptolemy of Alexandria compiled a new *zīj* in 120 A.D. by observing the motions of the stars. This *zīj* was used for 880 years. Al-Battānī, who realized that the observations of Ptolemy did not reflect the truth, compiled another *zīj*; and this *zīj* was used for about hundred years. Al-Zarqālī[7] saw that the motions of the Sun were intolerable discordant with al-Battānī's observations. The incompatibility between the two observations could not be reconciled. While this science was long forgotten, Alfonso,[8] the King of Spain, took an interest in astronomy, invited and convened master astronomers from all over the world and established an observatory in Toledo in 1251 spending 400,000 gold coins. The *zīj* which he had prepared was known as "Alfonso's *zīj*" and was used by Christians for about 200 years.

In 1461 the German scholars Peurbach[9] and Regiomontanus[10] found the mistakes in Alfonso's *Tables*. Although Regiomontanus began his observations in order to correct the *zīj*, he did not live long enough to finish his work. A few years later Nicolaus Copernicus, who was very successful and superior, found the mistakes in Alfonso's *Tables* and realizing that their foundation was unsound, he found a new solution in the year 1520.[11] He established this by depending upon the observations of astronomers at different times. The defect in question was the following: the eighth sphere moves with the ninth sphere by an oscillation *(raqqāṣiyya)* movement in 49,000 years. This motion is equal to two small circles moving from West to East in the depth of the ninth sphere. These two circles which come into being on the equinox of the ninth sphere have half a diameter of $9°$ each. This situation is against the observations of the majority. Later, Copernicus laid a new foundation and compiled a small *zīj* supposing that

---

[5] *OALT*, I, pp. 340-345.
[6] The text which is quoted here is essentially based on the manuscript in the Library of Kandilli Observatory, no. 403, fols. 1b-2a.
[7] The orthography of the name has been distorted during translation and transliteration. For his life see *Dictionary of Scientific Biography*, XIV, pp. 592b-595a.
[8] King Alfonso X of Castille (reg. 1252-1282). For his life see *Biographie Universelle*, I, pp. 619a-620b.
[9] Georg Peurbach (d. 1461). For his life see *Dictionary of Scientific Biography*, XV, pp. 473b-479a.
[10] Johannes Regiomontanus (d. 1476). For his life see *Dictionary of Scientific Biography*, XI, pp. 348b-352b; Salih Zeki, *Kāmūs-ı Riyāziyāt*, Library of Istanbul University, TY 914, VII, fol. 652; for more extensive information about him also see Lyyn Thorndike, *A History of Magic and Experimental Science*, V, New York, 1941, pp. 332-377.
[11] Given as 1425 in the Ottoman text.

the Earth is in motion. This *zīj* was used after him until the time of Tycho Brahe for approximately (hundred) years.[12]

Subsequently, Tycho Brahe, one of the notables of Dayne,[13] made observations with numerous excellent instruments and began to correct the *zīj* of Copernicus.[14] Depending upon his observations he would also write the ephemerides of the Sun, the Moon and the fixed stars. At this time, however, the campaign in Bohemia took place. He wanted to have the draft manuscript of his *zīj* printed, but his early death prevented him from the publication. Finally, a contemporary of Tycho from the province of Dacia[15] by the name of Longomontanus[16] compiled a *zīj* with few mistakes which was similar to the *zīj* of Tycho Brahe.

Afterwards, the scholar Kepler,[17] who worked in the company of the Spanish King Rudolf, compiled a *zīj* for all stars based on Tycho's observations and named it *Tabula Rudolphina*.[18] As he also mentioned, this *zīj* did not agree completely with the observations since the locations of the stars which Ptolemy observed were not the same as the locations in this *zīj*. The eclipses of the Sun and the Moon did not conform to this *zīj*, either. Later I, Ibrāhīm el-Zigetvarī, known as Tezkireci, translated the *zīj* which the scientist Durret compiled in thirty years of observation based on the *zīj* of Lansberge.[19] It was compiled by benefiting from the old Julian and the new Gregorian *zīj*es as well, accepting the Earth as motionless. It was difficult to use since it contained numerous printing errors. I showed the translation to Müneccimbaşı Müneccimek Mehmed Efendi.[20] After examining the work quite well and not having understood anything, he said: "Europeans have many vanities similar to this one." I smiled and showed him how to use it; he was very pleased to apply these [French] calendars to the tables of Ulugh Bey and other *zīj*es. When he later copied the translation, he bestowed upon us a benefaction as much as the treasures of Egypt and prayed much saying: "You saved me from suspicion. Now I have full

---

[12] It appears that there must be a number in the text which was rubbed out and is now illegible. Considering the death dates of Copernicus and Tycho, however, which were 1543 and 1601, respectively, the period during which this *zīj* has been used is approximately sixty years. Copernicus compiled his *zīj* in 1520. Tycho went to Prague in 1588 and conducted his observations there in the company of King Rudolph II. According to this calculation, the *zīj* of Copernicus was used more than sixty years. Depending on this evidence, we can say that the *zīj* of Copernicus has been used at least half a century.

[13] This name is spelled as "Dayne" in the text.

[14] The name appears as "Peronika" in the text.

[15] Longomontanus was born in the city of Longberg in Denmark.

[16] Severin, Christian (d. 1647). For his life see *Dictionary of Scientific Biography*, XII, pp. 332a-332b; *Larousse du XXe Siècle*, V, pp. 510c-511a, *Kāmūs-ı Riyāziyāt*, Library of Istanbul University, TY 916, IX, fol. 1326.

[17] In the text "Kepelaryus" is written as it is read.

[18] See *Dictionary of Scientific Biography*, VII, pp. 304b-306a.

[19] Philip Van Lansberge (d. 1632): Belgian astronomer and mathematician. For his life see *Dictionary of Scientific Biography*, VIII, pp. 27b-28b; *Biographie Universelle*, XXIII, pp. 468-474.

[20] He became *müneccimbaşı* (chief astronomer) in the year 1071/1661 and died in 1078/1668 while in office. He is also known as Şekîbî Mehmed Efendi. (See Mehmed Süreyya, *Sicill-i Osmanī (SO)*, IV, p. 177; Şeyhī, *Vekāyi el-Fuzelā*, Süleymaniye Library, Hacı Beşir Ağa, no. 479, fol. 528a). Salim Aydüz "Osmanlı Devleti'nde Müneccimbaşılık ve Müneccimbaşılar",

confidence in our *zīj*es." At that time the Nemçe (Austrian) campaign, known also as the Uyvar campaign,[21] took place. When we arrived at the winter quarters of Belgrade on our way, I worked out all the mean positions[22] of the *zīj* again according to the evidence and rearranged its shape on the same longitude. I expounded it in Arabic. This *zīj* was sexagesimal[23] and its mean positions were universal; I arranged it according to the signs of the zodiac. I turned the tables into a *zīj* hitherto unseen in shape and brevity, entitled it *Sajanjal al-aflāk fī ghāyat al-idrāk* and presented it to *Kādīasker* Ünsī Efendi[24] whose great support resulted in the compilation of this *zīj*. Again upon the encouragement of some friends, I arranged the *zīj* in question in this framework and rendered its explanation in Turkish. The ancients found many instruments and made observations in order to observe the motions of the stars correctly. But now an arc of altitude rectified to 90° and a pair of compasses would suffice for those who have a deep knowledge of the mathematical sciences. And for the proofs I compiled a different and new *risāla (Risāla-i gharība wa mubtakara)* which was easier than the *Almagest* in all its operations. I also compiled a work on calendars to be used by nations which came out to be more graceful and succinct than all calendars. The mentioned observer applied his *zīj* to the *zīj*es of all these observers and found it appropriate."

University of Istanbul, Institute of Social Sciences, Department of History of Science, Master's Thesis, Istanbul, 1993.

[21] This took place between the Ottomans and Austrians in the year 1663. The commander-in-chief on the Ottoman side was Köprülü Fazıl Ahmed Paşa (d. 1087/1676). (See *İslâm Ansiklopedisi (İA)*, VI, Istanbul, 1977, pp. 898-899).

[22] Tables of mean motions of the Sun, Moon and planets were standard in Islamic *zīj*es. See the article "zīdj" in *The Encyclopaedia of Islam*.

[23] *Sittīnī* (Sexagesimal): "A fraction whose denominator consists of a power of the number 60 is called a 'fraction of sixty'. The fractions of 2/60, 4/360, 25/21600 are all fractions of sixty... *Sittīnī* division is based on divisions by powers of the number 60. Accordingly an hour was divided into 60 minutes, a minute into 60 seconds, and a second into 60; again, similarly, the angle of 360° was equivalent to the same number of minutes divided each into 60 seconds." (See *Kamūs-ı Riyāziyāt*, Library of Istanbul University, TY 914, VII, fol. 760). "The astronomers usually used fractions with a denominator of sixty in their calculations similar to the Babylonians and Greeks in the ancient periods; e.g., 35/8 was written in the form of fractions divided into sixty as = 3 degrees 37 minutes 30 seconds (3° 37' 30"), namely = 3 + 37/60 + 3/3600" (see *İA*, V/I, Istanbul, 1964, p. 445).

[24] Ünsī Efendi, whose actual name was Abdüllatīf, was born in Kütahya. After serving in numerous positions as *kadı*, he was charged with the position of *kadıasker* (chief judge) when Köprülüzāde Fāzıl Ahmed Paşa was appointed *serasker* (commander in chief) during the Uyvar campaign in Ramadān of the year 1073/1663. Abdüllatīf, a poet and a writer in three languages (Turkish, Arabic, Persian), who used the pseudonym of Ünsī in his poems, passed away on the 16th Saturday of Jumāda I of the year 1075/1664. (For his life see Şeyhī, *Vekāyi el-Fuzelā*, fols. 241b-242b; Üşşākīzāde, *Zeyl-i Şakāik*, ed. Hans Joachim Kissling, Wiesbaden 1965, pp. 293-303; *SO*, III, p. 360). Muhammed ibn Abdullah ibn Ahmed ibn Kāsım el-Hüseynī el-Musavī, who is known by the nickname Kibrit, has a work entitled *al-Wasā'il al-Qudsīyya ila'l-Masā'il al-Unsīyya* which essentially deals with astrology. The author composed the work at the end of Dhilhijja of the year 1041/1632 in Istanbul. The present copy of the work was compiled from the copy which the author wrote. [Süleymaniye Library, Esad Ef., no. 1418/5 (fols. 28a-38a)]. Kibrit wrote this work upon the encouragement of Ebū Saʿīd Muhammed b. Esʿad. He presented it to Abdüllatīf Hüseyin b. Abbās, known as Ünsī, whom he refers to as "our Şeyh and Seyyid". This source gives Ünsī's real name and his father's name. Although in the biographical memoirs Ünsī's proper name is cited as Abdüllatīf, it appears that this was his nickname.

Later around the year 1094/1683, Cezmī Efendi (d. 1104/1692),[25] *kādī* of Belgrade, found a copy of *Sajanjal* which was probably given to Ünsī Efendi. He re-examined the work and prepared another edition of it. Cezmī Efendi briefly says the following in his introduction to the work:

"I learned this *zīj* from the experts and calculated its calendar. I explained the mean motions of the Sun, Moon and planets again. Nevertheless, I still had a suspicion about the accuracy of the work. With God's grace, I had attained the superiority of having become the sweeper of the library of the head of the most-respected masters *(Cārūbkash-i makhādim-i dhawī'l-takrīm).*[26] I took his permission to examine the worn-out book, and as I was scrutinizing the pages, I came across *al-Zīj al-Ḥākimī* which ʿAlī ibn Yūnus compiled for the Fatimid caliph al-Ḥākim in the 372 Yazdigird year[27] according to the longitude of 55°.[28] And I realised that this *zīj* was the source of *Zīj-i Frengī*, that the Europeans took over this *zīj* and applied it to the longitude of their countries, and that they only supplemented the subject of "the proof of non-uniform movement for the equinox" to this *zīj*. I came to the conclusion that the *Zīj* of Ibn Yūnus was the source of *Zīj-i Frengī*. Thus, I trusted in its correctness. In the meantime, I also obtained the copy of Müneccimek and made the necessary corrections. And I decided to make a neat copy of my work. The difference between this *zīj* and the calendars based on Ulugh Bey's *Zīj*, however, detained me from copying the book. At this time, the conjunction of the planets of Jupiter and Saturn

---

[25] He was Mehmed Abdülkerīm Efendi, the son of Azīz Efendizāde Yaḥyā Efendi who was the brother of the poet Bahāī Efendi, one of the *şeyhülislām*s of the reign of Mehmed IV. After serving in various positions, he was appointed *kadı* of Belgrade in Muḥarram of the year 1093/1682. He died on his way to Diyarbekir where he was appointed to the position of *kādī* in the year 1104/1692. (See Kādīasker Sālim Efendi, *Tezkire*, Istanbul 1315 (1897), pp. 200-202; İsmāil Belīğ, *Nukhbat al-Āthār li Dhayl Zubdat al-Ash'ār*, ed. Abdulkerim Abdulkadiroğlu, Ankara 1985, pp. 70-72; Uşşākīzāde, *Zeyl-i Şakāik*, pp. 664-665; *SO*, III, p. 354). See also *OALT*, I, 357.

[26] The person indicated here is Köprülü Fāzıl Ahmed Paşa. The library which is mentioned is the Köprülü Library, and the worn out book cited must be the work by ʿAlī ibn Yūnus (d. 400/1009) entitled *al-Zīj al-Kabīr al-Ḥākimī*. The author began to write his work upon the will of al-Azīz of the Fatimids around 380/990 and completed it shortly before his death during the reign of al-Ḥākim, the son of al-Azīz. Ibn Yūnus made his observations in the observatory at the mountain of Muqattam. These tables, which are inevitably far from complete, exist in pieces in Leiden, Oxford, Paris, Escurial, Berlin and Cairo (*IA*, V/II, Istanbul, 1967, p. 837a). Another copy of this *zīj* is in Hacı Selim Ağa Library, no. 728/2 (fols. 38b-181b). (See Ramazan Şeşen, ed. *Nawādir al-Makhṭūṭāt*, I, 205). No such book is extant in Köprülü Library today. It may be that this copy is the one which is in Hacı Selim Ağa Library at present. The record of possession on folio 38a of the copy indicates that this was probably the copy which came into Tezkireci Köse İbrāhīm Efendi's possession since it bears the record of possession of Mustafa Feyzī, the *masraf kātibi* of Hacı Ahmed Paşa of the Köprülü family. According to the register, the work came into Mustafa Feyzī's possession in 1162 (1749) in Egypt:

دخل إلى نوبة المذنب مصطفى فيضي كاتب مصرف صدر اسبق

الحاج احمد باشا يسر الله له ما يشاء سنة ١١٢٦ در مصر

[27] Yazdjard (Yazdgard) III, the last Sassanid ruler, was Shahriyār's son and the grandson of Khusraw Parwīz. He had a calendar prepared accepting as its beginning the date of his accession to the throne which was 16 June 632. The 372nd year of this calendar coincides with the 1004th year of the Christian era. About this calendar see *Biographie Universelle*, XXI, p. 175.

[28] This is the longitude of Cairo used by Ibn Yūnus. For Ibn Yūnus, see *Dictionary of Scientific Biography*, XIII, pp. 574-580.

which, according to the calculations of calendarists, would take place in 1094 H., was reckoned to take place on the last day of Jumāda II of the same year in the calendar prepared on the basis of Ulugh Bey's *Zīj*. The observations which were made showed that, however, this conjunction took place thirty-five days earlier. At this time it came out that the ephemeris which was prepared on the basis of Durret's *Zīj* and the conjunction of Jupiter and Saturn were in agreement. Thereupon I made a neat copy of the translation."

In the introduction of this edition, after touching upon the *zījes* of Tycho Brahe, Kepler and Lansberge, Cezmī Efendi deals with Durret's *Zīj* and, having explained the above-mentioned subjects, he completes his introduction. From this point of view, the introductions of the two editions are very different.

Besides, Cezmī Efendi supplemented a section to his work following the introduction under the name of "mu'āmarāt-ı kawākib", dealing with six subjects. Only the paragraph on the Sun entitled "mu'āmara-i āfitāb" exists in the manuscript copy of Tezkireci Köse. It appears that this paragraph was added by Cezmī Efendi.[29]

Apart from these, Cezmī Efendi has added to his work the subject of "the proof of non-uniform movement for the equinox" from Müneccimek's copy. The same subject, however, is treated in Tezkireci Köse's work as well with minor differences. In Cezmī Efendi's copy there is also a subject entitled "the manner of making a solar ephemeris" immediately preceding the tables. It appears that this part was supplemented by some other person. Accordingly, Cezmī Efendi's supplement consists of the new introduction and the subjects under the heading of "mu'āmarat-ı kawākib." Apparently Cezmī Efendi intended to appropriate the translation to himself without mentioning Tezkireci Köse's name. A note in the postscript of fol. 1 b of *Zīj-i Frengī's* manuscript copies of Kandilli and Hazine supports this:

"The present *zīj* originally belongs to the scholar by the name of Durret. The late Tezkireci Köse Ibrāhīm has translated it and says the following in his translation:

"The original of this *zīj* was compiled in line with the sexagesimal system. I arranged it according to the signs of the zodiac. And I gave it to the late Ünsī Efendi who was *kādī'l-muasker* during the period of the late Fāzıl Ahmed Paşa in the winter quarters of Belgrade. In the present copy, the account between the words 'lākin' and 'mu'āmara' are inserted later. It is possible that Cezmī Efendi set himself to adopt the translated work or to compile a kind of translation, and since one part of the work was opposed to the other part, he supposedly tried to conceal his fault by striving to mislead the reader in his argument."

In addition, the two copies of *Zīj-i Frengī* exhibit some differences. Furthermore, in different parts of *Sajanjal al-aflāk* there are notes of Cezmī Efendi, Feyzī and Mustafa Tālib (d. 1209/1794) related to the subject, and the

---

[29] After this come the *ta'līms* from among the *ta'līms* in the copy of Tezkireci Köse numbered 24, 23, 1, 2, 3, 4, 5, 6, 7, 8, 9 and 12. There are also two unnumbered *ta'līms* between 23 and 24 which exist in Tezkireci's copy as well. The other *ta'līms* do not exist in Cezmī Efendi's copy.

notes of various persons about the historical events between the years 1099-1106 H., that is, 1688-1694-95. It appears that the draft manuscripts have been copied without being organized in a coherent and consistent manner. For this reason, perhaps the *zīj* was not used for making ephemerides later. It should be compared with the copy of Durret's *Zīj*.

The Ottoman astronomers were aware of the progress made by the European astronomers during the Renaissance and the scientific revolution, as well as the new theory of Copernicus, Tycho Brahe's observations and other parallel studies; meanwhile their attitude regarding the first contacts with European astronomy, as in the above-given summary, can be evaluated as such: the Ottomans were able to follow the developments in Europe without a great gap in time. A book that was published in 1635 in Paris could be read with a critical view in Istanbul in the years 1660 or 1661 and would be first considered as a "European vanity" by some. The superiority of the West in the area of science was not immediately and definitely accepted. However, when the data given in this European *zīj* was found to be consistent with the datum of *zīj*es prepared by Muslim astronomers, the new European science was accepted and acknowledged.

Secondly, the Ottoman astronomers who believed in the use of the theory and applications of this new astronomy, did not accept the new information immediately and with enthusiasm; in their professional maturity, they were aware of the contributions of Muslim astronomers to this field throughout the Middle Ages and the European astronomers' debt to them.

Thirdly, the matter of the Sun's location at the centre of the universe and the Earth's motion - the basic elements of this new understanding of astronomy introduced by Copernicus which caused great controversy in Europe – was treated on the level of a secondary technical detail by the Ottoman astronomers. One of the reasons for this indifference may be the absence of a religious dogma which would cause disputes among Muslim astronomers on the subject of the geocentric and the heliocentric systems. Other factors could be that these debates in Europe were not transmitted to the Ottoman society and the absence of an environment that would foster such debates. Finally, this first translation indicates the aspects of Western astronomy which attracted the Ottoman astronomers' interest. In the following pages of this paper we shall observe the development of this trend.

Abū Bakr Efendi and the Translation of the *Atlas Major*

The second known work treating modern astronomy in Ottoman scientific literature was prepared by Abū Bakr ibn Behrām ibn ʿAbd Allāh al-Ḥanafī al-Dimashqī[30] under the name of *Nuṣrat al-Islām wa'l-surūr fī taḥrīr*

---

[30] Abū Bakr Efendi was born in Damascus, and for this reason he is known as al-Dimashqī. Having completed his education, he came to Istanbul and became attached to the Grand Vezir Köprülü Fāzıl Ahmed Paşa, then appointed as the assistant functionary *(mülāzım)* of Şeyh Mehmed Izzetī. Until the year 1099/1688, he held the position of *müderris* in various *medreses*. His rank was promoted to *hāmise-i Süleymaniye* in Ṣafar of this year. Again, in Jumāda I of this year, he was appointed to one of the *medreses* of Süleymaniye. In Jumāda I of the year 1101/1690, he was appointed *kadı* of Aleppo in place of Hocazāde Lütfullah. He was discharged from this position in Jumāda I of the year 1102/1691. El-Mevlā Idris was appointed in this place, and he died in Jumāda II of the same year. Through the mediation of Fāzıl Ahmed Paşa, he was

Atlas Mayor (The Victory of Islam and the Joy of Editing the *Atlas Major*). It was based on Janszoon Blaeu's work in Latin known as the *Atlas Major*. The word *taḥrīr* (edition) is used in the title of the work instead of "translation", which indicates that the book is a free translation and that there is room for al-Dimashqī's observations and thoughts in the text as well. Indeed, the author has abbreviated many passages of the text.

Janszoon Blaeu's geographical work of eleven volumes entitled *Atlas Major seu Cosmographia Blaeuiana qua solum coleum accuratissime describuntur*[31] was presented to Sultan Mehmed IV by Justin Collier, the ambassador of Holland in Istanbul in 1668. When Mehmed IV ordered the translation of the book, al-Dimashqī began his work in 1675 and completed the translation in 1685.

The first volume of the nine-volume copy of the translation of the *Atlas Major* in the Topkapı Palace Museum Library[32] contains general information on geography and cosmography as well as several maps; the volumes two to five are on European countries, volume six is on the African continent, volume seven on Italy, volume eight on China, and volume nine deals with America and the islands.

In the introduction of the work, after stating the importance of and the necessity for the science of astronomy, the translator Abū Bakr Efendi gives some information on the state of astronomy as a science in the Islamic world and on the views of the Europeans concerning this subject. He rejects the claims of European scholars to the effect that after Naṣīr-i Ṭūsī, Fakhr-i Rāzī, Niẓām (al-Nisāpūrī) and Ali Kuşçu, the science of astronomy declined among Muslims to such an extent that nobody knew its name. He points out that there were numerous scholars in Islamic lands well-informed in astronomy; in particular, his teachers with whom he studied mathematics were unmatched. He states that many people are studying the rational sciences- particularly astronomy and geometry- but they are more occupied with the theoretical aspect of these sciences and by no means know the application, for in the past someone who knew astronomy would also know geography and astronomical observation. He

---

acquainted with Sultan Mehmed IV. Upon the sultan's order, he was entrusted with translating *Atlas Major* (Şeyhī, *Vekāyi el-Fuzelā*, ed. Abdülkadir Özcan, Istanbul, 1989, II, p. 33; al-Murādī, *Silk al-Durar*, I, pp. 50-51). Although it is known that Abū Bakr Efendi wrote a work entitled *Jawalān al-Afkār fī Awālīm al-Aqtār*, where he treated the views of new astronomy extensively, no copy of it has been found (see *OALT*, I, 355-356; Adıvar, p. 155). al-Dimashqī, who was particularly knowledgeable in mathematics, had a good knowledge of Latin. From his words in the introduction of a copy of his work *Coğrafya-i Atlas (Atlas Geographicus)*, it appears that when he was working on this translation other translators assisted him. (See University of Bologna Library, nr. 3608). For Abū Bakr Efendi and his work see *SO*, I, p. 174 ff.; *OM*, III, p. 315, note 3; Franz Babinger, *Die Geschichtschreiber der Osmanen und ihre Werke;* trans. Coşkun Üçok, *Osmanlı Tarih Yazarları ve Eserleri*, Kültür ve Turizm Bakanlığı, Ankara, 1982, pp. 248-250.

[31] Wilhelm Janszoon Blaeu (d. 1638): Dutch cartographer. His *Atlas Major* was printed in Amsterdam in 1634 for the first time under the title of *Novus Atlas*. This work was printed in Amsterdam between the years 1635-1645 in four volumes under the title of *Tonneel des Aerdrycks*. Subsequently, it was reprinted in Amsterdam between the years 1662-1665 in volumes varying between nine and twelve in number under the titles of *Atlas Major, Le Grand Atlas*, and *Grooten Atlas*. After his death, his profession was carried on by his sons Joan and Cornelis. For his life see *Dictionary of Scientific Biography*, II, pp. 185-186.

says: "Although Kātib Çelebi [1609-1657] worked at this science a little, he could not produce a complete work." This, he says, was the reason the Sultan ordered him to prepare a work on this subject.

Again in the introduction, al-Dimashqī says that the sciences which describe the states of existence are divided into two: one of them is astronomy, whose subject matter is the state of the heavens; the other is geography, whose subject matter is the state of the Earth. He states that owing to this science, the infidels learned many things and, predominating the countries of the world, they excelled in disturbing Muslims.

The chapter of the translation entitled "The Centre of the Universe", where the systems of the universe are mentioned, briefly introduces the systems of Ptolemy, Copernicus, Tycho Brahe and Andreas Argoli. When compared to the Latin original, the information given is very brief. Blaeu states that the geocentric system is correct and the others which are contrary to the Bible ("Sacrae Scripturae" in the original text) are wrong. Al-Dimashqī, however, does not relate Blaeu's statement that the theory of Ptolemy has been considered appropriate by everyone for centuries and is still valid.

In the same chapter al-Dimashqī summarizes in one paragraph the three pages of information in the Latin original of the work on the four systems of Ptolemy, Copernicus, Tycho Brahe and Andreas Argoli.[33] Al-Dimashqī says the following on the views of the scholars of astronomy about the order of the universe:

"They have disputed much about the centre of the universe. That is to say, many of the old and the new scholars said: 'The centre of the universe is the Earth. It is stationary and motionless. The other elements and the planets are around the Earth.' This is what the Bible says. Pythagoras, Ptolemy and their followers held this view. (Illustrations 3 and 4 show the Ptolemaic system as represented in the original Latin edition of the *Atlas Major* and in the Turkish translation of this work.) Others said: 'The centre of the universe is the Sun. It is stationary and motionless. The other elements and the planets are moving...' (Illustrations 5 and 6 depict the Copernican system in the original Latin edition of the *Atlas Major* and in the Turkish translation.) According to another view, the Earth is at the centre of the universe. The Sun, the Moon and the planets are in another location. The proponents of yet another view say that the Earth is at the centre, following it are the Crystalline, the fixed stars, and after those come Saturn, Jupiter, Mars, the Moon, Air, Water and Fire. The Sun, Mercury and Venus are in another location. In short, the most accurate of these views is the first one. The others are wrong since they are contrary to the Bible."[34]

We conclude the following after examining this work:

Since no comprehensive work on geography had yet been produced in Ottoman Turkish, in the second half of the 17th century Sultan Mehmed IV ordered the translation of the *Atlas Major*, a more extensive work than Kātib

---

[32] Topkapı Palace Museum Library, Bağdat, nos. 325-333.
[33] Andreas Argoli (d. 1657): Italian astrologer and astronomer of the 17th century. Among his works, which have been printed many times, the well-known ones are *Astronomicorum libri Tres* and *Pandosion Sphaericum*. See *Dictionary of Scientific Biography*, I, pp. 244a-245a.
[34] Topkapı Palace Museum Library, Bağdat, no. 325, fol. 3a.

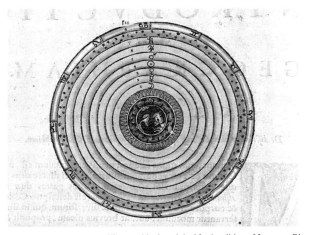

Illus. 3) The Ptolemaic system as illustrated in the original Latin edition of Janszoon Blaeu's work known as the *Atlas Major*. (Topkapı Palace Museum Library, Hazine 1423, p. 1).

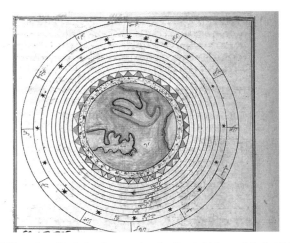

Illus. 4) Illustration of the Ptolemaic system from the *Nuṣrat al-Islām wa'l-Surūr fī Taḥrīr Atlas Mayor*, the Turkish translation of the *Atlas Major* by al-Dimashqī. (Topkapı Palace Museum Library, Bağdat 325, fol. 3a).

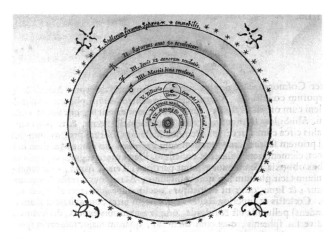

Illus. 5. Illustration of the Copernican system from the *Atlas Major* in Latin. (Topkapı Palace Museum Library, Hazine 1423, p. 2).

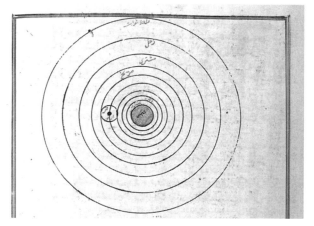

Illus. 6. The Copernican system as illustrated in the *Nuṣrat al-Islām wa'l-Surūr fī Taḥrīr Atlas Mayor*, al-Dimashqī's Turkish translation of the *Atlas Major*. (Topkapı Palace Museum Library, Bağdat 325, fol. 3b).

Çelebi's *Cihannümā* which was presented to him. As for the translator, it appears that in the introduction of the work which he would present to the sultan, he makes statements verifying and stressing the importance of the science of astronomy from different aspects, its necessity, and its compatibility with the Holy Qur'ān. It is especially noteworthy that the translator expresses his apprehension in the face of the claims of Europeans to the effect that Islamic science is lagging behind European science and that he answers the claims in question by pointing to the presence of the scholars also in the Islamic world who are well-informed in the sciences. At the same time, however, Abū Bakr Efendi admits that Islamic science regressed to the theoretical level since it receded from the information of geography and astronomical observation gradually and he states that Islamic astronomy fell behind the developments in Europe.

Müteferrika and the "New Astronomy"

The two 17th century works examined above introduce for the first time the concepts and the principles of the new astronomy in different ways, but in general very shortly and indirectly. Following these works, we come across a text which contains more extensive information on the subject in the first half of the 18th century. Ibrāhīm Müteferrika (d. 1158/1745),[35] Ottoman statesman, diplomat and a vanguard of reform policy, made a major contribution to the Ottoman cultural and intellectual life by establishing the first Ottoman printing press. Among the books which he printed was Kātib Çelebi's well-known work *Cihannümā*. (Kātib Çelebi began to write the first version of this in 1648; he began the second in 1654. The basis for the text published by the Müteferrika Press in 1732 was this second version.) The chapter which Ibrāhīm Müteferrika

---

[35] For Müteferrika's life see Niyazi Berkes, "Ibrahīm Müteferrika" in *EI²*, III, Leiden, 1979, pp. 996-998; *idem*, "İlk Türk Matbaa Kurucusunun Dini ve Fikri Kimliği", *Belleten* XXVI/104 (1962), pp. 715-737; Adıvar, pp. 166-174. Ibrāhīm Müteferrika was born in Transylvania (Erdel) of Christian parents probably between the years 1670 and 1674. Relying upon Ibrāhīm's brief autobiographical account in his *Risāle-i İslāmiye* (MS Istanbul, Süleymaniye Library, Esad Efendi 1187), Niyazi Berkes states that Ibrāhīm, who has been believed to be a Calvinist on the basis of the accounts of some Europeans, was in fact a Unitarian. Again, according to Berkes, contrary to the traditional account of Ibrāhīm's conversion to Islam, which states that he was taken prisoner by the Turks and was converted perforce, Ibrāhīm must have fled from the Catholic Habsburg rule and entered the Turkish service upon his will (see Berkes, "Ibrāhīm Müteferrika", in *EI²*, III, Leiden, 1979, pp. 996-997). He has been a theological student at a Unitarian college which was given over to the Calvinists in Transylvania. He then converted to Islam and entered the Turkish service. Whether or not Ibrāhīm Efendi studied at the *Enderun* (The inner section of the Sultan's palace) is not known, but after he entered the bureaucracy, he was promoted to the position of *müteferrika* serving as the "special counsellor and envoy of the Ottoman sultan". Ibrāhīm Efendi also took part in diplomatic negotiations, particularly with Austria and Russia. In the mid-16th century Unitarianism became widespread in Transylvania thanks to the religious freedom and toleration which resulted from the fact that the Ottomans controlled Hungary and supported the independence of Transylvania from the Catholic Habsburgs. The main principles of Unitarianism, however, aroused the hostility not only of the Catholics, but also of the Lutherans and the Calvinists. Indeed, Unitarians were violently opposed and persecuted because of their heretical beliefs. Moreover, Michael Servetus (d. 1553), the founder of Unitarianism, was burned by the Calvinists. (Berkes, "İlk Türk Matbaa Kurucusunun Dini ve Fikri Kimliği", pp. 724-725). As we have already mentioned in the above pages, because of his Christian background, Müteferrika feared that the great reaction which came forth in the Christian world against the system of Copernicus would also break out in the Islamic world. This attitude becomes clear from the hesitant and cautious statements in his supplement to Kātib Çelebi's *Cihannümā* regarding the new astronomy.

added to Kātib Çelebi's geographical work entitled *Cihannümā*[36] while he was printing this work in 1732, contains detailed explanations of the systems of the universe and the new astronomy. During the century following its publication, this supplement of Müteferrika preserved its quality of being the most extensive text in Turkish dealing with the "new astronomy". After one year, again with Müteferrika's translation of the astronomical work in Latin entitled *Atlas Coelestis*, a separate work dealing with the old and the new astronomy was introduced into Ottoman scientific literature.[37]

Müteferrika, in his supplement to *Cihannümā*, first states that all scholars of astronomy are like-minded on the question that the universe was constructed of concentric spheres one inside the other. After stating, however, that they have different approaches to the structure and the details of this subject, he divides the conceptions of the scholars of philosophy and astronomy on the structure of the universe into three. He attributes the first point of view to Aristotle and Ptolemy, the second to Pythagoras, Plato and Copernicus, and the third to Tycho Brahe (see illustration 7). He notes that European scholars qualify the first view as "old astronomy", and the second and the third views as "new astronomy".

Before he begins to explain these three points of view, Müteferrika points to the relationship between them and matters of religion and belief. In his view, it is a requirement of religion to believe that the universe is the work of the exalted creator. Once this is accepted, approving any one of the views concerning the shape, arrangement and the order of the universe does not depend on religion. Thus Müteferrika clearly states that the stance of religion on this subject will be impartial and whichever of these views one subscribes to, the religious aspect of the matter is merely related to the existence of the creator of the universe. But, on the other hand, he assumes a hesitant and extremely cautious attitude by saying that "the first view is held acceptable and superior by everyone, while the second and third views are not held in respect and are rejected". Nevertheless, he states that the explanation of these three views will be beneficial and deems it appropriate to affix this supplement to such an important work as *Cihannümā*. At the same time, he says that he has explained all three views because he considered it necessary for those who wish to know and hear that which everyone in the world knows and sees, and for those who pursue knowledge. He has also added illustrations to the book which would facilitate the "pictorial representation *(tasvīr)* of the state of the universe and the conception of the sphere of the world".

Following these cautious statements on the different views of the universe, Müteferrika examines each one individually. When he explains the geocentric view, which he calls "the conception of Aristotle, Ptolemy and their followers", he especially notes that it was accepted by Muslim philosophers.

Among the three views of the universe, the most space is devoted to the explanation and the discussion of the heliocentric system, and, here too,

---

[36] Kātib Çelebi, *Kitāb-ı Cihannümā li Kātib Çelebi*, Istanbul: Dār el-Tıbaat el-Āmire, 1145 H. (1732).
[37] *OCLT*, I, pp. 134-138.

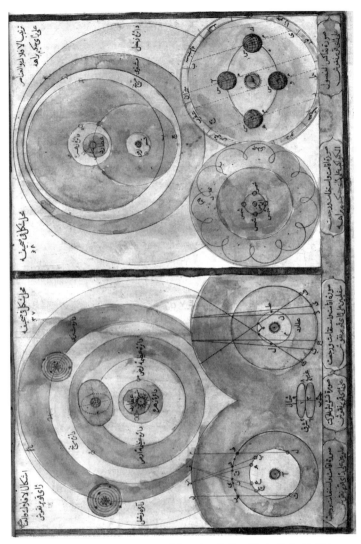

Illus. 7) Illustrations of the systems of Copernicus (left) and Tycho Brahe (right) in
Müteferrika's edition of the *Cihannümā*. (Kātib Çelebi, *Kitāb-ı Cihannümā li Kātib Çelebi*.
Istanbul: Dār el-Tıbaat el-Āmire, 1145/1732).

Müteferrika's cautious attitude is evident. Our author, after referring to Aristotle's view which he qualifies as appropriate and sound, describes this new view as unfortunate and useless. He cannot help but repeat that this view has no connection with religious beliefs. Meanwhile, he calls upon bold scholars to criticize this new view and to support Aristotle's view by asserting their counter-evidence on the subject. In this way, he says, their contributions will embellish his work.

Müteferrika, who carries on his contradictory statements, maintains that even Muslim scholars opposed this view in the past, but neither clarifies what this objection was nor identifies the scholars. He does not refrain from saying that though this viewpoint may be 'null and void', it attracted supporters at every age and time as numerous works were written on this subject reflecting the viewpoint of the older generation. Here, however, he falls short of explaining his own opinion amidst lengthy and vague utterances.

Müteferrika then presents the essentials and the illustrations of the new theory. He makes a comparison between the geocentric and heliocentric views and deals with the three classical criticisms put forth against the new system. The first criticism states that this new view is opposed to the statements in the sacred books, and he deals with the question of the Earth's motion here. He says that merely following the Copernican system, we must accept the movement of the Earth. On this matter, he asserts the philosophical arguments of Descartes and states that although it is written in the Bible that the Earth is always stationary, what is meant there is not immobility, but permanent existence. After he deals with the second criticism to the effect that the locations of the stars in the blue vault of heaven do not change relative to the Earth, he puts forth the influence of the Earth's state of motion on the bodies over the globe as the third criticism and responds to it, contrasting the physics of Galileo with that of Aristotle.

Müteferrika, on the other hand, takes over the historical development of the heliocentric system from a book by Edmond Pourchot.[38] Citing examples from ancient Greek philosophers who believe that the Earth rotates around its axis and also around a centre, he summarizes the views of the Pythagoreans and Aristarchus on this subject. He states that these views are contrary to the faith and the rites of Christianity, and furthermore, they are opposite to the understanding of ignorant people and that Latin poets have criticized them. He relates, however, that as a result of developments in astronomical instruments and new astronomical observations, this new view became current to a degree; even Cardinal Nicolaus Cusanus,[39] the celebrated Christian scholar, inclined to it; later, a Polish priest by the name of Copernicus sought to establish this view with his observations between the years 1500-1530 and since that day, this view became known by his name, i.e., "Copernican astronomy". He points out that

---

[38] Edmond Pourchot (d. 1734). See *Biographie Universelle*, XXXV, pp. 556-558.

[39] Cusa, Nicholas (Nicolaus Cusanus, Nicolaus Von Cusa) [d. 1464]. German philosopher and mathematician. The name appears as Kujeyani (Kuzyanus?) in the text. For his life see *Biographie Universelle*, X, pp. 582-584; *Dictionary of Scientific Biography*, III, pp. 512b-516b.

II

Descartes was also among the European scholars who later accepted this view, which was supported by his recently-developed philosophy.

After the heliocentric view, Müteferrika deals with Tycho Brahe's pattern of the universe as the third view. But he does not take much interest in it and indicates that it was put forth in order to reach an agreement between the two previously-mentioned views.

The most remarkable feature of Müteferrika's supplement to the *Cihannümā* is his cautious manner when explaining the heliocentric system. We have mentioned the reason for this hesitant and extremely cautious attitude of Müteferrika in a paper of general scope which was previously published on this subject (see p. 1, note 2 in this text.) He assumed this attitude since he anticipated a reaction in the Islamic world similar to the strong and violent opposition which came forth in the Christian world against the Copernican system. Müteferrika was well aware of this reaction in Christendom since he had once been a priest and then converted to Islam. In the pages where he introduces the heliocentric system, referring to Aristotle's view which he qualifies as appropriate and sound, he repeats that the acceptance of this view has no connection with religious belief and he calls upon bold scholars to criticize this new view and to support Aristotle's view. He states, without giving any example, that even Muslim scholars opposed the heliocentric view. But he does not specify what this objection was and who the opposing Muslim scholars were. He says that although this new view my be "null and void", it attracted supporters in every period since it was the view of the predecessors and that many works were written on it. Nevertheless, he always fails to explain his opinion since he uses dubious and contradictory statements. This cautious attitude of Müteferrika on the subject of the heliocentric system was to change to a great extent, however, when he translated the *Atlas Coelestis* from Latin shortly after the publication of the *Cihannümā*.[40]

The Translation of the *Atlas Coelestis*

Approximately one year after he printed the *Cihannümā* with his supplement on astronomy, Müteferrika translated the work by Andreas Cellarius[41] entitled *Atlas Coelestis* and first printed in 1708. Müteferrika completed the translation, which he began upon he order of Sultan Ahmed III, in the year 1733, and he entitled this translation containing astronomy alone *Mecmūatü Hey'et el-kadīme ve'l-cedīde,* meaning "Compendium of old and new astronomy".[42] Although Müteferrika had explained the concepts and the principles of new astronomy in the *Cihannümā,* printed in 1732, it is significant that he was ordered by the sultan the very next year to translate into Turkish a work devoted exactly to the same subject.[43]

Müteferrika states that his purpose in translating Cellarius' book is to explain the contents of the work as far as possible, to "modify and translate" the

---

[40] *OCLT,* I, p. 135.
[41] For Andreas Cellarius see *Biographie Universelle,* VII, p. 504b.
[42] The only manuscript copy of this work which is available is in the Library of the Military Museum, Istanbul, no. 5302 (74).
[43] *OCLT,* I, pp. 135-136.

19

Latin passages in the work into Turkish, and to summarize the views of the astronomers on the structure of the universe. He made some changes and additions when he translated the work. One of these changes may be deemed important from the viewpoint of the history of art, if not the history of science. This results from the fact that the illustrations in *Atlas Coelestis* have been reworked to accord with Turkish taste in painting and the "manner and style of Muslims". The author has added to his translation the information contained in the supplement to *Cihannümā* in summary form incorporating some additional illustrations. On the other hand, while three theories are treated in the *Cihannümā*, Müteferrika also refers to the view of Aratus[44] in this work. Stating that this view is accepted only by the ignoramuses of the Greek and Latin worlds and completely rejected by reasonable people today, he does not explain it and is satisfied with only presenting an illustration. He notes that he has included this view here only because it is contained in the *Atlas Coelestis*.

One of the remarkable points in Müteferrika's translation is his manner, now less prudent than his very cautious approach in the *Cihannümā*. The possible reason for this change in approach may be that, contrary to his fears, within the year following the publication of *Cihannümā*, he met with no opposition similar to the one in Europe on the part of the men of religion and state who read the book. Although Müteferrika mentions the connection of the subject with religion using his previous words and passages, he does not repeat them as frequently and we notice that now he is more assured.

Nevertheless, in this translation by Müteferrika the system of Ptolemy is influential when illustrations are in question. Although he explains all fourteen illustrations of Ptolemy's system one by one, he expains only the first of the two illustrations portraying the heliocentric system. Despite the fact that he is not negligent in expressing his appreciation for the system of Ptolemy, he does not begin his discussion here, as he did in the *Cihannümā*, by naming the new system "unfortunate and void" and calling everyone to criticize it. He is still prudent, however, when he completes his work stating that "at one time" this theory "attracted very many criticisms" and refers his readers to the answers in the *Cihannümā*.

Osman ibn Abdülmennān and the Translation
of *Geographia Generalis*

After the above-mentioned transfers of information by Müteferrika in the 1730s, we find the concepts of the new astronomy in a book of geography translated by Osman ibn Abdülmennān.[45] This book by Bernhard Varenius (d.

---

[44] Aratus of Soli (d. 240 B.C.): Greek poet and astronomer. His work in verse entitled *Phaenomena* gives an enumeration of the information in his time on earthly and heavenly bodies and of the signs which indicate the weather changes. For his life see *Biographie Universelle*, II, pp. 358-359; *Dictionary of Scientific Biography*, I, pp. 204b-205b.

[45] Osman ibn Abdülmennan gives his father's name as "Abdurrahman" and also as "Abdüssamed" in the other works which he translated *(Kitāb el-Nebāt* and *Hedīyet el-Mühtedī).* Although his death date is unknown, considering the year 1193 (1779) when he translated the very last work entitled *Hedīyet el-Mühtedī,* one can say that he was either alive at this date or passed away right afterwards. Since he also gives the Bosnian equivalents of the plant names in Latin in Matthioli's work, which he translated under the name of *Kitāb el-Nebāt,* it appears that he was of Bosnian origin. In his article entitled "Belgrad'daki Bayraklı Camii" published in the Xth issue

1676),[46] a pioneer of physical geography, was the *Geographia generalis in qua affectionnes generalles telluris explicantur* and was first printed in Amsterdam in the year 1650. Abdülmennän, the translator of *Belgrad Divānı*, also translated this book into Turkish as *Tercüme-i Kitāb-ı Coğrafya* in the year 1751 upon the order of Köprülüzāde Hacı Ahmed Paşa.[47]

The book is organized into an introduction *(mukaddime)*, six chapters *(bāb)*, and a conclusion *(hātime)*. The introduction of the work comprises two propositions *(mes'ele)*. In the first, the author mentions the place of the Earth in the universe and its shape; he then explains the necessity to draw maps with North at the top, and he makes a threefold classification of the Earth according to the latitudes and longitudes, lands and seas, and countries. The second proposition consists of explanations of some geographical terms in Latin. The conclusion of the work enumerates in some fifty pages the proofs of the sphericity of the Earth and includes a diagram of the geocentric system of Ptolemy.

The *Geographia Generalis* contains information on the structure of the universe and the sphericity of the Earth besides the general knowledge of geography. It is a work which was written according to the Ptolemaic system. However, after some reflection, in his translation Osman Efendi prefers the Copernican system although he remains loyal to the original text.

After comparing the geocentric and the heliocentric theories, by a pleasant analogy he explains that it is more reasonable for the Earth to rotate around the Sun than the Sun to rotate around the Earth: "If one wanted to roast meat on skewers, it is more logical to turn the skewered meat on the fire, instead of turning the fire around the meat".

An examination of this parable by Osman ibn Abdülmennän Efendi shows that he is not carried away with the hesitations, prudence and fastidiousness of Müteferrika and that he favors the heliocentric system. At the same time, as opposed to Abū Bakr Efendi, he does not cut the matter short and pass over the subject with obscure statements. After explaining his preference in a rational and logical way with a rather pleasant example, in the absence of an

of *Vakıflar Dergisi* (p. 386), M. Tayyib Okiç mentions some of the scholars and writers who grew up in Belgrade and discusses the significance of the fact that Osman Efendi gave the Bosnian equivalents of the names of plants in Turkish in the *Kitāb el-Nebāt*. For Osman Efendi's translation of "Materia Medica", one can consult the following works: M. Tayyib Okiç, "Natioli u turskom prevedu", *Gayret Mecmuası*, Sarajevo 1940, pp. 11-12; *idem*, "Hadis'te Tercüman", *İlâhiyat Fakültesi Dergisi*, XIV, p. 35. Besides his works entitled *Hediyet el-Mühtedī*, *Kitāb el-Nebāt* and *Tercüme-i Kitāb-ı Coğrafya*, Tercüman Osman Efendi also has the work *Dam a'im-i Kitāb-ı Ma'rifet el-Taktīr* on chemistry which he probably translated from German. At present, the only known copy of this last work is in Oriental Institute Library in Sarajevo, no. Y 608. This work is an expanded translation and the author of the original book is Berkhardos from Austria. For this work see Bedi N. Şehsüvaroğlu, *Eczâcılık Târihi Dersleri*, Istanbul 1970, pp. 306-307. For Osman Efendi also see Smail Balic, *Cultura Bosnjaka*, Wien, 1973, pp. 80-81; see also *OCLT*, I, 156-160.

[46] For his life and work see *Biographie Universelle*, XLVII, pp. 495-498; *Dictionnaire Universel d'Histoire et de Geographie*, p. 1929; *Dictionary of Scientific Biography*, XIII, pp. 583a-584a.

[47] For the translation of the work in Ottoman Turkish, besides the above-mentioned works, see *OM*, III, p. 188; Cengiz Orhonlu, "XVIII. Yüzyılda Osmanlılar'da Coğrafya ve Bartınlı İbrāhīm Hamdi Atlası", *İstanbul Üniversitesi Edebiyat Fakültesi Tarih Dergisi*, X, no. 14, September 1959, p. 116.

Islamic dogma, he presents the translation of the book in line with the preference of the author of the work. Indeed, one would not expect a translator who is not a professional geographer and astronomer to be more favorable to the new theory.

The Old and New Astronomy in the *Mārifetnāme* of Ibrāhīm Hakkı of Erzurum

The opinions on astronomy and the universe of Ibrāhīm Hakkı of Erzurum (d. 1194/1780), a versatile scholar and mystic of the 18th century, carry particular weight from the viewpoint of the history of science and culture. These are stated in his *magnum opus Mārifetnāme*, which he completed in 1757 and which was printed for the first time in 1825.[48] An examination of these views gives us an idea about the mentality of the author as well as offering us exceptional clues for understanding the changes taking place at the time in Ottoman intellectual life.[49]

The *Mārifetnāme* is full of thoughts and beliefs which are completely unrelated and often opposed to one another. At the same time, it contains long narrations and quotations without mentioning sources. Although the unity of expression among them is realized in the passages where the author begins saying "O dear one" ("Ey Azīz"), the remarkable feature of the work is that the author gives contradictory information on the same subject in different parts of the work. However, the work strikes one as most odd in that the author's own evaluations on the same subject are completely contradictory in different chapters. This leads us to assume that the author had some kind of intellectual problem or that he was forced to write in this way.

The *Mārifetnāme* is organized in an introduction, three "branches" of sciences, and a conclusion. In that section of the introduction entitled "Islamic astronomy" (pp. 2-21), the author first cites the Qur'ānic verses on the order of the creation of the universe and quotes the cosmological information on the creation of the beings in the universe from what he calls the commentaries on the Qur'ān and the books of tradition, without mentioning their titles. The introduction is intended to teach man the exaltation and the might of God by leading him to ponder the wisdom of the creation of mankind.

The first *fen* (branch) of science (pp. 24-158) treats substances, nonessential attributes and elements, as well as presenting the arithmetic and geometry necessary for the astronomy. This section both proves that the world is spherical and presents detailed information on the planets. Here, matters such as the plants, the animals, inanimate nature *(mevālid-i selāse)*, the four elements, the lines of latitudes and longitudes and the seven climates, are investigated thoroughly. Towards the end of the first branch, the author deals with the "new astronomy". The purpose of the first branch is to teach the secrets of the world

---

[48] The *Mārifetnāme* was printed eight times in the Ottoman period. It was first printed in Cairo in 1251 (1825), and again in the years 1255 (1839), 1257 (1841) and 1280 (1863). It was then printed in Istanbul in 1284 (1867) and again in 1294 (1877), 1310 (1892) and 1330 (1914). Afterwards, the latest edition of the work was reprinted several times in facsimile. For the editions see M. Seyfettin Özege, *Eski Harflerle Basılmış Türkçe Eserler Kataloğu*, I, Istanbul 1975, p. 1975, p. 1025. In this study we have referred to the 1330 (1914) edition in Istanbul, and the pages cited in the text refer to this edition.

[49] *OALT*, II, pp. 486-491.

which manifest God's unmatched power of creation, to communicate that "the universe is similar to a shell that contains man and man is the essence of the universe", and finally to ensure man's self-knowledge by helping him find ways of withdrawing completely from all beings except God. The first branch of science teaches the rational (philosophical) knowledge taken from the works of the scholars of physics, astronomy, geometry, arithmetic and astrology.

The second branch (pp. 158-157) is anatomy (the science of the dissection of bodies) and psychology (the science of the dissection of souls). This branch is intended to teach that the likes of the beings in the universe also exist in man and to make known that man is a small universe himself. The author presents this branch of science as "mir'āt el-ebdān" (the mirror of the bodies).

The third branch (pp. 257-278) is the knowledge of God through the improvement of belief and faith. It explains the ways of reaching the spiritual rank of the gnosis of God ("mārifetullāh mertebesi"). The author calls this branch of science "mir'āt el-kulūb" (the mirror of the hearts).

The conclusion (pp. 528-559) deals with the manners of behavior towards friends, relatives and neighbors, and its purpose is to make one popular in the eyes of friends and acquaintances. In the pages following the conclusion (pp. 559-561), the author treats the concept of "tevhīd" (divine unity) which facilitates for man the performance of acts of worship and devotion.

The information on astronomy and cosmography in the *Mārifetnāme* is to be found in different sections of the work. As we have also mentioned above, the information on the creation of the universe, which is presented under the title of "Islamic astronomy" in the introduction, is quoted from Qur'ānic commentaries and books of tradition. The second and the third chapters *(bāb)* of the first branch are devoted to the old and new astronomy. In these chapters the author proves that the world is spherical, gives detailed information on the Sun, Moon and planets, and also deals with the concepts of the new astronomy.

From the viewpoint of the sciences of cosmography and astronomy, Ibrāhīm Hakkı's perception of the universe appears as two different views in various parts of the work. As an example of these different views, we quote the diverse information in various parts of the work on the eclipses of the Moon and the Sun and the occurrence of earthquakes.

The work contains the following information on the movements of the Sun and the Moon from the East to the West and eclipses, which are considered to be a sign of doomsday (pp. 13-14):

"...God Almighty created a sea under and adjacent to the firmament... Afterwards, God Almighty created a carriage with 360 handles out of the gem of diamond for the Sun in this sea and also appointed an angel for holding each handle so that they bring the Sun in its carriage from the East to the West in the sea by pulling it. God Almighty also created a carriage of topaz with 300 handles and placed the Moon on it. He appointed an angel for holding each handle to bring the Moon in that sea from the East to the West... God Almighty fixed determined times for the solar and the lunar eclipses so that the men over the globe, seeing the changes in the Sun and the Moon, wake up and turn towards

Him entreating... At the time of the eclipse, the radiant Sun falls from its carriage and goes down the sky in the depths of the sea. If it falls completely, a full solar eclipse occurs, hence the light which covers the stars does not remain and only the greatest of them can be seen. If half of it falls in the sea that much of it is eclipsed... At the time of the eclipse, the beautiful Moon thus falls from its carriage into the depths of the sea either completely or in part, and the eclipse occurs during the time of the fall."

This information, which is stated as the words of the commentators on the Qur'ān *(mufassir)* and the traditionists *(muḥaddith),* is related neither to the Holy Qur'ān and the sound traditions - the two basic sources of Islamic religion - nor does it have any connection with the science of astronomy which developed in the golden age of Islamic civilization based on theory and observation. These explanations mixed with legend, written in a lyrical style and ornamented with metaphorical expressions, which do not have any connection with religion and science, address the taste of the uneducated masses. It appears that this material was compiled from "mawḍū'" (fictitious) traditions.

Ibrāhīm Hakkı depends on the cosmological literature on "Islamic folk astronomy", which al-Suyūṭī promoted, when he wrote his book on this subject which was taken as the major reference for generations to follow.[50] Al-Suyūṭī's (d. 911/1505) work entitled *al-Hay'at al-saniyya fi'l-hay'at al-sunniyya* was reorganized with some additions by Ibrāhīm ibn Abdurrahman el-Āmidī (alive in 1064/1654) under the title of *Risāla fi'l-hay'a ʿalā tarīq ahl al-sunna wa'l-jamāʿa* in the year 1064/1654. Naẓmizāde Hüseyin Murtazā b. Ali el-Baġdādī (d. 1134/1722) later translated Karamanī's work into Turkish.

On the other hand, contrary to the above explanations of the eclipses of the Moon and the Sun based on superstition, in another place the author presents a scientific explanation which he quotes from al-*Ghazālī's Tahāfut al-falāsifa*:[51]

---

[50] Cf. al-Suyūṭī, *al-Hay'at al-Saniyya fi'l-Hay'at al-Sunniyya* Arabic text, translated with a commentary (English), ed. Anton M. Heinen, Beirut 1982, p. 19, no. 35. The text, which is presented in Ibrāhīm Hakkı's *Mārifetnāme* (see pp. 13-14), is mentioned as quoted in the works of a majority of commentators and traditionists. It conforms with the text which Abū Muhammad ʿAbd Allāh b. Muḥammad b. Jaʿfar b. Ḥayyān al-Anṣārī al-Isfahānī (d. 369/979), known as Abū'l-Shaykh, and Ibn Marduwaih have transmitted on the authority of Abū'l-ʿIṣma Nūḥ b. Abī Maryam, Muqatil b. Ḥayyān, ʿIkrima, and Ibn ʿAbbās in their commentaries. In Ibrāhīm Hakkı's work the text appears in a more flowery language.
In our view, one should not consider on the same level the sources of the fictitious *(mawḍūʿ)* traditions in the various books which have been compiled in different centuries. In other words, since the sources of traditions are very diversified, it is impractical to categorize them in the same group and the same grade. For this reason, scholars have divided the books of tradition into four groups which consist of "ṣaḥīḥ" (sound), "ḥasan" (good) and "ḍaʿīf" (weak) traditions. (Cf. the work by Imām Aḥmad, better known as Shāh Walī Allāh al-Dihlawī, entitled *Hujjat Allāh al-Bāligha*, p. 105 ff., Cairo, 1322). The fourth group of these sources consists of useless books which were collected in the later centuries. These books were based on the accounts which were full of innovation and fancy, by people such as narrators, preachers, mystics and historians who were not authoritative. The books of Ibn Marduwaih, Ibn Shāhin and Abū'l-Shaykh belong to this group. Evidently, a person who has a knowledge of traditions would not rely upon the books in this group since they are the source of innovations and superstitions. (See Subhī al-Sālih, *Hadis İlimleri ve Hadis Istılâhları,* trans. Yaşar Kandemir, Ankara 1971, pp. 91-92.)
[51] *Tahāfut al-falāsifa,* ed. Süleyman Dünya, Cairo 1961, I, p. 78.

"... The eclipse of the Moon means the disappearance of the Moon's light as a result of the Earth's positioning between the Sun and the Moon... As for the solar eclipse, it results from the positioning of the Moon between the Earth and the Sun." Following this scientific statement, Ibrāhīm Hakkı puts forward with insight his ideas contradicting his previous explanations which were illogical and unscientific (p. 45):

"Whoever thinks that such things are related to religious matters does harm to religion since the proofs of arithmetic and geometry indicate that the mentioned events have taken place. If somebody tells a person who knows these subjects 'This is against the *şerīat*, that person does not suspect his knowledge but the *şerīat*. A person who wants to succour the *şerīat* in an unreasonable way has greater harm than the person who attacks the *şerīat* reasonably. As a matter of fact, 'a wise enemy is better than a foolish friend'."

There are also contradictory statements on the subject of earthquakes. According to the first unscientific explanation (p. 16), God Almighty charged a great angel with earthquakes and put in the streaks of the mountains under his order. When God Almighty wants to keep the people of a region away from sin and to protect them, upon God's order, that angel moves the streak of that region. This continues until the peoples of that place wake up as a result of the earthquake and turn towards God.

The second explanation (pp. 120-121) has a completely scientific nature:

"... If the vapor which forms and gets squeezed in the Earth reaches such a density that it can pass through the layers of the Earth and come out, or if the ground is too thick and hard for the vapor to get out, then that vapor roughly splits and tears the Earth pushing it and that region moves, this is called an earthquake."

The concepts of the new astronomy are introduced in the pages between 144-151 of the first branch of science in the *Mārifetnāme*. In this section Ibrāhīm Hakkı treats the acceptance of the new astronomy and the relation of the structure of the universe with religion.

Here the author relates that when the scholars who esteemed the heliocentric theory endeavored to support this view, naive people first directed an unpleasant allusion at them by way of reproach, but new astronomy became current as a result of the improved observations with the implements of astronomical observation which developed in the course of time. Moreover, he states that it is irrelevant to religion to consider whether the centre of the universe is the Earth or the Sun. In principle, one should believe that the universe is the work of the exalted creator, but the different beliefs concerning the shape of the universe do not concern religious matters. In this section, the information quoted on the planets which revolve in their orbits around the Sun at the centre, the period of their orbital movements and their satellites has a completely scientific nature.

Ibrāhīm Hakkı gives preference to the heliocentric system (see illustration 8) although he is not a professional scholar directly involved with mathematics and astronomy. He expresses his view clearly and more boldly than his predecessors without deeming it necessary to act cautiously like Müteferrika,

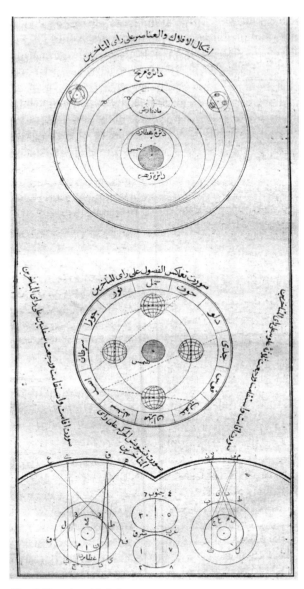

Illus. 8) Illustration of the heliocentric system in the *Mārifetnāme* of Ibrāhīm
Hakkı of Erzurum. (*Mārifetnāme*, Istanbul, 1330/1914).

who was his source. He says: "...it is much easier, more proper and reasonable for this spherical Earth, which is more suitable for movement owing to its small mass, to revolve around the great Sun once a year."

On the other hand, when the question of the relationship between the structure of the universe and religion is treated in the section entitled "Islamic astronomy", based on folk cosmological and superstitious explanations, the author states:

"... what has been written up to this point concerns the matters of religion which everyone should believe without any suspicion, for it would not be proper to compare the basic principles of the matters of religion with rational arguments inasmuch as the human intellect is deficient and incapable of perceiving them. Let us be content with explaining the shape of the universe this much here according to the Qur'ānic verses and the traditions of the Prophet."

A completely different mentality comes to light in this expression which is Ibrāhīm Hakkı's greatest contradiction.

Ibrāhīm Hakkı dwells on different sources in various parts of his *Mārifetnāme* when he explains his conception of the universe and some natural events, such as eclipses of the Sun and the Moon and the occurrence of earthquakes. One can group these sources mainly in three categories: 1- Religious (Qur'ānic verses and traditions of the Prophet and well-known religious references such as al-Ghazālī's *Tahāfut al-Falāsifa;* 2- Scientific books like Kātib Çelebi's *Cihannümā* (with Müteferrika's new additions); 3- Popular beliefs based on legends and superstitions (like al-Suyūṭī's book on folk cosmology based mainly on fiction rather than on astronomy). As it appears from an examination of these sources, Ibrāhīm Hakkı is addressing two separate groups of people. The first is the great mass of people with a rather low level of culture whom he calls "the common people" ("avām-ı nās"). When he addresses them, he does not see any harm in relating the superstitions and legends which were introduced to Islamic literature in certain periods and which became current in the epochs of regression, in his particular style beginning as "O dear one" ("Ey Azīz"). The second group of people addressed are "the intellectuals" ("havass-ı nās"), blessed with a certain level of culture. Besides these two groups of people, he felt that he was compelled to be cautious of some official *ulemā* who appeared to be narrow-minded, who were jealous of him, envied his popularity, personal influence and renown. It is most astonishing, however, to see that the author presents these opposed views in the same work at the same time, presents both of them in an equally enthusiastic way, and does not give preference to either one of the two contradictory thought patterns.

In order to understand the reasons why Ibrāhīm Hakkı presented such contradictory ideas in his work, one needs to briefly examine some events in his life. Ibrāhīm Hakkı completed his *Mārifetnāme*, on which he had been working for a few years, at the end of the year 1170/1757. It seems that the *Mārifetnāme* acquired a reputation even at the stage of its compilation. It is narrated that Kadızāde Muhammed Ārif b. Muhammed el-Erzurumī (d. 1173/1759-60)[52] *müftī*

---

[52] For Kadızāde's life see Baghdadī Ismail Pasha, *Hadiyyat al-Ārifīn (HA)*, II, p. 333; *OM*, I, p. 404; Kahhala, X, p. 115.

of Erzurum, was jealous of Ibrāhīm Hakkı, who was popular in the eye of a wide circle of people and who satirized him; therefore, the *müftī* took sides against him. According to an unconfirmed narrative, he assembled a council of some narrow-minded *mollās* and called Ibrāhīm Hakkı for interrogation. The *müftī* started to question Ibrāhīm Hakkı, saying: "Molla! You are said to talk about the Earth and the heavens and scribble things against the *şer-i şerīf*. How will it be"? According to the statement of the late Mesih İbrahimhakkıoğlu, a descendant of Ibrāhīm Hakkı, who cites this narrative, when Ibrāhīm Hakkı showed the *müftī* and his followers his *Mārifetnāme*, asking them which part of the book is against the *şerīat*, they could not find an opposition to the *şerīat* in the book and they all repented. Mesih İbrahimhakkıoğlu, who has heard this incident from his father or elder brother, relates it as a report which is considered reliable by everyone.[53]

Mehmed Nusret Efendi, in his *Tarihçe-i Erzurum,* gives Kadızāde Mehmed's biography and states that since he fanatically opposed the *ulemā* who were inclined to mysticism, Ibrāhīm Hakkı wrote a couplet describing his unpraiseworthy morals and physionomy.[54]

One may explain the reason for the contradictory statements in *Mārifetnāme* partly on the basis of the above-mentioned incident. Since Ibrāhīm Hakkı was previously informed about the fact that the *müftī* Kadızāde planned to interrogate him, he may have added the section on "Islamic astronomy" in the beginning of his work. The organization and contents of *Mārifetnāme* indicate that the section on "Islamic astronomy" was added to the work later by the author so that the *müftī* and his followers would not take sides against him. Indeed, contrary to the folk cosmological explanations in this part of his work, Ibrāhīm Hakkı explains natural events in a completely scientific way in the rest of his work.

As it appears from Mesih İbrahimhakkıoğlu's book, some people misunderstood Ibrāhīm Hakkı's ideas and behavior and were jealous of him for the esteem in which he was held. They therefore blamed him with putting forth ideas against the *şerīat* and faith in the opinion of the ignorant masses and tried to stain his reputation. For this reason, his two works entitled *Urwat al-Islām* and *Hay'at al-Islām,* both written in 1191/1777 three years before his death, depend on books of tradition and commentaries on the Qur'ān besides his own *Mārifetnāme.* Evidently his purpose was to protect himself against the unfounded accusations. As the title also indicates, his *Hay'at al-Islām,* written in a folk cosmological and legendary manner, is on "Islamic folk astronomy". In the introduction of this work Ibrāhīm Hakkı says: "I have abandoned reading the statements and books of philosophers on astronomy since they would spoil one's faith and weaken one's zeal in religious matters".[55]

Let us now examine the following letters of Ibrāhīm Hakkı which provide clues for understanding his regression from his views in *Mārifetnāme.* In a letter dated May 23, 1777 and addressed to his cousin Yusuf Nesim Efendi,

---

[53] Mesih İbrahimhakkıoğlu, *Erzurumlu İbrahim Hakkı,* Istanbul, 1973, p. 91.
[54] Mehmed Nusret Efendi [Müftizāde]. *Tarihçe-i Erzurum Yahut Hemşehrilere Armağan.* Istanbul: Ali Şükrü Matbaası, 1338 (1922), p. 101. Ibrāhīm Hakkı's couplet is quoted in this book as well.
[55] İbrahimhakkıoğlu, p. 154.

Ibrāhīm Hakkı advised his son Ahmedülhayr as his last injunction to read his poems with all his heart, but not to recite his couplets among people. Unfortunately, there is no evidence as to which couplets Ibrāhīm Hakkı meant here; we can speculate, however, that they were probably related to mysticism. He also enjoined his son to behave according to the words in *Urwat al-Islām* so that he would not go astray from the right path.[56] On this matter, M. İbrahimhakkıoğlu relates another incident. A student of Ibrāhīm Hakkı by the name of Derviş Halil, who read his teacher's mystical works, spread the word in Hasankalesi that Ibrāhīm Hakkı had written a book of mystery *(sır kitabı)*. Evidently, people suspected that İbrāhīm Hakkı had some secret ideas which would harm Islamic faith and they hassled his cousin Yusuf Nesim. Ibrāhīm Hakkı was aware that many scholars had been blamed with heresy on the basis of unsound evidence.

In order to protect himself against such unjust accusations, Ibrāhīm Hakkı sent a secret letter to his cousin Yusuf Nesim, together with his book *Urwat al-Islām*. This letter was dictated by Ibrāhīm Hakkı and written by his son Ismāil Fehim. Here, Ibrāhīm Hakkı advises Yusuf Nesim, saying: "My son, last year you stated in your letter: 'People annoy me saying that our Efendi wrote a book of mystery *(sır kitabı).'* Show the book entitled *Urwat al-Islām* to those who annoy you, not with this letter but with the other letter, saying: 'This book is the compilation of our Efendi after his *Mārifetnāme'.* Also read them the letter and say: 'Mischief-makers such as this habitual liar Halil have slandered our Efendi in order to gain undue esteem for themselves'. When you arrive at Hasankalesi, show this book and the stated letter to people and warn them, saying: 'Our Efendi has no work or deed against the *şerīat.* Some people, however, imputed words and acts to our Efendi which are against the *şerʿ-i şerīf.* God will put them in public disgrace since they try to defame our Efendi".[57]

At the end of Ibrāhīm Hakkı's words, his signature and seal were set for fear that someone might find the letter. In the remaining part of the letter, Ismāil Fehim addressed Yusuf Nesim, saying: "Our Efendi rejected his own couplets to such a degree that I saved his book *Insāniye,* which consists of his mystical poems, while he had thrown it into the fireplace. Our Efendi states that those who wish him good should forget and ignore his couplets. Instead, they should read the *Urwat al-Islām,* since the essential knowledge is based on the Qur'ān and the traditions. He has forgotten all other words."[58]

Until we have more extensive information, documents and studies on the life, time and works of Ibrāhīm Hakkı, the above-mentioned records may serve to hint at the reason for the presence of the contradictory parts of the *Mārifetnāme* and Ibrāhīm Hakkı's regression in his later works from his scientific ideas in the *Mārifetnāme.* On the other hand, we can say that since the *Mārifetnāme,* a popular book which addressed a great mass of people for a century, was printed eight times in the period 1825-1914, this made it possible

[56] *Ibid.,* p. 153.
[57] *Ibid.,* p. 161.
[58] *Ibid.,* p. 162.

for the heliocentric system to become known and spread outside the circle of the Ottoman astronomers and scholars.

Halifezāde Çınarī Ismāīl and his translation of Clairaut's and Cassini's *Astronomical Tables*

After Tezkireci Köse Ibrāhīm of Szigetvar translated the astronomical tables by the French astronomer Noel Durret in the 17th century, we see two new translations of astronomical tables in the second half of the 18th. Both of these tables were translated by Halifezāde Ismāīl Efendi (d. 1790),[59] an astronomer who was educated in Istanbul. Also known as Çınarī Ismāil Efendi, he produced numerous compilations and translations in the field of astronomy. His first Turkish translation of astronomical tables is the work titled *Rasad-ı Kamer*, also known as *Tercüme-i Zic-i Kılaro*.[60] He translated this *zīj*, which is based on the tables of the Moon prepared by the French astronomer Alexis-Claude Clairaut, in 1767. The *Théorie de la Lune* and *Tables de la Lune* of Clairaut (d. 1765) were first printed in St. Petersburg (1752) and Paris (1754), respectively. The second edition of *Théorie de la Lune* appeared in Paris in 1765, with new tables of the Moon based on new observations.[61]

Before the Ottoman astronomer translated these tables in 1767, the St. Petersburg Academy rewarded Clairaut in 1750, two years before the first edition of the work, for his theoretical studies on "the problem of the three bodies" in space mechanics, whereas in the Ottoman context, the translation of astronomical tables was deemed sufficient. This event gives us an important clue to the scientific approach of the Ottomans.

In his *Histoire des Mathématiques*, published in 1766, the French historian of mathematics Montucla states that during the years when these astronomical tables were translated, Sultan Mustafa III (d. 1774) ordered "the most recent and the most perfect" books on European astronomy from the Academy of Sciences in Paris.[62] Moreover, the Turkish pioneer in the history of Islamic astronomy and mathematics, Salih Zeki, states that according to his findings in the records of the Academy, the astronomical tables of Lalande were among the books which were brought from the Academy through the mediation of Baron de Tott.[63] We can speculate that among the books which were sent there were the astronomical tables of Clairaut and of Cassini, for the latter were translated by Çınarī Ismāil Efendi.

The second set of astronomical tables which Ismāil Efendi translated is entitled *Tables Astronomiques du Soleil, de la Lune, des Planètes, des Etoiles Fixes et des Satellites de Jupiter et de Saturne*, compiled by the French

---

[59] For his life see *Kāmūs-ı Riyāziyāt*, Istanbul, 1315/1897, I, pp. 327-330; *SO*, I, pp. 371-372; *OM*, III, pp. 259-260; Adıvar, pp. 199-201; Uzunçarşılı, *Osmanlı Tarihi*, IV/II, Ankara: Türk Tarih Kurumu Yayınları, 1983, p. 537; al-'Azzāwī, *Tārikh 'Ilm al-Falak fi'l-'Irāk*, p. 290; 'Abd Allāh al-Jubūrī, *al-Mustadrak 'ala'l-Kashshāf*, pp. 292-293; *OALT*, II, pp. 530-537.

[60] See *Kāmūs-ı Riyāziyāt*, Library of Istanbul University, TY 914, VII, fol. 743, Library of Istanbul University, TY 916, IX, fol. 1130; *OALT*, II, p. 533.

[61] For his life and works see *Biographie Universelle*, VIII, pp. 593-598; *Larousse du XXe siècle*, II, pp. 283c-284a; *Dictionary of Scientific Biography*, III, pp. 281a-286a.

[62] Adıvar, p. 199.

[63] *Ibid.*, citing *Kāmūs-ı Riyāziyāt*, Library of Istanbul University, TY 910, III, fol. 317.

astronomer Jacques Cassini (d. 1756)[64] and printed in Paris in 1740. He translated these astronomical tables into Turkish in 1772 under the title of *Tuhfe-i Behīc-i Rasīnī Tercüme-i Zīc-i Kasīnī.*

Although Salih Zeki and Adnan Adıvar assert that Cassini's astronomical tables were brought through the mediation of Yirmisekiz Mehmed Çelebi, whom Sultan Ahmed III sent as ambassador to the French King Louis XV towards the end of the year 1132/1720,[65] this is incorrect. Yirmisekiz Mehmed Çelebi, in his *Sefāretnāme*, when he mentions his visit to the Observatory of Paris in 1133/1721, reports the following:

"... When Cassini died before he completed his observations, his son became the director of the observatory and all the treasures were handed over to him. He wrote down the matters which his father considered to be contary to Ulugh Bey's tables and gave us this. His father's writings have not been printed yet, but his son wishes to print them after he completes the observations..."

From these statements it appears that Jacques Cassini wrote on some problems which his father found in disagreement with Ulugh Bey's tables and gave this text of his to Yirmisekiz Mehmed Çelebi. In fact, Jacques Cassini completed the observations of his father Jean Dominique Cassini and later published them in 1740 under the title of *Tables Astronomiques.*[66]

On the other hand, from the passage in the introduction of the Turkish translation of the astronomical tables, "... the new observation of the French astronomer Kassini which he wishes to publish in French in the 1740th year of the Christian era...", we see that Halifezāde did not translate "the matters contrary to the astronomical tables of Ulugh Bey", which were brought by Yirmisekiz Mehmed Çelebi, but rather the *Tables Astronomiques* printed in Paris in 1740.

Logarithms were first introduced in Ottoman scholarly circles also thanks to this translation of the astronomical tables by Halifezāde. Besides the logarithmic scales of the numbers from one to ten thousand, Halifezāde supplemented to this translation one scale each for the logarithm sinus and logarithm tangent of all the arcs from 0° to 45° minute by minute. All of the logarithms in these scales have five digits. In this translation Halifezāde uses the equivalents of *"nisba"*, *"nisāb-i jaybiyya"* and *"nisāb-i zılliyya"* for the terms "logarithm", "logarithm sinus" and "logarithm tangent", respectively.

The translation of these tables had important influences on the Ottoman practice of compiling ephemerides. Upon the order of Sultan Selim III, these began to be organized according to Cassini's *Astronomical Tables,* and Ulugh Bey's tables were abandoned in the course of time.

---

[64] For his life see *Biographie Universelle,* VII, pp. 301-302; *Dictionary of Scientific Biography.* III, pp. 104b-105a ff.; *Kāmūs-ı Riyāziyāt,* Library of Istanbul University, TY 916, IX, fols. 1128-1130.
[65] *Kāmūs-ı Riyāziyāt,* I, p. 316; Adıvar, p. 199.
[66] E. İhsanoğlu, "Tanzimat Öncesi ve Tanzimat Dönemi Osmanlı Bilim ve Eğitim Anlayışı", *150. Yılında Tanzimat,* ed. Hakkı Dursun Yıldız, Ankara: Türk Tarih Kurumu, 1992, pp. 335-395. See also *OALT,* II, p. 534.

Müneccimbaşı Hüseyin Hüsnī
and the translation of Lalande's *Tables*

The fourth translation of astronomical tables in Ottoman circles took place at the beginning of the 19th century. In 1814 Müneccimbaşı Hüseyin Hüsnī Efendi[67] translated the *Tables Astronomiques* (Paris, 1759) by the celebrated French astronomer Lalande (d. 1807)[68] into Arabic. We estimate that he must have translated this work into Turkish before the year 1826.

In the introduction of his translation, Hüseyin Hüsnī Efendi states that Lalande's astronomical observations rendered the astronomical tables of Ulugh Bey and Cassini outdated; these new astronomical tables will be a guide until Doomsday and no one will find any mistakes in them. He also mentions that the ephemerides based on these observations will be the most correct and the most perfect ones.

A report ("*arz ve ilām*") presented jointly to the Sultan in 1248 (1829) by Hekim Mustafa Behçet and Müneccimbaşı Hüseyin Efendi states that Cassini's astronomical tables caused significant mistakes in the organization of ephemerides; while in these tables the margin of error in the solar and the lunar eclipses is about twenty-five minutes, in Lalande's astronomical tables this margin falls to a few seconds, and therefore from now on Lalande's tables should be used. Thus, upon the order of the Sultan, Lalande's astronomical tables began to be used.[69]

As seen above, contrary to the multifaceted geography books translated from European languages the first works translated in the field of astronomy have a different characteristic. The first translation on the subject of new European astronomy, namely, Durret's work, and the subsequent ones all involved *zījes*. The translations of the *zījes* of Cassini[70] in 1186/1772 and of Lalande[71] during the reign of Mahmud II (1223-1255/1808-1839) followed the translation of Durret's *zīj* in 1071/1660. These activities indicate that the Ottoman astronomers followed the publications of Western literature on astronomical tables.

Thus, it appears that the Ottomans confined their interest to following the most recent developments in practical astronomy (i.e. the compilation of ephemerides) which was necessary rather than translating the principal sources and theoretical works of Western astronomy which brought about developments or changes in science, established its new foundations and laid the way for it to follow its course.

---

[67] For his life and works see *SO*, II, p. 224, 229; *OM*, III, pp. 260; Library of Kandilli Observatory, no. 183, fol. 13a; no. 323, fol. 3b; *Dhayl Fihris Dar al-Kutub*, I, p. 32; Adıvar, pp. 200-201; *OALT*, II, pp. 581-584.

[68] For the life and works of Joseph-Jérôme Lefrançais de Lalande see *Biographie Universelle*, XXIII, pp. 215-232; *Dictionary of Scientific Biography*, VII, pp. 579b-582a; *Kāmūs-ı Riyāziyāt*, Library of Istanbul University, TY 916, IX, fols. 1290-1293.

[69] *Takvim-i Vekāyi* (Ottoman official gazette), 6 B 1248 (29 November 1829), no. 46, p. 3, line 2.

[70] For his *zīj* see *Kāmūs-ı Riyāziyāt*, Library of Istanbul University, TY 914, VII, fol. 743.

[71] For his *zīj* see *Kāmūs-ı Riyāziyāt*, Library of Istanbul University, TY 916, IX, fols. 1290-1293.

II

Modern Astronomy in the Newly Established Educational
Institutions in Istanbul and Cairo

We now consider the introduction of modern astronomy to the Ottoman world from another perspective and investigate the place of the new astronomy in the new educational institutions which began to be established towards the end of the 18th century.

The Hendesehāne, known as Ecoles des Théories or Ecoles des Mathématiques in French, was the first independent institution devoted to modern military technical education in the Ottoman Empire. It was established in April 29 1775 in Tersāne-i Āmire and was later called Mühendishāne.

Within the framework of the "Nizām-ı Cedīd" movement which began during the reign of Sultan Selim III (1789-1807) a "new" Engineering School was established in 1793 under the name of "Mühendishāne-i Cedīde" for the purpose of providing technical training for bombardiers, sappers and gunners. The names of Mühendishāne-i Cedīde and Tersāne Mühendishānesi were changed into "Mühendishāne-i Berrī-i Hümāyūn" and "Mühendishāne-i Bahrī-i Hümāyūn", respectively by a regulation issued by Sultan Selim III in 1806.[72]

Hüseyin Rıfkı Tamanī[73] was one of the first teachers who taught this class in the Mühendishāne-i Berrī-i Hümāyun[74] and who served in the capacity of chief instructor (bashoca) in this school between the years 1806-1817. Although he has various works on mathematics which were printed several times, he does not have a separate work published in the field of astronomy. His student Ishak Efendi, however, examined and summarized those sections of his teacher's lecture notes on astronomy relating to geography and published them in the year 1831 under the title of el-Medhal fi'l Coğrafya.[75]

One clearly sees that this work was written according to the geocentric system from reading the numerous statements such as "Let it be known that the

[72] Mustafa Kaçar, "The Development in the Attitude of the Ottoman State Towards Science and Education and the Establishment of the Engineering Schools (Muhendishanes), Proceedings of the International Congress of History of Science (Liège, 20-26 July 1997) volume VI, Science, Technology and Industry in the Ottoman World, ed. E. İhsanoğlu, A. Djebbar and Feza Günergun, Brepols Publishers, Belgium 2000, pp. 81-90.
[73] Hüseyin Rıfkı Tamanī (d. 1817), who served in the capacity of chief instructor (bashoca) in Mühendishāne-i Berrī-i Hümāyun for a long period, was born in the town of Taman in Crimea. Besides his position as chief instructor, he also served in many official posts, became a vanguard in the introduction and the expansion of modern sciences in the Ottoman state, and translated many books of science into Turkish. For more detailed information see Mir'āt-ı Mühendishāne, pp. 27, 32-33; OM, III, pp. 261-262; Adıvar, pp. 206-207; Türk Ansiklopedisi, XIX, Ankara, 1971, p. 425; E. İhsanoğlu, Başhoca İshak Efendi (Türkiye'de Modern Bilimin Öncüsü), Kültür Bakanlığı Yayınları, Ankara 1990, pp. 9, 14-15. (The last mentioned book has an English summary); see ODMT, II, 423-430.
[74] Although some authors state that Hüseyin Rıfkı Efendi served as bashoca in the Mühendishāne during the 22 years from its establishment until the year 1232/1817, the documents which we have in hand indicate that in the year 1211 (1796), Hüseyin Rıfkı Efendi was the second khalīfa (assistant) in the Mühendishāne. (The rescript dated 4 Jumāda I 1211/November 5, 1796, Cevdet, Maārif, no. 4717). Although Çağatay Uluçay and Enver Kartekin state that Hüseyin Rıfkı Efendi was appointed bashoca in the year 1800 in their works (Ç. Uluçay and E. Kartekin, Yüksek Mühendis Okulu, Istanbul, 1958, p. 85), we have not come across a document indicating that he was appointed bashoca before 1221/1806. According to what we have been able to find, the first document recording Hüseyin Rıfkı Efendi as bashoca is dated 10 Jumāda I 1221 (26 July 1806). (Prime Ministry Archives in Istanbul, Cevdet Maārif, no. 5506).
[75] El-Medhal fi'l-Coğrafya, Dersaadet, Matbaa-i Āmire, 1247 (1831), 88 pp. and 8 plates.

II

universe in appearance (the physical universe) is a sphere and its centre is the Earth... The Sun and the Moon rotate around the globe and move about the signs of the zodiac." This book was compiled in order to teach the officers how to use maps. In practice, however, it does not make a great difference whether the astronomical concept upon which this work is based is connected with the geocentric system of Ptolemy or the heliocentric system of Copernicus. This indicates that the practical aspect of the education in the *Mühendishäne* had great weight and that the concepts of new astronomy were not the focus of attention, as well as showing clearly that the old tradition of science still continued.

On the other hand, it is remarkable that Tamanī, who translated and adapted Western scientific sources, carried on the old tradition regarding astronomy. It is also worthy of attention that his student Ishak Efendi, who prepared this book for publication, neither commented on nor supplemented any new information to this subject since, as we shall explain in the following pages, Ishak Efendi presented a long text of the concepts and the theoretical and technical details of new astronomy in his own work *Mecmua-i Ulūm-i Riyāziye*. He published this book between the years 1830 and 1834 when he organized his teacher's notes on astronomy.

Seyyid Ali Bey (d. 1846),[76] who in 1817 was appointed *başhoca* in *Mühendishāne-i Berrī-i Hümāyun* after Hüseyin Rıfkı Tamanī, has works on geometry and on fortifications. Besides these, he translated the *Fathiyya*, an Arabic astronomical work by Ali Kuşçu (d. 1474), the most renowned Ottoman astronomer of the 15th century into Turkish. Seyyid Ali Bey, in the introduction of the translation which he entitled *Mir'āt-ı Ālem*, explains the geocentric views of Aristotle and Ptolemy; the heliocentric views of Pythagoras, Plato and Copernicus; and Tycho Brahe's individual view. Of these he prefers the first, i.e., the geocentric view. When he explains the reason for his preference, he states that this view is widespread in Islamic countries, the astronomical tables prepared in order to organize ephemerides are based on it, and therefore it is in accord with his practice.

Seyyid Ali Bey mentions that before he translated the *Fathiyya*, he examined some books and treatises related to the subject carefully and chose Ali Kuşçu's *Fathiyya* among them since it was "concise and significant", and translated it for the reason that the sciences of philosophy and mathematics were esteemed in his period.

Although there were two expanded Turkish translations[77] of this classical work by Ali Kuşçu in the 16th century, remarkably Seyyid Ali Bey handled this work again in the 19th century and printed it in 1824 with some additions and abbreviations. This shows that Ali Kuşçu was held in great favor and the old tradition of astronomy still existed; at the same time it indicates that

---

[76] For his life see *Mir'āt-ı Mühendishāne*, pp. 51, 61-62, 66, 82; *OM*, III, p. 275; Adıvar, pp. 209, 220; *OALT*, II, pp. 587-588.
[77] The first translation was prepared by Seydī Ali Reis (d. 970/1562) and the other one by Abdullah Perviz Efendi (d. 987/1579) under the title of *Hulāsat el-Hey'e* and *Mirkāt el-Semā*, respectively. It is remarkable that Seyyid Ali Bey does not mention these two previously translated works.

Seyyid Ali Bey wanted to carry on the tradition of the school in which he was educated.

Ishak Efendi,[78] who in 1830 was appointed *başhoca* in the *Mühendishāne* upon the dismissal of Seyyid Ali Bey, devoted a rather large space of 250 pages to the subject of astronomy in the fourth volume of his most important work entitled *Mecmūa-i Ulūm-i Riyāziye*. Here, he emphasized the Copernican theory and presented a long and perhaps the most technical explanation of this system as yet in Ottoman scientific literature.

As in several earlier books of astronomy Ishak Efendi too introduced the three main systems of the universe (see illustration 9). After explaining the reason why the theories of Ptolemy and Tycho Brahe are no longer in demand, he narrates the historical development of the Copernican view. He states that one can explain many astronomical events more easily under the assumption that the Earth is in motion, without comparing his view much with the old systems. At the end of this explanation, Ishak Efendi cannot help acting prudently and states that "although errors are likely to happen", this view is more agreeable with *ilm-i hikmet*.

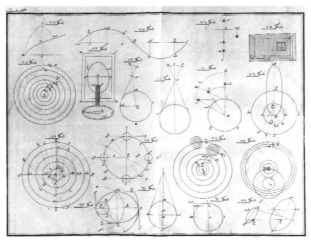

Illus. 9) Illustrations of the Ptolemaic system (picture 84), Tycho Brahe's system (picture 85) and Copernican system (picture 86) in the *Mecmūa-i Ulūm-i Riyāziye* by Ishak Efendi. (Vol. IV, Istanbul, 1834).

---

[78] For his life see *Mir'āt-ı Mühendishāne*, pp. 27, 34-42; *Kāmūs-ı Riyāziyāt*, I, pp. 299-301; *SO*, I, p. 328; *Ahvālnāme-i Müellifāt-ı Askeriye-i Osmāniye*, pp. 23-32; *OM*, III, pp. 254-255; Adıvar, pp. 219-221; *Türk Ansiklopedisi*, XX, Ankara, 1972, pp. 231-232; F. Reşit Unat, "Başhoca Ishak Efendi", *Belleten*, XXVIII, 1964, pp. 89-115; E. İhsanoğlu, *Açıklamalı Türk Kimya Eserleri Bibliyografyası*, Istanbul, 1985, pp. 10-11; *idem*, "Osmanlı Devleti'ne 19. Yüzyılda Bilimin Girişi ve Bilim-Din İlişkisi Hakkında Bir Değerlendirme Denemesi", pp. 95-96; E. İhsanoğlu and Feza Günergun, "Mecmûa-i Ulûm-i Riyâziye", *Türkiye Birinci Felsefe-Mantık-Bilim Tarihi Sempozyumu Tebliğleri*, Ankara 1986, pp. 19-21; E. İhsanoğlu, "Başhoca Ishak Efendi" *VI. Milletlerarası Türkoloji Kongresi* (unpublished paper), Istanbul 1988; *idem*, *Başhoca İshak Efendi*.

Throughout his four-volume work as well as in his other works, Ishak Efendi put forth and explained many new concepts of modern science in the Ottoman Empire in particular and perhaps in the Islamic world in general for the first time by way of translating and summarizing contemporary European sources. For example, he explained the new theories and laws of Descartes and Newton.

Ishak Efendi's technical survey of the contemporary knowledge in Europe was influential not only in Ottoman Turkey, but also in Cairo and other Islamic centres. The fourth volume of *Mecmūa-i Ulūm-i Riyāziye* was first printed in 1834 in Istanbul and also in Cairo in 1845. For many years, this work continued to be a principal source for those who wished to learn the many new sciences which developed in the West.[79]

While Ishak Efendi introduced the old and new astronomy in detail in his work published in Istanbul in 1834, Rifā'a al-Ṭahṭāwī (d. 1873),[80] a renowned intellectual and a translator of the period of Mehmed Ali Paşa, published a book of geography in Arabic entitled *al-Ta'rībāt al-Shāfiya li-Murīd al-Coghrāfya* in Cairo in the same year.

This work, which was the first book of modern geography published in Arabic in the Ottoman Empire, was compiled for the purpose of educating the students in the newly established schools in Cairo. It mentions the theories of new astronomy in a very brief and simple manner. On the other hand, the chapter on astronomy in the book entitled *Mecmūa-i Fenn el-Bahriye*, which was translated from European languages into Turkish for naval cadets and was printed in Bulak four years after Ṭahṭāwī's book, treats old and new astronomy very shortly. Here, however, the name Ptolemy is written as "Tolomya" instead of *Batlamyus* as it is known in the whole Islamic literature. This point is noteworthy since it shows that the translators did not take an interest in the old literature of science.

Among the books published in Cairo during the period of Mehmed Ali Paşa, we have not been able to find a separate book of astronomy. In that period, the source containing the most extensive information on new astronomy was again Ishak Efendi's book. It appears that the knowledge which was deemed satisfactory for Istanbul was also deemed satisfactory for Cairo, hence scholars there did not consider looking elsewhere.

The Transfer of Knowledge from Russian Science: Kudsī of Baku and *Asrār al-Malakūt*

After the quotations from and the translations of Western sources of astronomy in Latin and French between the 17th and the 19th centuries, we come across a work compiled using Russian sources. This was written by Abbāskulu Ağa b. Mirzā Muhammed Hān-ı Sanī (d. 1262/1846),[81] the

---

[79] *OMLT*, I, pp. 280-283.
[80] For his life and works see al-Ziriklī, *al-A'lām*, III, p. 29; Serkis, *Mu'jam al-Matbū'āt*, pp. 942-947; Jamāl al-Dīn al-Shayyāl, *Tārīkh al-Tarjama wa'l-Ḥarakat al-Thaqafiyya fī 'Aṣr Muḥammad 'Alī*, Cairo, 1951. See also *OCLT*, I, pp. 218-221.
[81] Abbāskulu Ağa, the son of Mirzā Mehmed Hān II, was one of the renowned poets, writers and historians of the Azerbaijani Turks in the 19th century. Abbāskulu Ağa, who used the

Azerbaijani literary scholar who used the pseudonym of Kudsī in his poems. Kudsī of Baku presented his work, entitled *Asrār al-Malakūt*[82] written in Persian and later translated into Arabic, to the Ottoman sultan Abdülmecid in person in the year 1846. The Grand Vezir Reşid Paşa charged Hayatizāde Seyyid Şeref Halil el-Elbistanī (d. 1267-68/1851)[83] with translating into Turkish this work which had won the sultan's esteem and aroused his interest so that it would be understood better. Besides translating Kudsī's work, el-Elbistanī supplemented some sections by making use of some previously printed works[84] on the subject and the work was published under the title of *Afkār al-Jabarūt fī Tarjamat Asrār al-Malakūt* in 1848.

The publication of this work shows the inclination of the authors to produce separate works on astronomy which proves that the new astronomy had gradually begun to arouse popular interest. The *Asrār al-Malakūt* was summarized from a book which Kudsī had previously compiled on geography. After presenting a short history of astronomy in the introduction, the author states that the Copernican theory is the most successful and the most correct one

pseudonym of Kudsī in his poems, can be counted as one of the most important poet-writer types which appeared among Caucasian Turks as a result of the union of the old Islamic culture with the views of the New European Civilization after the Russian invasion. He was born in the village of Emir-i Hāciyān near Baku in 1208/1794. He took service with General Yermoloff, the commander of the Caucasian-Russian armies, in 1820 at Tiflis. Abbāskulu Ağa, who learned Russian there, encountered European civilization through the works of Russian authors. He expanded his knowledge in such branches of science as history, geography and cosmography which he knew fundamentally. He traveled in many places in Caucasia, some regions of Anatolia and the Persian Azerbaijan on duty. He served in the company of General Paskevitch in the Turkish-Russian and the Russian-Persian wars as well as in peacetime. In 1833 he went on a journey, which lasted about two years, traveling in the countries of North Caucasia, the Don, Ukraine, Russia, Lithuania, Latvia and Poland where he met intellectuals and scholars. Upon his return, he settled in the city of Kuba and began to write various works. Abbāskulu Ağa's house in Kuba was a meeting place for the scholars and the poets of Kuba, and even for the thinkers and the artists who came from other places. Towards the end of his life, he came to Istanbul together with Hacı Molla Abdullah, the *kadı* of Kuba, for the purpose of going on a pilgrimage to Mecca. He met some scholars and poets of the time in Istanbul as well as presenting his work entitled *Asrār al-Malakūt*, which he once wrote in Persian and later translated into Arabic, to Sultan Abdülmecid together with an introduction addressed to the sultan. Hayatīzāde Seyyid Şeref Halil Efendi who translated this book upon the order of the Grand Vezir Reşid Paşa, states that Abbāskulu Ağa came to Istanbul on the 12th day of the Shawwāl of 1262 (3 October 1846) and that he himself even talked with Abbāskulu Ağa on various sciences and literature for about three hours in the social gathering of the great scholars of the time. Halil Efendi also praises Abbāskulu Ağa's knowledge and intelligence and states that according to his research, he passed away when he was on pilgrimage. Abbāskulu Ağa, died when he was 52 years old and was buried in the valley of Fatima between Mecca and Medina. (Fuat Köprülü, "Abbāskulu Ağa", *Türk Halk Edebiyatı Ansiklopedisi*, Istanbul 1935, no. 1). For Kudsī also see Muhammed Ali Terbiyet, *Dânişmendân-ı Azerbaycan*, first ed., Matbaa-i Meclis, Tehran, 1914, pp. 305-306; *Journal Asiatique*, 1925, CCVII, pp. 149-157.

[82] A manuscript copy of the Arabic text copied in 1313/1895 is in Library of the Iraq Museum, no. 19467. For a description of this work see Usāma Nāşir al-Nakshbandī-Zamya Muḥammad 'Abbās, *Makhṭūṭāt al-Falak wa'l-Tanjīm fī Maktabat al-Matḥaf al-Irāqī*, Baghdad, 1982, p. 99.

[83] Şeref Halil, the son of Hayātī Ahmed Efendi, was essentially occupied with languages. His principal work is a Turkish dictionary in four volumes entitled *Rawḍat al-Ashrāf fi'l-Muḍāf ilayh wa'l-Muḍāf*. For his other works see *OM*, I, p. 299; *OALT*, II, pp. 591-592.

[84] Hayātīzāde not only translated *Asrār al-Malakūt*, but expanded his translation with quotations from books of old astronomy, the commentaries on the Qur'ān, the books of tradition and other works, as it seemed appropriate, besides such books as İbrāhīm Hakkı's *Mārifetnāme*, Kâtib Çelebi's *Cihannümā*, Rifā'a al-Ṭahtāwī's *al-Ta'rībāt al-Shāfiyya*, *Mecmūa-i Fenn el-Bahriye* and İshak Efendi's *Mecmūa-i 'Ulūm-i Riyāziye* (*Afkār al-Jabarūt*, p. 5 ff.); *OALT*, II, p. 592.

for solving the problems which are dealt with. Kudsī, who wrote his work according to the conception of the new astronomy from beginning to end, gives preference to the Copernican system clearly and decisively, and explains his reason for preferring this system as follows:

"... Whether one prefers the Ptolemaic or the Copernican system, no inconvenience would ensue from the viewpoint of our religion because these matters are related to reason, not to religion. Since imitation *(taqlīd)* would not be lawful in matters pertaining to the intellect, we accept what the intellect prefers. Some Muslim scholars of new astronomy, who compared the Copernican view with the rules of reason and observation, defended its correctness on the basis of the Qur'ānic verses and the traditions of the Prophet. They were surprised to see how the Ptolemaic view, which did not conform to the principles of science and observation, continued to be well-known for such a long time. I realized that the Copernican view conforms to the clear and definite proofs deduced from geometry and moreover, to the Qur'ānic verses and the traditions of the Prophet; as for the Ptolemaic view, it is the opposite."

The conclusion of the work asserts that the conception of the universe which the new astronomy put forth is supported by the Qur'ānic verses and the traditions of the Prophet. Here, the author endeavors to reach an agreement between Religion and Science to an extent not hitherto encountered in any of the works mentioned above.

From the viewpoint of the history of culture and science, the most conspicuous and manifest feature of this work is the following: Neither the author nor the translator were professional astronomers, geographers, engineers *(mühendis)* or mathematicians; they were not technically educated in Western science; and they were brought up essentially in the classical oriental culture. In spite of this, merely thanks to their knowledge of languages, they worked at a new concept of a branch of Western science, accepted this new concept without hesitation, and described this system as the most rational one. Moreover, contrary to their predecessors, they not only pointed to its harmony with religion, but went further and stated that there were indications concerning the universe in accordance with it both in the Holy Qur'ān and in the traditions of the Prophet, the two basic sources of Islam. Therefore, we see that in mid-19th century the idea of the harmony between religion and science existed among enlightened men of religion and no conflict between religion and science had arisen yet.

Conclusion

In the above pages, we have examined as a case study the relationship between the Ottomans and Western astronomy during the broad time span of the 200 years from the middle of the 17th century until mid-19th century, based on manuscripts and printed works which have been dealt with above. Accordingly, we can point out this relationship in four stages:

1. Awareness and familiarity;
2. Utility and application;
3. Education as auxiliary intermediate courses imperative for the modernization of the state and the army; and

4. The harmonization of religion and science.

These stages, however, are not separated from each other by clear-cut lines and one cannot say that a given stage began only after the previous stage had ended. These stages are put forward as an attempt to explain the most characteristic aspects of this long and complicated "process". As we have seen, each stage has intricate parts, namely, a particular point which constitutes the principal subject of an advanced stage may have appeared as preliminary ideas and auxiliary themes in the previous stages. Since this point was also the main subject of that advanced stage, however, we have treated it under a separate heading.

Around mid-17th century, the Ottomans still considered themselves superior to the Western world. Moreover, since they had the scientific potential and institutions, namely, scientific and cultural autarchy to meet their own needs, they did not deem Western science to be necessary. This did not mean, however, that the Ottomans were far from or unaware of the scientific developments in the West.

Based on the first work which we have examined, entitled *Sajanjal al-aflāk fī ghāyat al-idrāk* and translated approximately between the years 1660 and 1664 by Tezkireci Köse Ibrāhīm, we can note the following observations about transfer of science:

The Ottomans were capable of following the developments in Europe without a great time lapse. Since the Ottoman astronomers had their own rich experience and were aware of the fact that Muslim astronomers had great contributions to astronomy during the Middle Ages, they did not accept this new European science right away. They, however, accepted it only after observing its compatibility with their own science.

While Europe was shaken by the heliocentric theory of Copernicus, the Ottoman astronomer Tezkireci Köse Ibrāhīm Efendi treated the basic concepts of this theory only on the level of a technical detail since the transition from the geocentric to the heliocentric system did not have any effect on practical calculations. Another reason for this approach may be that Ottoman society in this period was not conducive to debates on systems of the universe, probably as a result of the absence of religious dogmatism on this issue.

The subsequent works translated by the Ottomans were al-Dimashqī's 1685 translation of Janszoon Blaeu's *Atlas Major* and Osman ibn Abdülmennān's 1751 translation of Bernhard Varenius' work, which he entitled *Tercüme-i Kitāb-ı Coğrafya*. These translations show that Turkish scientists did not have any prejudice against the concepts of the new astronomy and particularly against heliocentricity. Since European writers, whose works these two translators rendered into Turkish were in favor of the geocentric system, they, too, transmitted the information in this manner. Among them, Osman ibn Abdülmennān, although he essentially followed the original text, preferred the heliocentric system by reasoning.

Thanks to the establishment of the printing house, works on the new astronomy which would address a broad mass of readers were also published. Ibrāhīm Müteferrika, known as the founder of the state financed printing house

II

in Ottoman society and who was originally a European convert to Islam, presented these new concepts in a very cautious manner in his supplement to Kātib Çelebi's *Cihannümā* which he printed in 1732. The reason for his prudence was his awareness of the strong reactions to these concepts on the part of the Church and some religious circles in Christendom.

Müteferrika translated Andreas Cellarius' *Atlas Coelestis* from Latin one year after he published *Cihannümā*. In this translation, he is much less prudent than he was in the *Cihannümā*, since there was no reaction in the Ottoman Empire similar to the one in Europe against such views.

Until this stage, the concepts of the new astronomy had been dealt with in professional scholarly circles of Istanbul and among astronomers. In the middle of the 18th century, such concepts were introduced to a new audience and presented to the people of Anatolia in a very interesting and contradictory style. Ibrāhīm Hakkı of Erzurum, a famous mystic of the 18th century, explained these concepts very clearly, presented them benefiting from the printed book of Müteferrika, and gave preference to the heliocentric system in a very manifest way. In another part of the book, however, he presented an understanding of astronomy based entirely upon Islamic folk cosmology and legendary explanations as well as fictitious traditions under the heading of "Islamic astronomy".

As it appears, Ibrāhīm Hakkı addressed both "the common people" and "the intellectuals" in his *Mārifetnāme* compiled in 1757. Besides these two groups, he had to be cautious of some official *ulemā* who envied his popularity and renown. In the above pages, we have mentioned the rumor according to which *müftī* Kadızāde Mehmed opposed Ibrāhīm Hakkı and blamed him with violating the *şerīat*. Most probably, in order to protect himself from such unjust accusations, Ibrāhim Hakkı included contradictory views in his work. For the same reason, he wrote his books *Urwat al-Islām* and *Hay'at al-Islām* in 1777, which are based on religious sources, as well as the section on "Islamic astronomy" in his *Mārifetnāme*. This regression from his scientific views in *Mārifetnāme*, which one observes in his later works, also appears to be a defense against some narrow-minded *ulemā*.

In this first stage of awareness and familiarity, the Ottomans received some information and preliminary observations about the scientific developments in Europe through their ambassadors who visited European capitals. Although in most cases they did not penetrate deeply into the matter, some of these ambassadors were closely interested in these developments on a technical level. For instance, Yirmisekiz Mehmed Çelebi visited the Observatory of Paris in 1721; and Hattī Mustafa Efendi visited the Observatory of Vienna in 1748. While Hattī Mustafa Efendi's visit was an ordinary diplomatic one, Yirmisekiz Mehmed Çelebi's visit took place in a much more different way. Merely due to his interest in astronomy, Yirmisekiz Mehmed Çelebi, who was not content with his first visit to the Observatory of Paris, paid a second visit. He had a detailed discussion with Cassini, the director of the observatory, and examined the modern astronomical instruments which were new to him. He also consulted Cassini about the observations which were contrary to Ulugh Bey's tables and received a written report from Cassini on these matters. Upon his return to Istanbul, Yirmisekiz Mehmed Çelebi communicated this report to various Ottoman astronomers.

Later, when the second stage of transfer began in the second half of the 18th century, Sultan Mustafa III ordered that "the most recent and the most perfect" books on European astronomy be brought from the Academy of Sciences in Paris. Hence, in this same period, Halifezāde Ismāil Efendi translated the work by the French astronomer Clairaut (d. 1765) under the title of Tercüme-i Zīc-i Kılaro in 1768 and another work by Cassini (d. 1756) under the title of Tuhfe-i Behīc-i Rasīnī Tercüme-i Zīc-i Kassini in 1772. Since the translation of these astronomical tables vitally effected Ottoman ephemeris compilation, Sultan Selim III ordered that ephemerides be organized in conformity with Cassini's astronomical tables, hence Ulugh Bey's tables fell out of use.

Cassini's tables were widely used, but the big margin of error in the calculations of these tables caused significant mistakes in the preparation of ephemerides. Therefore, the French astronomer Lalande's Tables Astronomiques were translated first into Arabic in 1814, and later into Turkish before the year 1826. Upon Sultan Mahmud II's order, ephemerides were organized according to these tables.

These translations of astronomical tables by the astronomers Durret, Clairaut, Cassini and Lalande between the 17th and the 19th centuries show that the Ottomans were informed about the literature on astronomical tables in the West.

The present stage of our research indicates that in these two stages, the principal sources and theoretical works in the West which brought about fundamental changes in astronomy escaped the attention of the Ottomans and they preferred to translate the works necessary for calendar making. This marks the practical aspect of Ottoman science which has been its primary characteristic.

Besides, we observe that in this period, the translation of some major geographical books was encouraged by the sultan or the grand vezir. On the other hand, no information was available regarding the translation of some main astronomical works which probably stood beyond the attention of the sultans. The fact that two printed works of Tycho Brahe are found in the Hazine section of Topkapı Palace Museum Library supports this remark.[85]

One can compare this "picture" during the first two centuries of the transfer of European astronomy to the Ottoman world, which we have treated as a case study, with the "picture" of the same branch of science in the first two centuries which witnessed the development of Muslim civilization. This comparison may lead us to an evaluation against the last period. In the first period, the activities of transfer and translation of scientific texts of pre-Islamic

---

[85] Tychonis Brahe, Dani, Astronomiæ instauratæ pregymnasmata by Dani Tychonis Brahe. –Prostat: Franco-furti Apud Godefridum Tompachium, 1610. (H 2804) and Tychonis Brahe, Dani, Mundi aethereirecentioribus phaenonenis, epistolarum astrono micarum, by Dani Tychonis Brahe, Prostat: Francofurti Apud Godefridum Tampachium, 1610. (H 2815).

science *('ulum al-awā'il)* resulted from meeting the same practical needs. Although in the first period the principal works on Indian and Greek astronomy such as *Siddhanta* and *Almagest* were translated, in the second, the basic European works by scientists such as Copernicus, Tycho Brahe, Kepler and Newton were not translated, indeed, no attempt was made to understand or interpret them. However, in this context we should not forget the fact that in this period the Ottomans had a rich tradition of science and it was not necessary to change it or substitute it for a new one immediately. The Ottomans deemed it necessary to transfer what they considered to be the most significant aspects of European science right away. This fact is important in indicating the character of the new period. At the same time, it is certain that the presence of an indigenous tradition of astronomy may have played a role in preventing the transfer of the science of new astronomy as a whole.

An interesting period of transition took place in the first three decades of the 19th century which characterized the third stage of transfer. It appears that Hüseyin Rıfkı Tamanī Efendi (d. 1817), *başhoca* in the Imperial School for Military Engineering with access to European sources, preferred to teach astronomy and geography on the basis of geocentricity during the first two decades of that century. He did not, however, teach these subjects in line with the concepts of modern astronomy. After Hüseyin Rıfkı Tamanī, Seyyid Ali Bey was appointed *başhoca* (1817-1830) in this school. Without referring to European sources, Seyyid Ali Bey turned towards classical Ottoman sources and translated the Arabic *Fatḥiyya* of Ali Kuşçu, the Ottoman astronomer of the 15th century, into Turkish. In the introduction of his translation, Seyyid Ali Bey gave preference to the geocentric view and stated that it was widely used in Islamic countries for preparing astronomical tables and organizing ephemerides.

An important categorical change took place with the appointment of Ishak Efendi as *başhoca* in the Imperial School for Military Engineering. Ishak Efendi is to be considered a vanguard of modern science in the Ottoman and Muslim world. He devoted 250 pages to modern astronomy in his four-volume *Mecmūa-i Ulūm-i Riyāziye.* In this work many of the modern sciences to the Ottoman and Muslim world were presented for the first time. It was printed in Istanbul in 1834 and reprinted in Cairo in 1845. Thanks to this book, the Ottoman world and the Muslim world, which followed the developments of the Ottomans, had a comprehensive survey on modern astronomy for the first time.

The last work which we have treated in this paper shows us the fourth stage in the transfer of science, i.e., the discussion of a harmony between religion and science. As a case in point, the work by Kudsī Efendi of Baku is interesting in that it has been devoted to this subject. His *Asrār al-Malakūt* was presented to Sultan Abdülmecid in 1846 by the author himself. It is based entirely upon the concepts of new astronomy and in favor of the Copernican system. In explaining the reasons for preferring the Copernican view, the author states that this view is in conformity with the definite proofs deduced from science by reasoning and with the Qur'ānic verses and traditions of the Prophet as well; the Ptolemaic view, however, is contrary to the rules of science and observation. In regard to the history of culture and science, the most striking characteristic of this work is that in the conclusion, the author makes great efforts to indicate the harmony between religion and science to a degree which was not seen in any of the earlier works. It can therefore be stated that in mid-19th century, religion and science were essentially considered compatible. One sees that Kudsī's book accomplished this compatibility, based upon a basic argument developed by al-Ghazālī, within the framework of the principles which Müteferrika put forward with very cautious and uneasy statements and Ibrāhīm Hakkı of Erzurum stated in the scientific part of his work in a very clear and manifest way. The harmony reached in Kudsī's book is more developed and far from contradictions in comparison with these previous works. Indeed, this book can be considered as a prototype for those ideas which were asserted later particularly in the 20th century in the Muslim world to the effect that modern science and Islam are compatible, and also for those who went one step further and asserted that many discoveries of modern science can be deduced from the principal sources of Islam and particularly from Qur'ānic verses.

Although the developments after the fourth stage, which took place until the beginning of the 20th century, are outside the historical framework which has been delineated in this study, the landmarks of the subject can be stated in the following way for those who intend to do further research on this matter:

1. The acquaintance of broader masses of people with modern science through modern educational institutions as a part of the curriculum.

2. The dissemination of this modern science among uneducated but literate masses through the emerging newspapers and periodicals.

3. Attempts to establish a "scientific institution", i.e., an observatory, to meet the needs of the modern state and society.

4. Finally, after three unsuccessful attempts, the establishment of the University of Istanbul ("Dārü'lfünūn") in 1900 in which Faculty of Science was founded for the purpose of teaching science *per se*. This final development added a new dimension to the transfer of Western science to the Ottoman world.

II

Note on transliteration

In this paper, works in Ottoman Turkish, proper names and terms in the Ottoman context have been transliterated in accordance with the official modern Turkish orthography. The lengthened vowels have been indicated. In transliterating the titles of books, proper names, terms and usages in Arabic and Persian, the system of transliteration in *The Encyclopaedia of Islam,* new edition *(EI²)* has been followed except that 'j' is used instead of 'dj' and 'q' instead of 'ķ' Some pronunciations of modern Turkish orthography are noted below:

c- Pronounced 'j' as in 'John'
ç- Pronounced 'ch' as in 'charity'
ğ- a soft guttural, pronounced almost like the 'gh' in 'ought'
ı- like the a in serial or io in cushion
ö- pronounced like the French 'eu', as in 'lieu'
ş- pronounced like the 'sh' in 'shovel'
ü- pronounced like the French 'u', as in 'lune'

Some well-known place-names (e.g. Cairo, Damascus) are spelled according to the anglicized forms.

Acknowledgements

The present article is the updated and reedited version of the paper first presented at the international symposium on "Modern Sciences and the Muslim World: Science and Technology Transfer from the West to the Muslim World from the Renaissance to the Beginning of the XXth Century" (Istanbul, 2-4 September 1987). The article was later published in *Transfer of Modern Science & Technology to the Muslim World,* ed. E. İhsanoğlu, IRCICA: İstanbul, 1992.

I wish to take this opportunity to thank Professor David A. King of Johann Wolfgang Goethe University, Frankfurt for his patience in reading the first printed version of this article and making valuable suggestions to improve the text. It was also an important occasion to draw some new conclusions and to update some of the footnotes including the research results which appeared since the first publication of this article.

It gives me pleasure to reiterate my gratitude to my colleagues who did not spare their valuable assistance during the preparation of the first text of this paper. I wish to express my thanks to IRCICA for its support in every respect during the long period in which this paper was prepared. On this occasion I cherish the memory of my dear student and colleague the late Cevat İzgi who assisted in the preparation of the first text. I wish to thank Dr. Semiramis Çavuşoğlu for checking the English text of this paper and Mr. Vedat Kaya for typing the text. I treasure the memory of the late Prof. Dr. Muammer Dizer, Director of Kandilli Observatory with whose assistance I had access to the books in the observatory's library. My thanks also go to Mr. Eli Nicolas of Institut Français d'Etudes Anatoliennes for his valuable assistance in translating the passages in Latin.

III

OTTOMAN SCIENCE :
THE LAST EPISODE IN ISLAMIC SCIENTIFIC TRADITION AND THE
BEGINNING OF EUROPEAN SCIENTIFIC TRADITION

Upon consideration of the scientific activities carried on throughout the six-century history of the Ottoman Empire (1299-1923), it may be argued that the history of Ottoman science witnessed several distinctive trends. Though the historical evolution of Ottoman science shared many features common to the history of scientific endeavor in other parts of the Muslim world (beyond the boundaries of the Ottoman Empire), there were also some important differences. In the process of this historical evolution, the Ottomans were considered " pioneers " in some areas of scientific endeavour.

Inspired by the medieval Islamic scientific tradition at the beginning of its history, Ottoman science quickly became influential in the old scientific and cultural centers of the Islamic world. However, when in the seventeenth century the Ottomans were introduced to European science, they became instrumental in spreading this new scientific tradition through the Muslim lands. Thus, the Ottomans, who in their day were the representative of the whole Islamic world, were able to combine Islamic scientific tradition with the newly emerging Western science. Around the turn of the nineteenth century, however, the Ottomans opted for the exclusive practice of European science over Islamic, and the Islamic scientific tradition gradually faded away as a new tradition emerged in accordance with modern Western scientific norms. This paper aims to draw a general picture of the history of Ottoman science, and show the role it played in bringing together two different civilizations. It also attempts to survey the scientific activities conducted by the Muslims that formed the majority of the population in the Ottoman Empire.

Scientific activities were, of course, also undertaken among the non-Muslim population of the Ottoman Empire, which was comprised of Jews and various Christian subjects, including Greeks, Armenians, Bulgarians, Serbs, Romanians, and Hungarians. The scholarly and scientific activities of these people,

III

which were recorded in languages other than Turkish, Arabic, or Persian have not been included in this paper. The main reason for this exclusion is that studies on the scientific activities of non-Muslim Ottoman subjects are as yet insufficient in number to allow us to offer a general overview of this aspect of the history of Ottoman science.

The first part of this survey deals with the Ottoman absorption of classical Islamic scientific tradition. This tradition came to the Ottomans by way of a pre-Ottoman Islamic scientific legacy — most importantly, the scientific legacy of the Seljuk Turks, who had ruled Anatolia and the Middle-East before the Ottomans. The second part will analyze modern Western scientific tradition, which was introduced to and developed within the Ottoman world as a result of the Ottomans' close relations with Europe in later centuries.

In the first centuries of the history of Islam and the first stages of the political and legal formation of Islamic civilization, teaching activities and education were conducted mainly in mosques. Translations made from Greek and Indian sources, which sought to introduce those sciences called *Ulumu'l-evail* (the sciences of the ancients, including philosophy, mathematics, medicine, astronomy, physics, chemistry, and so on) led to the creation of new institutions to house these new scientific activities. An early example is the *Beytü'l-hikme* which was founded in Baghdad. Institutions of this type, later known as *Darü'l-hikme* and *Darü'l-ilm*, dealt mainly with non-religious sciences and were founded under the patronage of the sovereigns.

In the eleventh century A.D., a new educational institution called the *medrese* was established in the Eastern part of the Islamic world, in Iraq, Iran, and Khorasan. The main focus of the formal education offered in the *medrese* was Islamic jurisprudence and related religious sciences. The *medreses* were supported by pious endowments (*waqf*s) that were established by individuals for the sake of God. The State did not directly interfere with education or curriculum in the *medrese*. The content of the curriculum and methods of instruction were based on the historical evolution of the teaching profession and the dynamic relationship between the various religious and intellectual trends.

In the Great Seljuk period, the *medrese* (with its traditionally adjoining dervish monastery) became the most popular form educational institution. Due to the efforts of Nizamü'l-Mülk (1015-1092), the Grand Vizier of Sultan Alparslan (and later of his son, Sultan Melikshah,) *medreses* became widespread. Nizamü'l-Mülk founded a " Nizamiye Medrese " in every prominent city in the Seljuk Empire, the most famous among these being the Baghdad Nizamiye Medrese, founded in 1065.

According to G. Makdisi, the institutional education offered within the medieval Islamic lands was concerned solely with " religious " sciences. The legal structure of the *medrese* system was developed by Islamic jurists (*fuqaha*). Consequently, *fiqh* (Islamic law) studies were given priority in the *medrese* system, and rational sciences were not included in the curriculum. Higher officials, who would go on to be appointed to religious, juridical, and political posts within the Empire, were educated and trained according to the *sunni* doctrine in these *medreses*. Before long, the establishment of *medreses* became widely popular in the *sunni* community, and, as a result of the State's expanding organization, these *medreses* clustered in specific centers.

Foundation of new *medreses* continued at full throttle in the smaller Muslim states that eventually branched off of the Great Seljuk Empire. During the reign of Nureddin Zengi (d. 1174) and Selahaddin Eyyubi (d. 1193), members of the royal family and prominent statesmen began to establish *medreses*, hospitals, and dervish monasteries in different cities. In this period, the *medreses* even spread to small towns, where many charitable and educational institutions were founded.

Islamic culture and science enjoyed its most glorious period under the Seljuks, and during the subsequent period when smaller states budded off of crumbling Seljuk Empire. The writings of many famous scholars, works of art, buildings, and other architectural monuments that have come down to us attest to the glory of this period, and preserve the memory of its magnificence and power even into the present day. During the era of the Seljuks and their followers, cities such as Baghdad, Merv, Isfahan, Nishapur, Mosul, Damascus, Cairo, Aleppo, Amid (Diyarbakir), Konya, Kayseri, and Malatya all became a scholarly centers thanks to the establishment in those places of cultural, educational, and medical institutions.

After Cengiz Khan's (d. 1227) invasion, cultural and scientific activities in the Islamic world came to a standstill. However, learning was to regain momentum a century later during the Ilkhanid period, specifically during the reigns of Hulagu (d. 1265) and Gazan Khan (d. 1304), and later during the Muzafferid period. The observatory founded in the town of Meraga in Azerbaijan and the attached school of mathematics and astronomy, together with the cultural activities in the adjacent regions of Iran and Transoxania were behind this revival of learning. Scientific and scholarly activities were greatly influenced by this new tradition. In Gazan Khan's period Tabriz replaced Meraga as the leading center of learning. *Medreses* in the cities of Shiraz, Kirman, and Yazd also made significant advances during the Muzafferid period.

The second great Central Asian emperor, Timur (d. 1405), reigned over a vast realm extending from the borders of China to the Ukraine, and occupied India, Iran, Iraq, Azerbaijan, Turkistan, southern Russia, Anatolia, Syria, and Egypt between 1370 and 1405. During his reign, Timur patronized *medreses*, libraries, and other institutions of learning, as well as funded the development of cultural institutions in Turkistan, Khorasan, and Transoxania. Indeed, his patronage of such institutions of learning was extensive by comparison with

III

14

other contemporary rulers. Consequently, during the reigns of Timur and his son Shahruh (d. 1447), Samarkand and Herat became famous centers of learning and culture, attracting scholars, artists, and students from all over the Islamic world.

Timur conquered the Ottoman lands briefly ; but after his death, the peace established between the Ottomans and the Timurids during the 43-year reign of Timur's son Shahruh was instrumental in drawing these two *sunni* empires closer to one another in matters of politics and religion. The political cooperation between the Timurid and Ottoman states also led to fruitful cultural exchange. Ottoman students went to study in Samarkand and Herat, and scholars and artists emigrated from Central Asia to the Ottoman State. It is very likely that the close associations with the Timurid lands cultivated by the Ottoman students who went to study there played an ongoing role in this exchange.

Those scholars who were patronized by the Timurid sultans were greatly disturbed by the conflict and political instability in the Timurid Empire following the deaths of Shahruh and his elder son Ulug Beg (d. 1449) — the latter ruler also a great scholar. These upheavals ultimately forced scholars to emigrate from Khorasan, Transoxania, and Azerbaijan to more peaceful lands, where scholarly and scientific activities were better appreciated. Only the Ottoman lands and India, which was then under the rule of another Turkish dynasty, the Mughals, had the conditions these scholars required to pursue their work. While some poets and physicians went to India, many prominent mathematicians and astronomers preferred to come to the Ottoman lands.

OTTOMAN INSTITUTIONS OF LEARNING AND SCIENCE

In the Ottoman Empire, the most important institution of religious, scholarly, and cultural activities was the *medrese*. When new lands came under Ottoman rule, the Ottomans immediately founded numerous *medreses* all over these lands, as well as sponsoring those that had already been established by previous rulers.

In the Ottoman period, the basic structure of *medrese* education remained in keeping with prior tradition, but the methods and the objectives of teaching and the scope of the curriculum with regard to the sciences underwent several important changes. The first Ottoman *medrese* was established in Iznik (Nicaea) in 1331 by Orhan Bey (1326-1362), the second Ottoman sultan. Thereafter, members of the Ottoman dynasty, statesmen, and prominent men of learning continued to found *medreses* throughout the vast expanse of the Empire. The continuity and development of these institutions was ensured by financially secure pious endowments (*waqf*s).

Table 1: Ottoman *medreses* in important centers and regions

	XIVth century	XVth century	XVIth century	Undefined period	Total
Iznik	4				4
Bursa	19	11	6		36
Edirne	1(Darüsshifa)	20	10		31
Istanbul		23	113	6	142
Anatolia	12	31	32	13	88
The Balkans	4	12	18	5	39
Syria			3		3
Hijaz			6		6
Yemen			1		1
Total	40	97	189	24	350

Table 2 : Distribution of *medreses* according to the reigning Sultans

Orhan Gazi (1326-1359)	10
I. Murad (1359-1389)	7
I. Bayezid (1389-1402)	23
Çelebi Mehmed (1402-1421)	7
II. Murad (1421-1451)	38
II. Mehmed (1451-1481)	30
II. Bayezid (1481-1512)	33
I. Selim (1512-1520)	8
I. Süleyman (1520-1566)	106
II. Selim (1566-1574)	17
III. Murad (1574-1595)	42
III. Mehmed (1595-1603)	5
Medreses in undefined periods	24
Total	350

Table 3 : Rumeli *medreses* in the Ottoman period

Regions	*Medreses*
Greece	189
Bulgaria	144
Albania	28
Bosnia-Herzegovina, Croatia, Montenegro	105
Kosova-Macedonia-Serbia, Slovenia, Voivodinia	134
Romania	9
Hungary	56
Total	665

16

In a short period of time, a large number of students were educated in the emerging Ottoman *medreses*, and an active scholarly environment was formed on the Ottoman soil, thanks to the instruction offered in the *medreses* by great men of learning. The tradition of sending the Ottoman scholars abroad and inviting scholars from other countries to Anatolia also continued. These travels for scholarly purposes are further evidence of the scientific dynamism of this period.

Active cultural and educational relations were established between Ottoman Anatolia and its Muslim neighbors : Iraq, Iran, and the Persian heartland of Transoxania in the east, and the Arab countries in the south, especially Syria and Egypt. Men of learning trained in these places were welcomed in the Ottoman world and their works were used as textbooks in the *medreses*.

Shortly after Byzantine Constantinople's final surrender to Sultan Mehmed II (" Fatih " or " The Conqueror ") in the year 1453, the Sultan built the first major religious and educational complex of Ottoman Istanbul. This was the Fatih *külliyesi* (Fatih complex), consisting of a mosque and courtyard surrounded by an elementary school (*mektep*), colleges of higher learning, a hospital, public kitchens, and auxillary buildings. The construction of this *külliye* set an example for the Sultan's successors and high-ranking members of the ruling class to follow. The colleges of the Fatih Complex (known as the *Sahn-i Semân medreses*) included 16 *medreses* (constructed adjacent to the mosque) : eight middle-level schools (*tetimme*) and eight higher level schools (*âlî*). This was an important breakthrough in the development of the *medrese* system in the Muslim world. Graduates of the Fatih *medreses* went on to serve as teachers, bureaucrats, *kadi*s (judges), *kazasker*s (highest ranking judges) and *mufti*s in the Ottoman Empire.

As a result of a general climate of political stability and economic prosperity, distinguished scholars and artists from all over the Islamic world assembled in the new Ottoman imperial capital of Istanbul. In addition to the favorable political conditions, the changes made in the traditional *medrese* system during the reign of Sultan Mehmed II fired the progress of Ottoman scientific education and production. With the foundation of the Fatih *medreses*, the Islamic world experienced an unprecedented wave of scientific progress. This was due in part to a broadening of the sope of *medrese* education, opening the door to the rational sciences, including subjects such as logic, mathematics, astronomy, and the natural sciences, in addition to continued instruction in religious sciences. For the first time — at least as regards the Ottoman *medrese* system — the charter of these colleges officially stipulated that the teachers should have some knowledge of *ulum-i akliye* (rational sciences) in addition to the traditional religious sciences. It is thus implied that the teaching of rational sciences was slowly being included within the formal *medrese* education. The increased interest in rational sciences also led to an increased production of scholarly works.

The institutions that were established under the patronage of Sultan Mehmed II and his son and successor Sultan Beyazid II were instrumental in the progress and development of Ottoman science in the second half of the fifteenth century. The continuing patronage given to scholars and the cultivation of a scholarly environment in the sixteenth century, during the reigns of sultans Selim I, Süleyman I, Selim II and Murad III, resulted in an active period in the history of Islamic science in Istanbul. Original contributions were made in, for example, Taqi al-Din's work in the field of astronomy. We shall return to this later.

When Egypt was conquered and annexed to the Ottoman Empire in 1517, many *sunni* lands in the Middle East and North Africa came with it into the fold of the Ottoman Empire. By the sixteenth century, when the Ottomans had become a world power, the Ottoman world was virtually synonymous with the Islamic world.

Establishment of the Süleymaniye Complex by Sultan Süleyman the Magnificent (r. 1520-1566) marked the final stage of development in the Ottoman *medrese* system. The charter of the Süleymaniye *medreses* made specific reference to the teaching of rational sciences in the *medreses* established by Sultan Mehmed II, and it, too, stipulated that these sciences should be taught during the course of formal education offered in these colleges. A specialized *medrese* named *Dâruttib* (medical college) was added to the conventional four *medreses*. Thus, in addition to the *shifâhânes* (hospitals) where medical students were trained in the traditional hands-on way, an independent college was now established in the Süleymaniye Complex to give medical education. This college functioned until the middle of nineteenth century when European-style medical schools were opened. Other specialized *medreses* established by the Ottomans were the *Darülhadîs* and the *Darülkurrâ*, the former being the highest institution in the hierarchy of the *medreses*. As the pious foundations which provided the financial resources of the *medreses* grew rich, scholarly and educational life developed to even greater heights.

In addition to formal scholarly education on offer in the *medreses*, there were other trade and professional institutions where sciences, including medical sciences and astronomy, were taught and practiced by the master-apprentice method. These scientific educational institutions were typically housed in the mansions of distinguished men of society, or in the *shifâhânes* (hospitals), *muvakkithâne* (timekeeper's office), or the office of the *müneccimbashi* (chief astronomer).

The institutions in Ottoman society which provided health services and medical education were called *dârüsshifâ, shifâhâne, or bîmâristan*. The Seljuks of Anatolia had built *dârüsshifâ*s in Konya, Sivas, and Kayseri. Similarly, the Ottomans built several *dârüsshifâ*s in major cities such as Bursa, Edirne, and Istanbul. Some European sources report that there were a large number of *dârüsshifâ*s in Istanbul in the sixteenth and seventeenth centuries. These were not constructed independently but rather were part of a *külliye* (complex). The most important Ottoman *dârüsshifâ*s were the following : the Fatih *dârüsshifâ* in Istanbul (1470), the Beyazid II *dârüsshifâ* in Edirne (1481), the Süleymaniye *dârüsshifâ* in Istanbul (1550), the Haseki *dârüsshifâ* in Istanbul (1550), and the Hafsa Sultan *dârüsshifâ* (1522-23), built by the wife of Sultan Selim I in her name in the city of Manisa. Physicians were trained and patients treated in these *dârüsshifâ*s, some of which continued functioning right up to the middle of the nineteenth century, when European-style hospitals began to take over.

*Muvakkithane*s, or offices of the timekeepers, were public buildings located in the courtyards of mosques and *mescid*s in almost every town. They became widespread especially after the conquest of Istanbul by the Ottomans. They were run by the *muvakkit*, the person responsible for keeping the time. In particular, the *muvakkit* was responsible for keeping track of the correct prayer hours. Instruments used for timekeeping in the fifteenth and sixteenth centuries included the *rubu tahtasi* (quadrant), the *usturlâb* (astrolabe), the sextant, the octant, the hourglass, the sundial, the mechanical clock, and the chronometer. Depending on the timekeeper's expertise, the *muvakkithane*s might also function as the centers where mathematics and astronomy were taught.

In addition to the above-mentioned institutions, all of which were financed by *waqf* endowments, two institutions, the offices of the *hekimbashi* (Chief Physician) and the *müneccimbashi* (Chief Astronomer) were set up within the official Ottoman bureaucracy.

Office of *hekimbashi* (Chief Physician) was vested with full responsibility for the Sultan's medical needs, as well as those of the imperial family and the palace staff. However, he was more than just a private physician to the palace. In addition, the administration of all medical institutions and practitioners in the State, including physicians, surgeons, ophthalmologists, and pharmacists was placed under the direction of the office of the *hekimbashi*. The best-trained physicians were chosen from among the learned classes for this office. Some of the physicians who took the post of Chief Physician were among the most celebrated scholars of their time.

Also under the Office of the Chief Physician was the office of the *müneccimbashi* (Chief Astronomer), who was selected from among the religious intellectuals that graduated from the *medreses*. The institution of *müneccimbashi* was established in the late fifteenth or early sixteenth century. The office was entrusted with the preparation of official calendars, prayer and fasting timetables, and horoscopes for prominent statesmen and members of the imperial family. Until the year 1800, Ottoman calendars were made according to the *Zij* of Ulug Beg ; from that date onward Jacques Cassini's *ephemerides* were used.

The Chief Astronomer (or one of the other senior astronomers in the office) determined the auspicious dates and hours for important events, including imperial accessions, wars, imperial births, weddings, the launching of ships, and so on. The Chief Astronomer also kept track of unusual astronomical events (and the earthly cataclysms to which they were believed to be related), such as the passage of comets, earthquakes, fires, and solar and lunar eclipses. The Office of the Chief Astronomer forwarded information concerning such events to the palace together with their professional interpretations. The administration of the *muvakkithane* also came under the duties of the Chief Astronomer. The famous observatory founded in Istanbul in the late sixteenth century was administered by the Chief Astronomer Taqi al-Din Rasid (d. 1585). A total of thirty-seven individuals held the post of Chief Astronomer over the course of the Empire's history. The post of Chief Astronomer was retained all the way up until the Empire came to its final end ; it was abolished in 1924. In its place, the *bashmuvakkitlik* (office of the Chief Timekeeper) was established in 1927.

FOUNDATION OF THE ISTANBUL OBSERVATORY

Chief Astronomer Taqi al-Din was born to a family of Turkish descent in Damascus. He came to Istanbul from Egypt in 1570, and was appointed *müneccimbashi* by Sultan Selim II (1566-1574). Shortly after Sultan Murad III's (1574-1595) accession to the throne, he undertook the construction of an observatory in Istanbul. This was an elaborate structure which included, in addition to the observatory itself, residential quarters, offices for the astronomers, and a library. It was planned as one of the largest observatories in the Islamic world, and equipped with the best instruments of the time. Taqi al-Din's Istanbul observatory was comparable to Tycho Brahe's (1546-1601) Uranienborg Observatory, built in 1576. Indeed, there is a striking similarity between the instruments used by Tycho Brahe and those of Taqi al-Din.

Taqi al-Din had fifteen assistants in the Istanbul observatory. From his *zij* titled *Sidrat Muntaha'l-Afkâr*, we understand that he started making observations at the new observatory in the year 1573. It is generally agreed that the observatory was demolished on 4 Dhilhijja 987 in the Islamic calender, corresponding to January 22, 1580. Therefore, we may suppose that Taqi al-Din's observations were carried out between 1573 and 1580.

Taqi al-Din developed a new method of calculation to determine the latitudes and longitudes of the stars by using Venus and the two stars near the ecliptic, known as Aldebaran and Spica Virginis. By using the method called " three points observation " he calculated the eccentricity of the orbit of the Sun, and yearly mean motion of the apogee. According to Copernicus the eccentricity is 1p 56' ; according to Tycho Brahe it is 2p 9', and according to Taqi al-Din it is 2p 0' 34" 6''' 53'''' 41''''' 8''''''. As compared to modern calculation, Taqi al-Din's is the most accurate value. According to Copernicus, the annual motion of the apogee is 24" ; to Tycho Brahe it is 45", and to Taqi al-Din it is 63". Its real value is 61". As far as world astronomy is concerned,

20

Taqi al-Din's results can be said to be the most precise in the calculation of solar parameters.

As well as producing new observations and methods of calculation, Taqi al-Din invented a number of new astronomical instruments. The instruments in use in his observatory included the following : the armilary sphere, parallactic ruler, and the astrolabe, which were originally invented by Ptolemy ; the azimuthal and mural quadrants, which were invented by Muslim astronomers ; a *mushabbaha bi'l-manatic* (the sextant, an instrument with cords used to determine the equinoxes), built by Taqi al-Din, which very much resembled an instrument of the same type invented by Tycho Brahe. Taqi al-Din also built a wooden quadrant to measure azimuths and elevations, and mechanical clocks to measure the true ascension of the stars. The latter was one of the most important discoveries in the field of practical astronomy in the sixteenth century, because the clocks previously in use were never accurate enough. To quote Taqi al-Din, " We constructed a mechanical clock with three dials which show the hours, minutes, and seconds. We divided each minute into five seconds ".

Based on his own observations, Taqi al-Din prepared a *zij* for Sultan Murad III entitled *Sidrat Muntaha'l-Afkâr fi Malakut al-Falak al-Davvar - Zij-i Shahinshahi*. This *zij* contained the tables of the sun, but these were incomplete. His work titled *Jaridat al-Durar wa Kharidat al-Fikar* contained the tables of the moon, which were not based on his own observations. It must be noted that Taqi al-Din's observations were exact and he used the decimal fractions instead of the sexagesimal fractions in his astronomical calculations. He also prepared trigonometry tables according to decimal fractions.

Taqi al-Din, who read the works of Muslim scholars, produced a critique of early astronomical works in addition to reporting his own findings. Without doubt his works are representative of the most advanced stage of Ottoman science. Within a short time of its foundation, the Istanbul observatory was contributing the latest advances in the Islamic tradition of astronomy. Unfortunately, the observatory was demolished on account of false religious assertions made by envious statesmen — a dispute in which Taqi al-Din was caught between rival sides. The demolition of the observatory marks the beginning of the decline of a classical Ottoman scientific tradition.

OTTOMAN SCIENTIFIC LITERATURE IN THE CLASSICAL PERIOD

The Ottoman scientific literature in the classical period was produced mainly within the milieu of the *medrese*. Original works and translations were produced in the fields of religious sciences, mathematics, astronomy and medicine, in addition to which a great number of textbooks were compiled by scholars. These works were written in Arabic, Turkish, and, to a lesser extent, Persian. (These were the three languages — known as the *elsine-i selâse* —

III

which the Ottoman scholars mastered). At first, Ottoman scientific literature was written mostly in Arabic, the basic and common language of Islamic civilization. However, in the fourteenth and fifteenth centuries, a movement was begun toward the production of Turkish translations of Arabic works. This activity was encouraged by those administrators who did not know Arabic, and in time Turkish was used to disseminate information to the public in their own vernacular language. Thus a concerted effort was made to translate works in every field of Islamic learning and practice (from encyclopedic manuals on medicine and drugs, geography, astronomy and the interpretation of dreams, to treatises on music and dictionaries) into simple and clear Turkish, and in this way to introduce Islamic culture to a wider public. From the eighteenth century onward, the majority of scientific works were written in Turkish, and upon the establishment of the first printing press in Istanbul in 1727, Ottoman Turkish became the most frequently used language in the transmission of knowledge about the modern sciences in the Ottoman Empire.

Bursali Kadizâde-i Rumî (d. 844/1440), the first prominent Ottoman scholar, made important contributions to the development of an Ottoman scientific tradition and literature. He started his scholarly career in Anatolia, where he wrote his first book. He then moved and settled in Samarkand. Among his works were *Sharh-Mulakhkhas fi'l Hay'a* (a commentary on the *Compendium on Astronomy* of Chaghminy) and *Sharh Ashkal al-Ta'sis* (a commentary on *The Fundamental Theorems of Geometry* of Samarkandî), both written in Arabic. He was made the chief instructor of the Samarkand *medrese* and the director of the observatory founded by Ulug Beg (d. 1449) in Samarkand. He was also the co-author of *Zij-i Gurgani (Zij-i Ulug Beg)*, the famous astronomical tables of Ulug Beg, written in Persian. He simplified the calculation of the sine of a one-degree arc in his work *Risala fi Istikhraj Jaybi Daraja Wahida (Treatise on the Calculation of the Sine of a One-Degree Arc)*. Kadizâde-i Rumî's works made a significant impact on the Ottoman Empire, and his influence continued to expand through the works of Ali Kushcu (d. 1474) and Fathullah al-Shirvânî (d. 1486), two students of his who came to the Ottoman Empire from Turkestan.

In the preface of his work *Sharh Ashkal al-Ta'sis*, Kadizâde-i Rumî opined that the philosophers who pondered the creation and the secrets of the universe, the jurists (*faqih*s) who passed *fetvâ*s on religious matters, state officials, and *kadis* (judges) should all have knowledge of geometry. Thus, he emphasized the importance of studying mathematics for the pursuit of philosophical, religious, and worldly matters. This way of thinking reflects the general character of Ottoman science in the classical period — which lasted until the Ottomans began to adopt Western science. However, in the period of modernization, it became apparent that the Western concept of man's domination of nature through science and technology was foreign to Ottoman scholars, who came out of an Islamic scientific legacy and belief system.

III

Among other astronomy books read in this period were *Urjuza fi Manazil al-Kamar wa Tulu'iha* (*Poem on the Mansions of the Moon and their Rising*) and *Manzuma fi Silk al-Nujum* (*Poem on the Orbits of the Stars*), both written in Arabic by 'Abd al-Wahhab b. Jamal al-Din b. Yusuf al-Mardani'. Two books by Nasir al-Din al-Tusî, the founder of the Maragha school, titled *Risala fi'l-Takwim* (*Treatise on the Calendar*) and *Si Fasl fi'l Takwim* (*Thirty Sections on the Calendar*) were translated from Persian into Turkish. Ahmed-i Dâ'î (d. *ca.* 825/1421) is known to be the translator of the second work.

In addition to Samarkand, Egypt was another important source for Ottoman science in the classical period. Haci Pasha (Celâleddin Hidir) (d. 1413 or 1417), a well-known physician educated in Egypt, wrote two books in Arabic, *Shifa' al-Askam wa Dawa'al-Alam* (*Treatment of Illnesses and the Remedy for Pains*) and *Kitab al-Ta'alim fi'l-Tibb* (*Book on the Teaching of Medicine*), both of which played an important role in the progress of Ottoman medicine. This author had many other works in both Turkish and Arabic.

In the field of medicine, the works of Sabuncuoglu Sherafeddin (d. *ca.* 1468) were particularly important in the development of Ottoman medical literature. His first book on surgery written in Turkish was titled *Jarrahiyat al-Khaniyya* (*Treatise on Surgery of the Sultans*) was an all-in-one handbook for every branch of medical science. The book was a translation of Abu'l Kasim Zahrawi's *al-Tasrif*. However, it was enriched by miniatures depicting surgical interventions. A chapter on the cauterization of boils and another on the treatment of herpes was added together with a list of drugs mentioned in the book. As the book includes the first miniatures depicting surgical operations, it is well-known in the history of Islamic medicine. In addition to classical Islamic medical information, the work reports on the author's own experiences and shows evidence of influence from Turco-Mongolian and far Eastern medical traditions. Sabuncuoglu's influence was felt outside the boundaries of the Ottoman Empire — particularly in Safavid Iran — through his student Ghyas Ibn Mohammed Isfahanî.

Sultan Mehmed the Conqueror, who is famous among the Ottoman sultans for his patronage of scholars, displayed an avid interest in classical Greek sources and the contributions of European culture to scholarship. He ordered the Greek scholar Georgios Amirutzes of Trabzon and his son to translate the *Geography* of Ptolemy into Arabic and to draw a world map based on it. Mehmed had first taken an interest in European culture when he was the crown prince living in the Manisa palace. In 1445, the Italian humanist Ciriaco d'Ancona and other Italian scholars resident in the Manisa palace taught Mehmed Roman and European history. Later, while Patriarch Gennadios prepared for Mehmed his work on the Christian faith, the *Itikadname* (*Book of Belief*), Francesco Berlinghieri and Roberto Valtorio presented him with their respective works *Geographia* and *De Re Militari*. These works are still extant in the collections of imperial manuscripts housed in the Topkapi Palace Museum.

Among the most important examples of Mehmed II's patronage of Muslim scholars and encouragement of scholarly writing, we may cite his request to the prominent Ottoman intellectuals Hodjazade and Ala al-Din al-Tusi that each produce a work comparing Al Ghazzali's *Tahafut al-Falasifa* (*The Incoherence of the Philosophers*), a critique of peripatetic philosophers regarding metaphysical matters, with *Tahafut al-Tahafut* (*The Incoherence of Incoherence*), which was Ibn Rushd's reply to this critique.

However, undoubtedly the most notable scientist of the Conqueror's time was Ali Kushcu (whose real name was Muhammed b. Ali), a representative of the Samarkand tradition. Ali Kushcu wrote twelve books on mathematics and astronomy, including a commentary on the *Zij-i Ulug Beg* written in Persian. His other two works in Persian are *Risala fi'l-Hay'a* (*Treatise on Astronomy*) and *Risala fi'l-Hisab* (*Treatise on Arithmetic*). He rewrote these two works in Arabic with some additions — and under new titles, as *al-Fathiyya* (*Commemoration of the Conquest*) and *al-Muhammadiyya* (*Dedication to Sultan Mehmed*). Both of these works were taught in the Ottoman *medreses*.

A noteworthy scholar of the Bayezid II period was Molla Lutfî (1481-1512). Molla Lutfî wrote a treatise in Arabic about the classification of the sciences, entitled *Mawdu'at al-'Ulum* (*Subjects of the Sciences*), and compiled a book on geometry titled *Tad'if al-Madhbah* (*Duplication of the Cube*). Part of the latter was translated from Greek.

Mîrîm Çelebi (d. 1525), the grandson of Ali Kushçu and Kadizâde-i Rumî, was another famous astronomer and mathematician of the early sixteenth century. He contributed to the development of Ottoman scientific tradition particularly in the fields of mathematics and astronomy, and was famous for his commentary on the *Zij* of Ulug Beg.

Interesting contributions to Ottoman scientific literature were made by emigrant Muslim and Jewish scholars from Andalusia. Ilya b. Abram al-Yahudi is one such scholar. Al-Yahudi settled in Istanbul during the reign of Sultan Bayazid II. There, after converting to Islam, he changed his name to 'Abd al-Salam al-Muhtadi al-Muhammedi and authored several medical and astronomical works in Arabic. In a book dedicated to Sultan Bayezid II, which he first wrote in Hebrew then translated into Arabic in 1503, al-Yahudi introduced an astronomical instrument of his own invention called *al-dabid*, claiming that it was superior to the *dhat al-halak* (armillary sphere) invented by Ptolemy. This treatise illuminates an aspect of Ottoman scientific literature which is not yet well known.

The production of scientific literature advanced considerably during the reign of Sultan Süleyman the Magnificent. Two major books written in Turkish on mathematics are known from this period, the *Jamal al Kuttab wa Kamal al-*

24

Hussab (*Beauty of the Scribes and Perfection of the Accountants*) and *'Umdat al-Hisab* (*Treatise on Arithmetic*), both by Nasuh al-Silahî al-Matrakî (d. 971/ 1564).

With their continuously expanding territorial empire and concurrent quest to dominate navally in the Mediterranean, Black Sea, and Red Sea, and Indian Ocean, the Ottomans had a clear interest in excelling in geography. Both the geographical works of classical Islam and the more contemporary European works were very useful in this respect. The first source of Ottoman knowledge about geography came from the Samarkand tradition. Meanwhile, by supplementing existing geographical knowledge with their own fresh observations, the Ottoman geographers produced original works.

During the sixteenth century, noteworthy contributions to Ottoman geography were made by the prominent naval captain Pîrî Reis, whose earliest surviving map was drawn in 1513. It is a *portolano*, without lines of latitude or longitude, but with careful delineations of coasts and islands. This was once a large world map ; however, only a part of it has survived. This fragment depicts southwestern Europe, northwestern Africa, and the eastern coasts of South and Central America. It is considered most astonishing for its early delineation of the eastern coastline of the Americas. The map was based on Pîrî Reis' experience as a sailor, as well as on other Islamic and European source maps — including, Pîrî Reis claimed, some of Columbus' own early maps of the Americas.

Pîrî Reis presented yet another world map to Sultan Süleyman the Magnificent in 1528. Of this map, only the part which contains the North Atlantic Ocean and the newly discovered areas of North and Central Americas has survived. Pîrî Reis also wrote the famous *Kitâb-i Bahriye* (*Book of the Sea*) which he presented to Sultan Süleyman the Magnificent in 1525. This important work consists of drawings and maps of the Mediterranean and Aegean coasts, and gives extensive information about navigation and nautical astronomy. In writing this book, he made use of both classical Islamic and contemporary geographical sources, as well as his own observations.

Another Ottoman naval officer, Admiral Seydî Ali Reis (d. 1562), a prominent figure in the field of maritime geography, wrote a noteworthy book in Turkish titled *al-Muhit* (*The Ocean*). His work contains his observations concerning the Indian Ocean, and astronomical and geographical information necessary for long sea voyages.

Nasuh al-Matrakî's *Beyan-i Menâzil-i Sefer-i Irakeyn* (*Description of the Places on the Way to Iraq During the Two Campaigns*) is a good example of descriptive geography.

Tarih-i Hind-i Garbî (*History of Western India*) is yet another geographical work produced in the sixteenth century. It contains information about geographical discoveries in the New World. This anonymous work based on Spanish and Italian sources was presented to Sultan Murad III in 1583. The work is in three parts. The third part relates to the travels and adventures of Columbus, Balboa, Magellan, Cortés, and Pizarro in the first sixty year after the discovery of America, from 1492 until 1552. Works such as this (and Pîrî Reis') demonstrate that the Ottomans were aware of the geographical discoveries made by the Europeans.

Geographical and historical ties with Europe were crucial in making the Ottoman Empire the first non-Western environment where Western science and technology spread. Geographical proximity allowed the Ottomans to learn about European innovations and discoveries. From the fifteenth century on, the Ottomans began to adopt European technologies, expecially those that concerned firearms, cartography, and mining. They also had access to Renaissance advances in astronomy and medicine through their contacts with Jewish scholars who emigrated to the Ottoman lands. However, Ottoman interest in European advances was selective, partly because of their feelings of moral and cultural superiority over Europe, and partly because of the self-sufficiency of their economic and educational system. They therefore did not keep close track of the scientific and intellectual developments of the Renaissance and the Scientific Revolution as they were unfolding. It is anachronistic to claim, as some modern historians do, that the Ottomans failed to realize that these developments would challenge them in the future.

Starting in the sixteenth century with the arrival of some European physicians and the spread of various diseases from the West, new medical ideas and methods of treatment and prophylaxis were introduced into the Ottoman Empire. In the seventeenth century, the medical doctrines of Paracelsus and his followers were seen in Ottoman medical literature under the names of *tibb-i cedid* (new medicine) and *tibb-i kimyaî* (chemical medicine). One of the most famous followers of this trend was Salih b. Nasrullah (d. 1669). In his work titled *Nuzhatu'l-abdan*, Salih mentioned the new medicines described in various European sources, and disclosed the composition of their remedies. Al-Izniki (18th century) likewise cited Arab, Persian, Greek, and European physicians in his *Kitab-i Kunuz-i Hayat al-Insan Kavanin-i Ettibba-yi Feylesofan*, and presented information on the new medicines as well as the old known ones. Ömer Shifaî (d. 1742), too, in his work *al-Cevher al-Ferid,* claimed that he had translated the book from a European language into Turkish containing remedies taken from the works of Latin physicians. Thus, new information on the composition and use of European medicines was disseminated and put into practice alongside traditional medicines until the beginning of the nineteenth century.

The famous Ottoman scholar and bibliographer Kâtip Çelebi (d. 1658), also known by the name of Haci Halife, was one of the first intellectuals to notice the widening gap between scientific advancement in Europe and the Ottoman world. He was able to approach both classical Islamic culture and contempo-

26

rary European culture in an analytical frame of mind. He wrote books in Arabic and Turkish on a variety of subjects. With the help of a European convert, he translated the *Chronic* of Johanna Carion from Latin and titled it *Tarih-i Firengî Tercümesi* (*Translation of European History*). This book was first published in German in 1632, with numerous later editions and translations. He also compiled *Tarih-i Konstantiniye ve Kayasira* (*History of Constantinople and the Emperor*) which is also called *Ravnak al-Saltana* (*Splendor of the Sultanate*). Kâtip Çelebi's source for this work was evidently one of the reprints of the *Corpus universae historiae praesertim Byzantinae* (Paris, 1567), in this case one titled *Historia rerum in Oriente gestarum ab exordio mundi et orbe condito ad nostra haec usque tempora* (Frankfurt, 1587). The bulk of this book consisted of the histories written by Johannes Zonaras, Nicetas Acominatus, Nicephorus Gregoras, and Laonicus Chalcocondyles.

In the field of geography, Kâtip Çelebi translated the *Atlas Minor* of Mercator and A.S. Hondius under the title *Lawami' al-Nur fi Zulmat Atlas Minur* (*Flashes of Light in the Darkness of Atlas Minor*). Later, in his work *Mizan al-Hakk fi Ikhtiyar al-Ahakk* (*The Balance of Truth and the Choice of the Truest*), Kâtip Çelebi critiqued the intellectual life of his day.

Starting in the seventeenth century, there was a massive influx of translations of Western scientific works to the Ottoman Empire. Now we shall attempt to follow the introduction of modern scientific concepts as they were absorbed into the Ottoman scientific milieu. As far as we could establish, the first work of astronomy translated to Ottoman Turkish from a European language was the *Ephemerides Celestium Richelianae ex Lansbergii Tabulis* (Paris, 1641), the astronomical tables by the French astronomer Noel Duret (d. *ca.* 1650). Ottoman astronomer Tezkereci Köse Ibrahim Efendi (of Zigetvar) translated this book in 1660 under the title *Secencel el-Eflak fi Gayet el-Idrak*. This translation was also the first book in Ottoman literature to mention Copernicus and his heliocentric system. The initial reaction of the Chief Astronomer to this notion was to declare the book a " European vanity " ; but after learning its use and checking it against Ulug Beg's *Zij* (astronomical tables), he realised its value and rewarded the translator. His first reaction, however, typifies Ottoman reluctance to acknowledge European scientific superiority.

Ottoman astronomers of the classical period viewed heliocentricity, which aroused such controversy in Europe, as an alternative technical detail and did not make it a subject of any great polemics. One likely explanation for this nonchalance is that the heliocentric theory of the universe did not conflict with any religious dogmas. Subsequently, most astronomy books translated from European languages dealt with astronomical tables and did not touch upon the subject of heliocentricity again.

Among the translations completed in the second half of the seventeenth and early eighteenth centuries was a major book on modern geography by Abu

Bakr b. Behram el-Dimashki (d. 1691), based on Janszoon Blaeu's *Atlas Major* (1685).

Ibrahim Müteferrika translated Andreas Cellarius' *Atlas coelestis* (1708) from Latin, completing the translation in 1733 and calling it *Mecmuatü'l-hey'eti'il-kadime ve'l-cedide*, or " collection of old and new astronomy ". In 1751, Osman b. Abdulmennan translated Bernhard Varenius' work *Geographia Generalis* from Latin, calling it *Tercüme-i Kitab-i Cografya* (*Translation of Geographia Generalis*). Alongside these new translations, classical Ottoman astronomy, geography, and related scientific activities continued within their traditional framework.

A survey of Ottoman astronomical literature thus shows that after overcoming their feelings of superiority, Ottoman scholars readily accepted new information, techniques and concepts from Europe. The Ottoman administration looked favorably on such influences, and the religious scholars (*ulema*) were not particularly hostile — as evidenced in their attitude toward the theory of heliocentricity. There was no obvious conflict between religion and Western science at this stage.

TOWARDS MODERNISATION

As the seventeenth century drew to a close, the Ottomans, faced with European military and economic superiority, recognized the need to reform classical institutions. Yet they seemed unable to adapt to the changing conditions. They failed to understand modern economic problems, and remained bound to the traditional formulae defining the Near-Eastern political state. The weakness of the Ottoman army became apparent when it was defeated at Vienna in July 1683. Following this defeat, the Ottomans grudgingly began to acknowlege the superiority of the West in certain fields. The Treaty of Carlowitz (1699) impeded further Ottoman territorial advances in Europe, and it became painfully clear that Ottoman power was waning and Europe gaining the upper hand. Thus the Ottoman Empire embarked on a new era, one where leaders would look more attentively to the West as they tried to address some of Empire's problems.

Since the foundation of the Empire, the Ottomans had been keenly interested in European military and technological novelties, which they had selectively adopted for themselves. However, at the beginning of the eighteenth century, Ottoman-European relations entered a new, more intense, period. Confronted with the swift advances in European technology and the defeat of the Ottoman armies by the Europeans, the Ottomans became convinced that in order to master the techniques of modern warfare, they needed to open new channels of communication and technology transfer.

Then, during the first quarter of the eighteenth century, the Ottomans won decisive victories over the Persian and Russian armies in the east, as well as the Venetian armies in the west. Following these victories, a relatively long period of peace known as the *Lâle Devri* (Tulip Period) began. This period is most closely associated with the reign of Sultan Ahmed III (r. 1703-1730) and his Grand Vizier Nevshehirli Damat Ibrahim Pasha. It lasted until the Patrona Halil rebellion in 1730. During this period, the Ottomans were intensely interested in European ways of life and innovations. A new era began in Ottoman cultural life : novelties such as the printing house and pumping equipment for the firemen were imported from Europe. Meanwhile, close contacts between the high officials of the Ottoman state and European ambassadors were established.

Ottoman administrators followed new military developments closely after the year 1730. They trained officers and privates in modern techniques of warfare and equipped the army with the same weapons used by the European armies. In this way, they tried to restore the military balance with Europe. With this objective in mind, they established institutions of military education where new information and technological developments were taught. These schools stood in contrast to the traditional style of military training based on the master-apprenticeship relationship — an important new development in the Ottoman educational system as well as the military.

The first efforts to give modern technical training in the Ottoman military began in the *Ulufeli Humbaraci Ocagi* (The Corps of Bombardiers). The Corps was organized in 1735, with the efforts of Claude Alexander Comte de Bonneval (1674-1747), a general of French origin who had requested asylum from the Ottomans in 1729 and converted to Islam, taking the name of Ahmed. He was later known as Humbaraci Ahmed Pasha. The fact that he took refuge in Istanbul was a blessing for the Ottomans, who wished to make military reforms and could benefit from his expertise.

The Corps was divided into three subdivisions or *oda*, each consisting of twenty-five officers of different ranks. Some of these had administrative duties, while others handled military, educational, and medical affairs. The Corps was under the direct supervision of the commander-in-chief of the military and the Grand Vizier. The salaries of the officers and soldiers of the Corps of Bombardiers were paid from a special fund. A candidacy system was introduced and the salaries were fixed according to rank. The retirement pension of the bombardiers was also guaranteed.

The teaching staff of the Corps consisted of Ottoman as well as French and Scottish teachers. Mehmed Said Efendi was the instructor in geometry, Istanbullu Ibrahim Hodja taught arithmetic, geometry, drawing and the measurement of altitude, and Cenk Mimarbashi Selim Aga taught fortification, artillery, and mechanics. Classical Ottoman books on mathematics and European books were used concurrently. Thus, for the first time, mathematical sciences were included in the curriculum of a military institution and a new system of military training was introduced to Ottoman education. Before the foundation of the Corps, mathematical sciences were taught either in the *medrese* or in private tutoring. The military school thus opened a third venue.

After Ahmed Bonneval Pasha's death in 1747, the Corps of Bombardiers was administered for a short time by his adopted son Süleyman Aga, but it was soon dissolved. Nonetheless, Ottoman administrators redoubled their efforts to train army officers in the new military techniques, and in the 1770s new attempts were made to establish more enduring military-technical schools.

THE IMPERIAL SCHOOL OF MILITARY ENGINEERING

The foundation of the *Hendesehane* (literally " house of geometry ") within the Imperial Maritime Arsenal on April 29, 1775, represented a significant step forward in Ottoman military education. Referred to as the *Ecole de Théorie et de Mathématiques* in French documents, the *Hendesehane* was founded under the supervision of Baron de Tott, a French officer of Hungarian origin who came to Istanbul in 1770 and later entered the Sultan's service as a military adviser. The goal of the *Hendesehane* was to provide the imperial naval fleet (*donanma-i humayun*) with officers trained in the sciences, particularly geometry and geography. The French technician Sr. Kermovan and the Scotsman Campbell Mustafa taught mathematics at the *Hendesehane* until September, 1775, then Sr. Kermovan left Istanbul and Baron de Tott lost interest in this subject. With a regulation dated 1776, the *Hendesehane* became the first Ottoman institution where mathematics and the art of fortification were taught based on European sources, theories, and methods. After Baron de Tott left Istanbul in 1776, Cezayirli Seyyid Hasan, Second Captain in the Ottoman Imperial Fleet, was appointed to the post of *hodja* (professor) in the *Hendesehane*. Soon afterward, this institution was reorganised to conform to the traditional Ottoman bureaucratic structure.

After 1781, the *Hendesehane* was also called the *Mühendishane* (literally meaning " house of geometricians or engineers "), and ten students were trained there at a time. During the Grand Vizier Halil Hamid Pasha's term of office (1784), two military engineers, Lafitte-Clavé and Monnier, who were sent by the French Government to assist in reforming the Ottoman army and strengthening the fortifications, also taught in the *Mühendishane*, training officers in artillery, navigation and fortification.

As in the case of the Corps of Bombardiers (1735), Ottoman teachers were graduates of the *medreses* (and thus of the *ulema* class). Until the end of the eighteenth century both classical Ottoman books on science and foreign (mainly French) texts were used, particularly in the teaching of mathematics, astronomy, firearms, techniques of fortification, warfare, and navigation. After 1788, when the French left the Ottoman Empire, the *medrese* teachers took over responsibility for all courses at the *Mühendishane*.

As part of a series of new military reforms undertaken by Sultan Selim III (r. 1789-1807), a new *Mühendishane (Mühendishane-i Cedide)* was established in 1793 within the barracks of the Corps of Bombardiers. Here cannoniers, bombardiers, and sappers were to be trained. Courses commenced at the new *Mühendishane* in 1794 with a new generation of Ottoman engineer-teachers, among whom we may note the First Chief Instructor Hüseyin Rifki Tamanî, who had learned the art of fortification from French engineers like Lafitte-Clavé and Monnier in the old *Mühendishane*. The organization of this new institution was modeled on the other *Mühendishane*, and the teaching staff consisted of a professor (*hodja*), four assistant professors (*halifes*), and other functionaries. Here, members of the Corps of Bombardiers and Sappers were taught geometry, trigonometry, measurement of elevation, and surveying.

Between the years 1801 and 1802, a number of students selected from the Corps of Bombardiers, Sappers, and Architects were admitted to the *Mühendishane-i Cedide*. The teaching staff was composed of a professor and five assistant professors, and the students numbered about one hundred. In 1806, upon the issuing of an imperial decree of Sultan Selim III, a new regulation was prepared for this institution. Thereafter, it was called *Mühendishane-i Berri-i Hümayun* (The Imperial School of Military Engineering), a name which it retained for quite a long time.

The new regulation of 1806 reflected both European and Ottoman influences. The European practice of taking successive classes (four classes with four teachers, one of them being the chief instructor) was introduced for the first time in Ottoman educational system, while the practice of moving up from one class to a higher class (which was only possible if there was a vacancy in the senior class) continued in accordance with the old Ottoman administrative tradition. This Ottoman system of advancement was called *silsile* (chain).

In 1793, the first *Mühendishane*, located in the Imperial Arsenal, was transformed into a school where shipbuilding, navigation, surveying, and geography would be taught. The French naval engineer J. Balthasar Le Brun was put in charge of the school's administration. After he returned to France, he was succeeded by the Ottoman naval officers whom he had trained. Again by imperial decree of Sultan Selim III, this school officially became known as the *Mühendishane-i Bahri-i Humayun* (The Imperial School of Naval Engineering) and served the Empire for many years under this name.

NEW MILITARY SCHOOLS OF THE NINETEENTH CENTURY

Institutions of Modern Medical Education

Until the beginning of the nineteenth century, the *Süleymaniye Tip Medresesi* (Süleymaniye *Medrese* of Medicine), which was founded by Sultan Süleyman the Magnificent as part of the Süleymaniye Complex in the sixteenth century, had been the only *medrese* devoted to medical education. However, this should not be taken to mean that it was the only Ottoman institution for training physicians. Many physicians were trained in the hospital, or *darüsshifa*, themselves. In addition to these, there were some non-Muslim Ottoman subjects who had studied medicine in Europe, and Jewish physicians from Spain who had taken refuge in the Ottoman Empire.

Many state institutions like *Tophane-i Amire* (Imperial Gun Foundry), *Tersane-i Amire* (Imperial Maritime Arsenal) and other military organizations had their private physicians and surgeons. Although we know little about Ottoman medical training in the classical period, we believe that it was in most cases based on a master-apprentice tradition. In the eighteenth and nineteenth centuries, as education and professional life were modernized, medicine was among the fields that were given top priority. As in many other fields, the first attempts to introduce modern medical practices were made in the military. In order to train physicians and surgeons for the Imperial Maritime Arsenal (*Tersane*), a school called the " *Tersane* School of Medicine " was opened in January, 1806. The aim of this school was to teach modern medicine, and in so doing to increase the number of Muslim physicians in the Empire. The courses were to be given in French or Italian — the languages most commonly used by the Ottoman Levantine physicians and pharmacists who were in contact with Europe — and would use European textbooks.

Although this school was closed only two years later, in 1808, the significance of its establishment within the Imperial Maritime Arsenal must be noted. As we have already seen above, the Arsenal was an important conduit through which modern science and technology were intoduced into the Ottoman Empire. Three major streams of early modernizing influence may be identified in connection with the Imperial Maritime Arsenal : the first was shipbuilding — a shipbuilding based on modern methods and using new technologies, which were first introduced in the Ottoman Empire at the Arsenal. Second was the teaching of mathematics, astronomy, and engineering at the *Hendesehane* and *Mühendishane*, the two schools located within the Arsenal. Third was the introduction of modern medicine and medical education at the Arsenal medical school.

Toward the beginning of the nineteenth century, two personalities were particularly influential in shaping Ottoman medical education. The first was Shanizade Mehmed Ataullah Efendi (d. 1826), a scholar of many interests and a valuable encylopedist, trained in the sciences as well as in European languages. In his well-known five-volume medical work entitled *Hamse-i Shanizade*, modern medicine and anatomy were presented to Ottoman readers in a comprehensive form for the first time.

The second scholar was Mustafa Behçet Efendi (d. 1834), considered the founder of modern medical education in Turkey. The *Tiphane-i Amire* (Imperial School of Medicine) was founded in 1827 while Behçet Efendi held the post of Imperial Chief Physician. Like the earlier Arsenal medical school, the goal of this new medical school was primarily to train physicians and surgeons for the army. In 1831, a separate school for training surgeons, the *Cerrahhane-i Amire* (Imperial School of Surgery) was opened at the Gülhane gardens, adjacent to the Sultan's Topkapi Palace. In 1836, the medical school was transferred next door to the school of surgery, and the curriculum was redesigned under the direction of the French surgeon Sat de Gallière. In 1838, these two schools (medicine and surgery) were combined in a single institution named *Mekteb-i Tibbiye-i Adliye-i Shahane* (The Imperial School for Medical Sciences) and moved to the Galatasaray district of Istanbul, where they would be housed in a single building. C. Ambroise Bernard, a young physician from Austria, was appointed director to the school in the same year. Under Dr. Bernard, medical education in the Ottoman Empire entered a period of stability, both in terms of content and method. French became the language of instruction. The practice of promoting students depending on the vacancies in the senior classes (the above-mentioned " chain " system) was abolished and program of study limited to five years, at the end of which graduates received diplomas certifying completion of the course of study. This European style of academic promotion was also adopted in the Imperial School of Military Engineering, where institutional certification of learning and graduation replaced the more personalistic process of classical system, known as *icazet*. In that system, a student was given a license, or *icazet*, by his master when the latter deemed him prepared. The master was conceived of as a personal authority in a chain of transmission of (religious) learning and scholarly authority reaching back to the Prophet, the first source of divine knowledge.

Following the proclamation of the Imperial *Tanzimat* Rescript in 1839, non-Muslim subjects were allowed to enroll in the Imperial School of Medicine for the first time. With this change the number of Muslim students gradually decreased. Non-Muslim students proved more successful in their studies — and particularly in mastering French, because of their affinity with European culture. As a way of redressing this imbalance, Cemaleddin Efendi, the Superintendent of the Imperial School of Medical Sciences, set up a special class called the " distinguished class " (*mümtaz sinif*) where special emphasis was given to the teaching of Turkish, Arabic, and Persian languages. This class became the nucleus of the civilian school of medicine that was founded in 1867, and paved the way for the spread of modern civilian medical education among the Ottomans. Under the leadership of Kirimli Aziz Bey and his colleagues, who were all graduates of this special class, a *Mekteb-i Tibbiye-i Mülkiye* (Civilian School of Medical Sciences) was opened as a part of the Imperial School of Medical Sciences. Here Turkish was the language of instruction. Thus, civilian medical education became independent from military medical education. However, shorty after, in 1870 a decision taken by the *Darüshshura-i Askerî* (Medical Council of the Ministry of Defence) decreed that instruction in the Imperial School for Medical Sciences, as well, would thenceforth be given in Turkish. This decision provoked heated debates between some Turkish physicians and their non-Muslim Francophone colleagues. The Turkish physicians prevailed, and from that point on medicine in the Ottoman Empire was taught exclusively in Turkish. It should be noted that this decision applied to Ottoman provincial medical schools as well as those in Istanbul. Thus, for example, even in the Ottoman Medical School in Damascus, where the local language was Arabic, the language of instruction was Turkish.

As the number of students in the civilian medical school increased, the school moved to a larger campus in 1873. In 1909, both the military and civilian medical schools were moved again, this time to a new and larger building in the Haydarpasha district of Istanbul. Then, in 1915, the civilian school of medicine was annexed to the *Istanbul Dârülfünun* (university) as the Faculty of Medicine. This faculty at the *Dârülfünun* became the first among all subsequently-established Turkish faculties of medicine. This building is still in use, and since 1980 houses the Marmara University Faculty of Medicine.

The Imperial Military School

In 1826, Sultan Mahmud II (r. 1808-1839) abolished the Janissary Corps, which had until then been one of the cornerstones of the Ottoman army. In its place he formed a new corps called *Asakir-i Mansure-i Muhammediye* (Mohammed's victorious soldiers). In 1831, plans were drawn up for a military school that would train officers for this new force and educate them in modern techniques of warfare. Until then, officers had been trained in the Imperial School of Military Engineering.

The Ottomans founded this school on the model of the French *Ecole Militaire*. The school, which had a capacity of 400 students, opened its doors in 1834 under the name of *Mekteb-i Harbiye-i Shahane* (Imperial Military School) and was headed by Namik Pasha, a general who was educated in Europe and well versed in European languages. The organization of the Imperial Military School differed from the schools of engineering and medicine in several respects : it consisted of eight classrooms where different courses could be held separately. Teachers from both the schools of military and naval engineering collaborated in preparing a curriculum that would improve the general level of education. As part of this program, students and officers would be sent to Vienna and Paris for training — in part so that the demand for teachers in the new Imperial Military School might be met.

In 1838, Emin Pasha was appointed as the superintendent of the Imperial Military School. He divided the school into two sections. The first was a senior level called *Mekteb-i Fünun-i Harbiye* (School of Military Arts) which had a four-year program of training, and the second was junior level program of three years called *Mekteb-i Fünun-i Idadiye* (Preparatory School for Military Arts). During Emin Pasha's tenure, the number of the teaching staff was increased and instruction was modernized by a combination of European teachers and young Ottoman officers educated in Europe. In 1848, Kimyager Dervish Pasha, who succeded Emin Pasha as superintendent of the school, prepared a new set of regulations inspired by those at the French *Ecole de Saint Cyr*. These regulations re-organized the Imperial School of Military Arts, making it conform completely to the European model. A school of veterinary sciences *(Baytar Mektebi)* to meet the army's need for veterinarians was established in 1846 as part of the Imperial School of Military Arts.

The Emergence of Secular Civilian Institutions in the Ottoman Educational Life

At the outset of modernization in Ottoman scientific and educational life, during the pre-*Tanzimat* period (the end of the eighteenth and beginning of the nineteenth centuries), there were two kinds of educational institution in the Ottoman Empire : the *medrese* and the institutions of military and technical education (engineering, medical, and military schools).

Institutions of military education were established and developed within the framework of military reforms that began in the eighteenth century, as described above. The modern education provided by these institutions was different from the classical Ottoman style of the *medreses*, and led to the emergence of a new understanding of education. As previously mentioned, this transformation was first encountered in the schools of engineering, where there was a shift from the classical Islamic tradition to the Western tradition. For a while, the two approaches coexisted, but from the second half of the nineteenth century onward, a conflict began to emerge between " the old " and " the new " — often conceived of as a conflict between " religion " and "science ".

The schools of engineering are often seen as the first centers of Western-style education. Yet in some ways, they were actually examples of an Ottoman-Western synthesis. The main purpose of the institutions of military and technical education was described as follows : *" Mühendishanelere fünun-i berriye ve bahriyeden hendese, hesap ve cografya fenlerinin intishari ve Devlet-i Aliyye'ye ehemm ve elzem olan sanayi-i harbiyenin talim ve teallümü ve kuvveden fiile ihraci "*. Briefly, in the language of the time, the objective was to train the *mütefennin zabit* (military officers educated in the military sciences). The statesmen and scholars of the period believed that a regular army equipped with modern technology could save the country. This was the reason why they turned to the West : in order to train and equip the officers with modern sciences and technology.

The Imperial *Tanzimat* Rescript (1839) did not explicitly describe any specific goals with regard to education and learning. However, before long it was clear to all that reforms in the educational system as a whole were a necessary step in modernization. In January of 1845, when Sultan Abdülmecid visited the *Meclis-i Vâlâ* (Supreme Council of Juridical Ordinances), he pointed to the dire necessity of fighting ignorance in all aspects of life. In particular, he saw education as a means to improve the public works of the Empire and thereby prosperity of its people. He ordered that priority should be given to the education of the people.

Thus for the first time in Ottoman history, the State started considering the subject of " educating the public " — including the establishment of a Central Office to coordinate such a project. In 1845, a Temporary Council of Education planned a system of education based on European (and particularly French) models. The model called for three stages of education, *primary, intermediate*, and *high*. The *intermediate* was accepted by the Council as a school where young school boys would be trained for the *darülfünûn* (university), thus securing their place in the Ottoman educational system. As this intermediate training proved effective, the number of intermediate schools increased from four to ten between the years 1847 and 1852.

At this time, Ottoman administrators found that they had to recruit newly-trained bureaucrats and officials for the new departments created by reform in the Central Administration. Throughout the nineteenth century, trained civilians were needed in the fields of agriculture, animal breeding, forestry, mining, engineering, industry, law, and the fine arts. As a first step, the State devised a system that would help fill these posts. The result was modern schools, modeled on those in Europe, whose graduates would then be employed in the government services. The second step was finding the teaching staff that would train this new generation. A program was developed to solve this problem. This program dealt with middle-level schools such as the *Rüshdiye, Idadi*, and *Sultani,* establishments where male teachers who had been trained in the *Darulmuallimin* (teacher training school for men, established in 1846) and the *Darulmuallimat* (teacher training school for women, established in 1869) would teach. European teachers or Ottomans who had been trained in Europe were employed for teaching of the subjects listed above.

Sending Students to Europe for Training

The practice of sending Ottoman students to Europe for training began during the reign of Sultan Mahmud II. In 1830, four Ottoman students named Hüseyin, Ahmed, Abdullatif, and Edhem, who were studying in the *Enderun* (Palace School), were sent to France to study military subjects under the patronage of Serasker Hüsrev Pasha. Their expenses were met by the Imperial Treasury (*hazine*). In 1839, thirty-six students were selected from the military and engineering schools and sent to London, Paris, and Vienna to learn European technology. On their return, they would be employed in the Ottoman military factories and workshops where heavy industry and technology were used

III

— i.e. the arsenal, gunpowder works, rifle and cartridge factories, and the foundry.

Following the proclamation of the *Tanzimat*, eligible students who wished to receive training in the civilian services were also sent to Europe with funding from the state. Because of the influx of non-Muslims into the new educational institutions at this time (as discussed above), in the year 1840 a significant number of non-Muslims were among the students sent to Europe. The ratio of the Muslim to non-Muslim students varied in different periods. Between 1848 and 1856, about fifty students went to Paris, this number rose to sixty-one between 1856 and 1864. An Ottoman school by the name of *Mekteb-i Osmanî* was opened in Paris in 1857 to educate the Ottoman students in the arts and sciences and thus enable them to follow the classes offered in French schools. Until 1864, all Ottoman students in Paris attended this school. Between 1864 and 1876, a total of ninety-three students were sent to Paris ; forty-two of these went to study science, and fifty-one to receive apprenticeship training in various technical vocations. On January 13, 1870, twenty students from the *Mekteb-i Sanayi* (The School of Arts and Crafts) went to Paris to study in various branches of arts. Thus, the number of students that were sent to Europe gradually increased over the course of the nineteenth century. In addition to military education, training was given in the professions and in arts needed for civilian purposes.

Darülfünûn (University)

In the history of Ottoman education, the *Darülfünûn* (which can best be translated into European terminology as " university ") stands as a unique institution in civilian higher education. It had no counterpart in the classical Ottoman system, since it was quite different from the *medrese*. The idea of a *darülfünûn* emerged during the *Tanzimat* period in response to the need to educate a greater number of students in a non-military form of higher education. Towards the middle of the nineteenth century the first attempt was made to found a *darülfünûn*.

The Council of Public Instruction stated that the aim of the *Darülfünûn* was to train future employees and cultivate enlightened civil servants (*münevver bendegân*), who would go on to serve the State in the best possible way. This training included the study of modern sciences. In November 1846, a contract was signed with G. Fossati, a Swiss architect of Italian descent, to build the imposing three-story *Darülfünûn* building in the Ayasofya quarter of Istanbul. The building would have 125 rooms and would resemble large European universities. Unfortunately, construction dragged on for many years. Yet in 1863, by the order of Grand Vizier Keçecizade Fuad Pasha, a program of public lectures was started in the incomplete building. When the building was finally completed in 1865 it was deemed too big for the intended purpose ! It was promptly turned over to the Ministry of Finance, and a smaller building was planned for the *Darülfünûn*. In order not to interrupt the ongoing lectures, the government rented temporary quarters in the mansion of Nuri Pasha in the Çemberlitas district. (This building was sadly destroyed in the Hocapasa fire). In 1869, the new, smaller building for the *Darülfünûn* was completed and classes began.

That same year, a Regulation on Public Instruction was carried out with the aim of organizing the entire system of education, including the elementary schools. Courses, examinations, teachers and finances were all taken into consideration in this series of decisions, which defined the second attempt to found the *Darülfünûn*. The university was officially named the *Darülfünûn-i Osmani*, and its internal organization was strongly influenced by the French model. According to the Regulation, it was to be divided into three departments : philosophy-literature, natural sciences-mathematics, and law. Required for graduation were three years of coursework and one year for the preparation of a thesis, totaling four years. Students sixteen years of age and in possession of a preparatory degree or its equivalent could enroll. Each department had a detailed curriculum culminating in a graduation thesis based on the student's own research. Teaching certificates were also granted. A museum, a library and a laboratory were to be opened at the *Darülfünûn*. The courses to be offered were inspired by their French counterparts. Arabic and Persian as eastern languages, and French, Greek, and Latin as Western languages would be offered in the department of philosophy and literature. Both Islamic and Roman law were to be taught in the department of law. The teaching of disciplines springing from both Islamic and Western cultures in these two departments reflects the desire of Ottoman intellectuals of the *Tanzimat* period to achieve a new " Ottoman cultural synthesis ".

Registration at the *Darülfünûn* began in 1869. An examination was administered to some one thousand applicants, of whom only 450 were accepted. The *Darülfünûn* opened with a grand ceremony, with Grand Vizier Ali Pasha, Minister of Education Safvet Pasha, and other dignitaries were present. Tahsin Efendi (Hodja Tahsin) was appointed Director of the *Darülfünûn*. He was appointed because it was believed that he could establish a harmonious balance between Islam and the West, between the old and the new. Indeed, Tahsin Efendi, having been educated in the traditional *medrese* system, then later taking on the duties of director at the Ottoman school in Paris, was well situated to find common ground between Islamic and Western cultures. Rather than stumbling on the contradictions between the two, Tahsin Efendi would put forward a new convergent view that would satisfy the *Tanzimat* dignitaries, who were in search of a synthesis between the two cultures.

However, the conditions were not yet conducive to opening the university for classes, due to a lack of funds, teachers, textbooks, and so on. Under these conditions, the regulations of the *Darülfünûn-i Osmanî* could not be fully implemented, and the programs of education in the three departments could not be realized as envisaged. Thus the same curriculum was followed in every department, and all students attended the same lessons. For these reasons the second attempt to establish the university did not bear the expected fruit.

In 1873, Minister of Education Safvet Pasha gave the task of setting up the *Darülfünûn* to Sawas Pasha, an Ottoman Greek who was then the Director of the *Mekteb-i Sultânî* (Imperial Lycée of Galatasaray). The condition under which Sawas Pasha had to work was that the third attempt at the *Darülfünûn* should not be a burden on the state treasury. The idea was to establish the *Darülfünûn* on the foundations of the *Mekteb-i Sultânî*, which had been functioning since 1868. Thus, an institution of higher education was founded based on an institution of secondary education. The new *Darülfünûn* would be called *Darülfünûn-i Sultânî*, and would consist of schools of law, sciences, and the arts. In official accounts, these three departments were referred to as the " Higher Schools " (*Mekâtib-i Aliye*).

The *Darülfünûn-i Sultânî*, which opened for classes in the 1874-1875 academic year, consisted of the schools of Letters, Law, and — instead of the originally intended school of science — Civil Engineering (*Mühendisin-i Mülkiye Mektebi*). During the first academic year, the name of *Mühendisin-i Mülkiye Mektebi* was changed to *Turuk u Maabir Mektebi* (School of Public Works, the Turkish translation of the French *L'Ecole des Ponts et Chaussées*).

Students who studied in the *Darülfünûn-i Sultânî* for four years would prepare and defend a scholarly thesis and graduate with the title of " Doctor ". Law students would be employed by the Ministry of Justice and engineers by the Ministry of Public Works. Graduates of the Faculty of Letters would be employed as teachers of literature. Students who did not prepare a dissertation would take an examination designed to be easier than the doctoral examination. Such non-dissertation graduates of the faculties of Letters, Law, and Public Works would be appointed as junior teachers, secondary lawyers, and engine drivers.

During the 1874-1875 academic year, the numbers of students who attended the schools of Law and Public Works were twenty-one and twenty-six respectively. All succeeded in the examinations that were given at the end of the academic year. We have no known documents telling us whether the School of Letters had started giving instruction by this time or not. In 1881, the schools of Law and Civil Engineering were attached to the Ministries of Justice and Public Works respectively, and continued to function successfully as independent schools.

At the end of the nineteenth century, the foundation of new schools of higher education where students could specialize in the fields of civil service, medicine, law, commerce, industry, engineering, and architecture were completed to answer the requirements of the State for qualified personnel.

However, on 14 February, 1895, the Grand Vizier Said Pasha submitted a petition to Sultan Abdülhamid II requesting that in addition to higher institutions where professional training was given, a *darülfünûn* should be established on the model of American and European universities, with five faculties (*darülicâze*). Here the students would receive scholarly education, not merely for the sake of becoming civil servants of the state bureaucracy, but for the pursuit of scholarly academic study in various fields.

After three failed attempts, the beginning of the twentieth century saw the successful foundation of this new *Darülfünûn* in 1900. This new institution would house several departments in a campus under one administration. Thus the university by the name of *Darülfünûn-i Shahane* was established. It was the product of fifty years' experience, and was further supported by a now-sufficient number of secondary schools, well-educated students, and firmly established faculties of law and medicine and other institutions of higher learning. The *Darülfünûn-i Shahane* was the first, most important, and most influential of all modern conventional Turkish universities.

Learned Societies

In the classical period, informal gatherings of scholars, poets, and intellectuals were part of the Ottoman cultural life. Such groups usually met in the mansions of distinguished persons and the learned to discuss issues of intellectual and scholarly interest. However, these informal meetings never led to the establishment of formal societies. Societies similar to those in Europe were started only in the mid-nineteenth century, following the *Tanzimat* reforms in the fields of education and culture. From then on, many learned and professional societies were founded by Ottomans to advance the goal of professional solidarity and learning among their members.

The first attempt at such a society was made in 1851 by high level statesmen who founded the *Encümen-i Danish* (Learned Society). This society, the first of its kind in the Islamic world, was in some respects similar to the Paris *Académie Française*. It became the model for Ottoman learned societies, and was soon followed by others founded by the foreigners living in Istanbul. The first of these was the *Société Orientale de Constantinople* founded in 1852. This was followed by the *Société de Médicine de Constantinople*, established by foreign physicians from allied countries who gathered in Istanbul during the Crimean war. This society, which received the title " Shahane " (Imperial) by the Sultan's decree, survived until the end of the Empire, and continues today under the name of *Türk Tıp Dernegi* (Turkish Medical Society).

In 1861, a group of Ottoman intellectuals under the leadership of Münif Pasha established the *Cemiyet-i Ilmiye-i Osmaniye*, which would function in accordance with the State's educational and cultural policy. Its monthly *Mecmua-i Fünûn*, which contained news of modern European culture and

III

science, was the first of its kind published by the Ottomans. The activities of this society were discontinued due to lack of funds.

Approximately ten years later, Hodja Tahsin Efendi and his friends founded the *Cemiyet-i Ilmiye*, with the objective of reviving notion of the Islamic-European synthesis. Though this organization only existed for a year, between1879-1880, it published a journal entitled *Mecmua-i Ulum,* of which only seven issues came out.

With the proclamation of the second *Meshrutiyet* (constitutional government) in 1908, chemists, engineers, architects, dentists, veterinary surgeons, pharmacists, and agriculturists all established their own societies to build solidarity and pursue professional interests. During the nineteenth century, the societies established to strengthen the solidarity among physicians and pharmacists were of long duration and proved very influential. However, learned societies founded for the introduction and dissemination of modern culture were small in number and soon petered out.

New Scientific Establishments

Parallelling expanding education in modern sciences such as medicine, astronomy, botany, zoology in the nineteenth century, a growing need was felt for institutions where these sciences were applied. The Department of Quarantine established in 1831 was one of the first institutions to implement modern medical practices — which they did particularly with regard to the annual pilgrimage to Mecca. In 1862, quarantine hospitals were built in Istanbul and other important cities along the roads to Mecca in Anatolia, Rumelia, and the Arab provinces to prevent the spread of infectious diseases. The Ottomans, who at this time were particularly concerned with public health, closely followed developments in Europe in the fields of microbiology and vaccination. When Pasteur discovered the rabies vaccine in 1885, a lecture was given in Istanbul on the subject, and a delegation of physicians was sent to Paris to learn more. They presented Pasteur with the official decoration of the Ottoman Empire together with a prize of 10,000 FF from Sultan Abdul Hamid II. After the delegation's return, with the collaboration of Ottoman and European physicians, a rabies laboratory (*Da'ul-Kelb Ameliyathânesi*) was founded. Later, a bacteriology laboratory was established to combat cholera epidemics. These institutions played an important role in the prevention of rabies, cholera, and small pox epidemics by promoting vaccination.

Another modern scientific institution established in the second half of the nineteenth century was the *Rasathâne-i Amire* (The Imperial Observatory). Affiliated to the Ministry of Education, the Observatory was built on European models and directed by the Frenchman M. Coumbary. Though called " the observatory ", this structure functioned more as a meteorology station than a place for making astronomical observations. Under M. Coumbary's direction, it sent out weather reports from the big cities and exchanged reports with stations in Europe. This office continued to function under Turkish administrators after the departure of Coumbary. In 1910, after the appointment of Fatin Efendi [Gökmen] (who was first educated at the *medrese* then at the *Darülfünûn* Faculty of Science) as Director, the observatory started making astronomical observations as well as preparing weather reports. The observatory became known as the *Kandilli Rasathânesi* (Kandilli Observatory) after it was moved to the district of Kandilli in Istanbul. Today, it is attached to Bogaziçi University and continues to carry out astronomical observations and seismic measurements.

Adoption of the metric system

The enterprising attitude of the Ottomans that we have seen in many other areas was again encountered in the field of metrology and standardization. With a law issued in 1869, the Ottoman State adopted the meter, gram and liter as the official units of length, weight and volume. Following the passage of the law, various measures were taken to facilitate the acceptance, use and dissemination of the metric system. Thus began a gradual transition from Ottoman to metric weights and measures. Due to the longstanding familiarity of the general public with the traditional system, and the abuse of the shopkeepers, the compulsory use of the metric system was posponed at regular intervals. However, the new system was regulary used in government offices and by certain professionals such as pharmacists and physicians. Greenwich mean time was later introduced to complete metrological integration of the Ottoman Empire and European countries. Following its adoption by communication and transportation companies and military and government offices in 1910s, mean time also came to use in daily life. The use of GMT was made obligatory with a law issued in 1926. Thus, as with the Ottoman and metric weights and measures, *alla turca* and *alla franca* times were used concurrently in the Ottoman Empire for many years.

The Emergence of Modern Turkish Scientific Literature

The nineteenth century scientific literature clearly shows that modern science co-existed with old Turko-Islamic scientific traditions in a number of fields. Examples can be found in works where geocentric and heliocentric systems of the universe were introduced together. Similar cases can be traced in medical works. Eighteenth-century Ottoman medical works included practical medical knowledge taken from Europe alongside old concepts such as the concept of four bodily humors (*ahlat-i arba'a*).

Toward the end of the eighteenth century, teachers at the Imperial School of Engineering, which had been established to teach modern sciences to military officers, started to translate or compile books from European scientific literature. Generally, the instructors benefited from the textbooks that were used in

42

the European military technical schools. At the turn of the century, among the first scientific publications in the Ottoman Empire were about ten books on mathematics, geography, and engineering compiled and translated by Hüseyin Rifki Tamanî (d. 1817). Rifki Tamanî's student — and later his successor as the chief instructor of the *Mühendishane* — Ishak Efendi (d. 1836) published thirteen volumes based on Western (particularly French) sources. Among these, *Mecmua-i Ulum-i Riyaziye* (Compendium of Mathematical Sciences), in four volumes, is of special importance, as it is the first attempt to prepare a comprehensive textbook on various sciences written in the language of any Muslim nation. The text treated the subjects of mathematics, physics, chemistry, astronomy, biology, botany, and mineralogy. Ishak Efendi's efforts to find Turkish equivalents for new scientific terms, and his role in disseminating information on modern sciences extended beyond the borders of Ottoman Turkey.

Instructors at the *Mühendishane* and graduates and instructors at the *Mekteb-i Tibbiye-i Shahane* (Imperial School of Medical Sciences), which was reformed and reopened in 1838, also contributed to the translation of European scientific books. As modern education became widespread and civilian education was reorganized after the proclamation of the *Tanzimat* (1839), new scientific and technical books were printed. By the middle of the nineteenth century, the number of printed books on modern science and technology, as well as the variety of subjects they addressed, had increased considerably. During the pre-*Tanzimat* period, between the establishment of the first Turkish printing press in 1727 and the proclamation of the *Tanzimat* in 1839, some thirty books on science were printed ; that figure reached approximately 250 during the *Tanzimat* period (1840-1876). There was an increase in the number of books that were printed in the field of mathematics and medicine, but a decrease in the fields of geography, military sciences, engineering, astronomy, and navigation.

After the *Tanzimat,* different subjects were taken on. For example, Dervish Pasha published the first book on chemistry in Turkish, titled *Usul-i Kimya* (*Elements of Chemistry*, Istanbul, 1848). The first book on modern zoology and botany, titled *Ilm-i Hayvanat ve Nebatat* (*Zoology and Botany*, Istanbul, 1865), was translated from the French by the Chief Physician Salih Efendi.

During the first three decades of the *Tanzimat* period, about four books were printed every year, but in the following seven years (1870-1876), this number went up to about eighteen books a year. This dramatic increase is indicative of the increasing interest in modern sciences in Ottoman society.

The adoption of Turkish as the language of instruction in the Imperial School of Medicine was instrumental in developing a Turkish medical literature. This resulted in the translation and publication of several books on medicine, particularly after 1870. The first of these was *Lugat-i Tibbiye* (*Medical Dictionary*, Istanbul, 1873).

The decrease noted above in the number books published in geography, engineering, and military sciences indicated a shift of interest from military to civilian areas. This shift is confirmed if one examines the prefaces of the books compiled on the same subjects in the nineteenth century before and after the *Tanzimat*. In his work *Mecmua-i Ulum-i Riyaziye,* Ishak Efendi mentions the importance of chemistry in the war industry. On the other hand, Kirimli Aziz Bey, in his work *Kimya-i Tibbi (Medical Chemistry,* Istanbul, 1868-1871) pointed out that chemistry, like medicine, was the basis of several industries and technologies of non-military character.

Some nineteenth-century books in the field of mathematics were original contributions rather than translations. Among them *Linear Algebra* written in English by Vidinli Hüseyin Tevfik Pasha is noteworthy as a valuable contribution to the development of linear algebra. This book, which is concerned with three-dimensional linear algebra and its application to geometry, was printed twice in Istanbul between 1882 and 1892.

During the eighteenth and nineteenth centuries, the primary languages of Ottoman scientific literature were Turkish and Arabic. Although there were a few works in Persian, these were rather rare, representing less than 1% of production. However, the relative share of Arabic and Turkish works varied between manuscripts and printed works. While almost all the printed books about modern science and technology produced during these two centuries in Istanbul were in Turkish, a significant number of manuscripts were still written in Arabic.

During the eighteenth century, the total number of books on astronomy produced in the Ottoman world, both in manuscript and print form, were 331. Of these, 221 were written in Arabic, 101 in Turkish, 2 in Persian and the remaining 7 in a mixture of these three languages. In the nineteenth century, the total number was 263, of which 137 were written in Arabic, 123 in Turkish, and 3 in the mixed languages. These figures clearly show the increase in the use of Turkish — even in the context of a decrease in the total number of books. This shift is also indicative of the increase in the number of schools and educational institutions during the nineteenth century, and the promptness with which the resulting transition from the Eastern manuscript tradition to the modern printed book was made.

Most of the astronomy books written in the Ottoman Empire outside the Arabic-speaking lands were in Turkish. Although we do not have statistical figures for other sciences, we may assume that the same pattern holds for them. We may thus conclude that in the eighteenth and nineteenth century scientific manuscripts were produced in both in Arabic and Turkish, while most of the increasing number of printed works were in Turkish.

Our information on research-oriented activities at Ottoman institutions is limited. However, in the second decade of the twentieth century, research activities were included in the objectives of the *Darülfünûn.*

In the second half of the nineteenth century there are some significant — yet not well documented — research-driven works produced in Europe by Ottoman scholars who pursued their studies abroad. After these scholars' return from Europe, they took up positions in the modern educational institutions founded all over the Ottoman Empire.

At the beginning of the twentieth century, Ottoman Turkish was a well-developed scientific language with an elaborate terminology that was used to write about many modern sciences. There were sufficient textbooks and, to a lesser degree, authentic publications in Ottoman. This was a long journey from the fourteenth century, when the first scientific Turkish literature was produced by the Ottoman Turks who settled in Anatolia.

CONCLUSION

In this paper, scientific developments and changes in the Ottoman State over the course of six centuries have been examined, while scientific activities in both institutional and private settings have been studied from different viewpoints. This paper, which may be considered as a summary of the history of Ottoman science, divides Ottoman science into two eras. First, the classical Ottoman science that was established and developed under the influence of pre-Ottoman Islamic traditions, and second, the Western scientific tradition who expanded with the European political and military influence. In reviewing these two eras, the decisive influence exerted by pre-Ottoman Islamic scientific traditions on Ottoman scientific life is emphasized. However, instead of just statically imitating their classical inheritance, the Ottomans made original and new contributions to the classical Islamic science in many areas, ultimately carrying the tradition to a higher level.

The Ottoman sultans, while establishing new *medreses* across the vast Empire in a very short span of time, allowed existing *medreses* to continue to function. During the first centuries of the Empire, scholars who came from other Islamic lands were employed as instructors in Ottoman *medreses*. Soon, the *medreses* improved their programs of instruction by restructuring curricula, setting up new individual systems and organizing their staff. We must note that the attention and patronage given to the rational sciences during Sultan Mehmed II the Conqueror's reign was key to these subjects' subsequent inclusion in the traditional *medrese* curriculum. The Süleymaniye Complex, founded by Sultan Süleyman the Magnificent, was the last and best link in the development of the classical *medrese* system. The fact that *Daruttip* which was established within this complex was completely devoted to medical education is particularly noteworthy.

In addition to the *medreses*, scientific education was also given in institutions such as the *muvakkithane* and *shifahane*, where applied medicine, mathematics, and astronomy were practiced. Both traditional and new instruments of astronomy were used together In the Istanbul Observatory, which was founded towards the end of the sixteenth century by Taqi al-Din Rasid. Here, Islamic astronomical observations and studies continued. Short but original studies were also conducted and immediately published as books. The classical Ottoman scientific literature was generally produced within the *medrese* milieu. Numerous books were written by authors in Arabic and Turkish, and in lesser numbers in Persian. These books were primarily on religious subjects, but also included efforts in the fields of mathematics, medicine, and astronomy.

One of our most important observations is that books written by scholars of Muslim or Jewish origin who came from non-Ottoman lands (including, importantly, Andalusia) have a significant place in Ottoman scientific literature. When seen within the framework of classical Ottoman science, the patronage given by sultans and statesmen to *medreses* and scholars, and their evaluation of original works are all brought into focus. The Ottomans, who were aware of scientific and geographical advances in Europe, were beginning to take a keen interest in Western science and technology, particularly towards the end of the seventeenth century. In order to learn new technologies of warfare, they switched from their old policy of limited, selective adoption of scientific advances to one of more intensive borrowing and imitation.

At the beginning of the eighteenth century, the Ottomans became more interested in European science and technology. With the assistance of European experts, important steps were taken in transforming the army. After concentrated efforts to introduce modern military and technical education, new institutions were established. This started with the establishment of the *Ulufeli Humbaraci Ocagi* and continued with the opening of the *Mühendishane* and other military schools in the nineteenth century. Among the military schools, institutions of modern medicine took pride of place. In addition to the military and technical schools, modern civilian schools were also founded, thus spreading the modernization of scientific education among a wider segment of the population. To fill in the teaching staff positions in these institutions, a large number of students were sent to Europe for training. One of our most important observations is that the *Darülfünûn-i Shahane*, which was finally opened in 1900 after three unsuccessful attempts, was the pioneer in the foundation of the present-day schools of higher education in Turkey.

In addition to the official institutions of education, Ottoman intellectuals founded numerous professional and learned societies and made great efforts in the production of a modern Turkish scientific literature. They published a large number of books, mainly translations, and including some dictionaries. The majority of the works published in the eighteenth and nineteenth centuries were in Turkish, which indicates that Turkish had by then been accepted as a scientific language.

Though European science and technology was followed closely by the Ottomans, the modern scientific approach was never wholly accepted. Notably, the understanding of science as based on research was lacking. Due to the lack of this fundamental understanding, Ottomans were not able to conduct systematic research studies parallel to those seen emerging in Russia and Japan. Yet in spite of this setback, Ottoman scholars and intellectuals were able to produce many scholarly works in the sciences, and continued to develop and work on Ottoman Turkish as a scientific language. At the beginning of the twentieth century, ideas in a wide range of scientific subjects could easily be expressed in Ottoman Turkish.

The cultural and scientific heritage that was accumulated during the Ottoman period constitutes the scientific and cultural background of the numerous successor states to that Empire in the Balkans and the Middle East, but particularly that of the Republic of Turkey.

BIBLIOGRAPHY

A. Adivar, *Osmanli Türklerinde Ilim,* 5th ed., ed. A. Kazancigil and S. Tekeli. Istanbul, Remzi Kitabevi, 1991.
S. Aydüz, *Osmanli Devleti'nde Müneccimbasilik ve Müneccimbasilar,* Unpublished M.A. thesis, Istanbul University, Faculty of Letters, Dept. of the History of Science, 1993 (supervisor : E. Ihsanoglu).
S. Aydüz, " *Osmanli Devleti'nde Müneccimbasilik* ", F. Günergun (ed.), *Osmanli Bilimi Arastirmalari II (Studies in Ottoman Science II),* Istanbul, Istanbul University, 1998, 159-207.
S. Çavusoglu, *The Kadizadeli Movement : An Attempt of Sheri'at-Minded Reform in the Ottoman Empire,* Unpublished Ph.D. diss., Princeton University, Near Eastern Studies Department, 1990 (Supervisor : C. Kafadar).
T.C. Goodrich, *The Ottoman Turks and the New World, A Study of Tarih-i Hind-i Garbi and Sixteenth-Century Ottoman Americana,* Wiesbaden, Otto Harrassowitz, 1990.
F. Günergun, " Du Zira au Mètre : Une Transformation Métrologique dans l'Empire Ottoman ", Patrick Petitjean, C. Jami and A.-M. Moulin (eds), *Science and Empires,* Dordrecht, Boston, London, Kluwer Academic Publishers, 1992, 103-110 (*Boston Studies in the Philosophy of Science,* vol. 136).
F. Günergun, " Metric System in Turkey : Transition Period (1881-1934) ", *Journal of the Japan-Netherlands Institute,* vol. 6 (W.G.J. Remmelink (ed.), Papers of the Third Conference on the Transfer of Science and Technology between Europe and Asia since Vasco da Gama (1498-1998)), Tokyo, 1996, 243-256.
F. Günergun, " Standardization in Ottoman Turkey ", F. Günergun and S. Kuriyama (eds), *Introduction of Modern Science and Technology to Turkey and Japan,* Kyoto, International Research Center for Japanese Studies, 1998, 205-225.
E. Ihsanoglu, *Bashoca Ishak Efendi : Türkiye'de Modern Bilimin Öncüsü (Chief Instructor Ishak Efendi : Pioneer of Modern Science in Turkey),* Ankara, Kültür Bakanligi, 1989.
E. Ihsanoglu, " Some Remarks on Ottoman Science and its Relation with European Science & Technology up to the end of the Eighteenth Century ", *Journal of the Japan-Netherlands Institute,* vol. 3 (W.G.J. Remmelink (ed.),

III

Papers of the First Conference on the Transfer of Science and Technology between Europe and Asia since Vasco da Gama (1498-1998), Tokyo 1991, 45-73.

E. Ihsanoglu, " Ottoman Science in the Classical Period and Early Contacts with European Science and Technology ", E. Ihsanoglu (ed.), *Transfer of Modern Science & Technology to the Muslim World*, Istanbul, The Research Centre for Islamic History, Art and Culture, 1992, 1-48.

E. Ihsanoglu, " Introduction of Western Science to the Ottoman World : A Case Study of Modern Astronomy (1660-1860) ", E. Ihsanoglu (ed.), *Transfer of Modern Science & Technology to the Muslim World*, Istanbul, The Research Centre for Islamic History, Art and Culture, 1992, 67-120.

E. Ihsanoglu, " Ottomans and European Science ", P. Petitjean, C. Jami and A.-M. Moulin (eds), *Science and Empires*, Dordrecht, Boston, London, Kluwer Academic Publishers, 1992, 37-48 (*Boston Studies in the Philosophy of Science*, vol. 136).

E. Ihsanoglu, " Tanzimat Öncesi ve Tanzimat Dönemi Osmanli Bilim ve Egitim Anlayishi ", *150. Yilinda Tanzimat,* ed. Hakki Dursun Yildiz, Ankara, Türk Tarih Kurumu, 1992, 335-395.

E. Ihsanoglu, " Bashoca Ishak Efendi, Pioneer of Modern Science in Turkey ", *Decision Making and Change in The Ottoman Empire*, Missouri, The Thomas Jefferson University Press at Northeast Missouri State University, 1993, 157-168.

E. Ihsanoglu, F. Günergun, " Tip Egitiminin Türkçelesmesi Meselesinde Bazi Tesbitler ", A. Terzioglu (ed.), *Türk Tip Tarihi Yilligi-Acta Turcica Historiae Medicinae I*, Istanbul, 1994, 127-134.

E. Ihsanoglu, " Ottoman Science ", H. Selin (ed.), *Encyclopaedia of the History of Science, Technology, and Medicine in Non-Western Cultures*, London, Kluwer Academic Publishers, 1997, 799-805.

E. Ihsanoglu, R. Sesen, *et al.*, *Osmanli Astronomi Literatürü Tarihi (History of Astronomy Literature During the Ottoman Period)*, 2 vols, ed. and foreword E. Ihsanoglu, Istanbul, The Research Centre for Islamic History, Art and Culture, 1997.

E. Ihsanoglu, " Modernization Efforts in Science, Technology and Industry in the Ottoman Empire (18th and 19th Centuries) ", F. Günergun, S. Kuriyama (eds), *The Introduction of Modern Science and Technology to Turkey and Japan*, Kyoto, International Research Center for Japanese Studies, 1998, 15-35.

E. Ihsanoglu, " Changes in Ottoman Educational Life and Efforts towards Modernization in the 18th and 19th Centuries ", F. Günergun, S. Kuriyama (eds), *The Introduction of Modern Science and Technology to Turkey and Japan*, Kyoto, International Research Center for Japanese Studies, 1998, 119-136.

E. Ihsanoglu, " Osmanli Egitim ve Bilim Müesseleri ", E. Ihsanoglu (ed.), *Osmanli Devleti ve Medeniyeti Tarihi (History of the Ottoman State and Civilisation)*, vol. 2, Istanbul, The Research Centre for Islamic History, Art and Culture, 1998, 223-359.

E. Ihsanoglu, " Osmanli Bilimi Literatürü ", E. Ihsanoglu (ed.), *Osmanli Devleti ve Medeniyeti Tarihi (History of the Ottoman State and Civilisation)*, vol. 2, Istanbul, The Research Centre for Islamic History, Art and Culture, 1998, 363-444.

H. Inalcik, *The Ottoman Empire : The Classical Age, 1300-1600*, Trans. N. Itzkowitz and C. Imber, London, Weidenfeld and Nicolson, 1973.

III

M. Kaçar, " Osmanli Imparatorlugunda Askeri Sahada Yenilesme Döneminin Baslangici ", F. Günergun (ed.), *Osmanli Bilimi Arastirmalari (Studies in Ottoman Science)*, Istanbul, Istanbul University, 1995, 227-238.

M. Kaçar, *Osmanli Devletinde Mühendishanelerin Kurulusu ve Bilim ve Egitim Anlayisindaki Degismeler* (*The Foundation of " Mühendishane" in the Ottoman State and the Changes in the Scientific and Educational Life*), Unpublished Ph.D. diss. Istanbul University, Faculty of Letters, Dept. of History of Science, 1996 (supervisor : E. Ihsanoglu).

M. Kaçar, " Osmanli Imparatorlugu'nda Askeri Teknik Egitimde Modernlesme Çalismalari ve Mühendishanelerin Kurulusu (1808'e kadar)", F. Günergun, *Osmanli Bilimi Arastirmalari II* (*Studies in Ottoman Science II*), Istanbul, Istanbul University, 1998, 69-137.

E. Kâhya, *El-'Itaqi, The Treatise on Anatomy of Human Body*, Islamabad, National Hijra Council, One Hundred Great Books of Islamic Civilization series n° 85 (a), 1990 (Original Turkish text with English translation).

I. Miroglu, " Istanbul Rasathanesine Ait Belgeler ", *Tarih Enstitüsü Dergisi*, vol. 3, Istanbul University, Faculty of Letters, 1973, 75-82.

G. Russell, " The Owl and the Pussycat : The Process of Cultural Transmission in Anatomical Illustration ", E. Ihsanoglu (ed.), *Transfer of Modern Science & Technology to the Muslim World*, Istanbul, The Research Centre for Islamic, History, Art and Culture, 1992, 180-212.

N. Sari, Z. Bedizel, " The Paracelsusian Influence on Ottoman Medicine in the Seventeenth and Eighteenth Centuries ", E. Ihsanoglu (ed.), *Transfer of Modern Science & Technology to the Muslim World*, Istanbul, The Research Centre for Islamic, History, Art and Culture, 1992, 157-179.

A. Sayili, *The Observatory in Islam*. Ankara, Turkish Historical Society, 1960.

S. Sherefeddin, *Cerrahiyyetü'l-Haniyye,* 2 vols, ed. I. Uzel, Ankara, Turkish Historical Society, 1992.

S. Tekeli, " Onaltinci Yüzyil Trigonometri Çalismalari Üzerine bir Arastirma, Copernicus ve Takiyüddin (Trigonometry in the Sixteenth Century, Copernicus and Taqi al Din) ", *Erdem*, vol. 2/4, Ankara 1986, 219-272.

S. Tekeli, " Osmanlilarin Astronomi Tarihindeki En Önemli Yüzyili ", *Fatih'ten Günümüze Astronomi*, Prof. Dr. Nüzhet Gökdogan Sempozyumu, Istanbul Üniversitesi'nin Kurulusunun 540. Yildönümü, Istanbul University, 1994, 69-85.

S. Tekeli, " Taqi al-Din ", H. Selin (ed.), *Encyclopaedia of the History of Science, Technology, and Medicine in Non-Western Cultures*, Dordrecht, Boston, London, Kluwer Academic Publishers, 1997, 934-935.

A. Terzioglu, *Moses Hamons Kompendium der Zahnheilkunde aus dem Anfang des 16. Jahrhunderts*, München, 1977.

A. Terzioglu, " Bîmâristan, Islâm dünyasinda klasik hastahanelerin genel adi ", *Türkiye Diyanet Vakfi Islâm Ansiklopedisi*, vol. 6, Istanbul, 1992, 163-178.

I. Uzel, " Dentistry in the Early Turkish Medical Manuscripts ", Ph.D. diss. Istanbul University, Istanbul Medical Faculty, 1979, (supervisor : A. Terzioglu).

A.S. Ünver, " Osmanli Türkleri Ilim Tarihinde Muvakkithaneler ", *Atatürk Konferanslari 1971-72*, vol. 5, 1975, 217-257.

T. Zorlu, *Süleymaniye Tip Medresesi*, Unpublished M.A. thesis, Istanbul University, Faculty of Letters, Dept. of History of Science, 1998 (supervisor : E. Ihsanoglu).

IV

Başhoca Ishak Efendi
Pioneer of Modern Science in Turkey

AT THE END of the eighteenth and the beginning of the nineteenth century, military technical schools like the Mühendishane-i Bahri-i Hümayun and Mühendishane-i Berri-i Hümayun were founded for the teaching of modern sciences. These schools were developed within the framework of the reforms started in the Ottoman state. In addition to the sciences that were a continuation and repetition of those that were taught in accordance with the traditional curriculum, the new sciences that were developed in the West, and with which Ottoman scientists were not fully familiar, were introduced and taught in these schools.

Enough emphasis has not been given to the scholars who were instrumental in the transfer of the new sciences into the educational system and made the transition from the Eastern to the Western sciences possible. Ishak Efendi is one of the most important and prominent figures among them. The swift and widespread translations made by him and the new methods and organization he brought into the Mühendishane-i Berri-i Hümayun's educational system can be considered as the basis of intensive reforms made in the Ottoman sciences and educational system after the Tanzimat period.

ISHAK EFENDI'S ORIGIN AND LIFE

Much has been written and many speculations have been made about the origin and life of Ishak Efendi, but an extensive research based on archival sources has not been published so far. Before discussing the newly found archival documents, we briefly touch upon previous publications and viewpoints regarding Ishak Efendi's origin.

In these writings controversial information is given about his origin. Some writers say that he was the son of a Jew from Yanya (Janina) who converted to Islam, while the majority of the writers say that Ishak Efendi was a convert himself.[1] There are writers like Bursalı Mehmet Tahir Bey and Fuat Köprülü who claim that he was a Turk from Karlova.[2]

In *Mirat-ı Mühendishane-i Berri-i Hümayun,* author Esad Efendi gives detailed information about him and writes that Ishak Efendi was the son of a Jew who had converted to Islam. The family was from the township of Narda in Yanya and lived in the Celalipaşa district where the Jews resided. This was the generally accepted information.[3]

Avram Galante, in his works about the Turkish Jews, based his information directly on the *Mirat* and added, without showing any reference, that Ishak Efendi was known as *Tersane Hahamı* (Rabbi of the Dockyards).[4] Faik Reşit Unat surmises that Sultanzade Ishak Bey and Ishak Efendi could be the same person.[5] In none of these publications is there information on where and how he received his education. Only Salih Zeki Bey correctly guessed that he furthered his education at the Mühendishane.[6]

In light of our research made in the Ottoman archives, we will present new information about Ishak Efendi's origin and life from 1806 to his death in 1836 and analyze his contribution to the introduction of modern sciences into the Ottoman State.

The information we found in the archival documents clears all doubts about his origin. In the 1806 school records, his name appears as "Yanyavî Ishak,"[7] proving that he definitely comes from the town of Yanya. In the school lists of 1813 and 1815, he is registered as Ishak b. Abdullah.[8] This information confirms that Ishak Efendi himself was a convert to Islam. Having "Abdullah" as his father's name eliminates Esad Efendi's supposition that he is the son of a convert.

Thus, we can state that Ishak Efendi was born in the township of Narda in Yanya and belonged to a Jewish family. We could not find any information regarding his birth in the Turkish sources. M. Franco refers to Esad Efendi's work and mentions the year 1774 as his birth date,[9] but no such information is given in the *Mirat-ı Mühendishane*.

Ishak Efendi and his younger brother Esad Efendi had lost their father at an early stage of their lives. The two brothers were educated in different fields. Esad Efendi became the *defterdar* of Rumeli Ordu-i Hümayun and his son, Üsküdarlı Raşid Paşa, was trained in Mühendishane-i Berri-i Hümayun. As to Ishak Efendi, he had two sons, Sami and Bahai. While Ishak was a student at the Mühendishane, Sami assisted his father in his teaching and attended some of his courses. Sami had been an interpreter at the Divan and the Mühendishane for some time; he died soon after his father. Bahai Efendi died at a very young age.[10] No information was available in the documents studied concerning the daughter of Ishak Efendi, who was married to Halil Esrar Efendi.

We did not come across any information about Ishak Efendi's education prior to the Mühendishane. It is understood that he learned Hebrew from his family and Greek in the village where he was born. His knowledge of Turkish, Arabic, and Persian showed that he had a Medrese or similar education in his childhood or youth after his conversion to Islam. His teaching the *Isaogci* book in his Logic courses, his derivation of new

Ottoman terminology from Arabic roots, and his knowledge of this language are also proofs of the education he received at the Medrese and of his vast knowledge of Arabic linguistics.

In 1806 he was a *mülazım* (assistant) in the Third Grade at the Mühendishane. Then in 1812 and 1813 he was made the third assistant and second assistant respectively and finally in 1815 became a First Grade *şakird* (student).[11] It must be stated here that in those days the Mühendishane started with the Fourth Grade and ended with the First Grade. If the student's performance was satisfactory, he would be promoted to the next grade. The number of students in school was limited. Sometimes the student had to wait for several years before he could be promoted to the next grade. In the meantime, he could work in government offices, then leave work to continue his education.

Hüseyin Rıfkı Tamani,[12] chief instructor at the Mühendishane, realized that Ishak Efendi was an outstanding student; he was intelligent, industrious, and knowledgeable. When Tamani was entrusted with repairs of the holy places in Medina in 1816, he brought Ishak Efendi along with him as his assistant.

On his return from Medina in 1822, Ishak Efendi was enrolled in the Mühendishane again as a senior class student.[13] At that time, the Imperial Court Dragoman (Divan-i Hümayun Tercümanı) was Yahya Naci Efendi, who gave French lessons at the translation office while at the same time, teaching at the Mühendishane. The extra duty given to Yahya Naci was detrimental to his lessons at the Mühendishane; therefore Sultan Mahmud II ordered a replacement for the dragoman. Ishak Efendi was proposed for this post, whereupon the sultan ordered that his ability and knowledge be verified.[14]

Ishak Efendi was assigned to this post after the death of Yahya Naci Efendi.[15] He continued working at the court, without any impairment to his career at the Mühendishane until 1829. As related by historian Ahmed Lütfi Efendi, he was sent to the Balkan countries to repair military fortifications because he was out of favor with Reisül-Küttab Pertev Efendi.[16]

On his return to Istanbul in 1830, he was appointed to the Mühendishane as chief instructor,[17] to restore the order and raise the level of education, with the provision that if he did not succeed he would be penalized. He started making many changes in the school administration and dismissed the incompetent teachers. In the meantime he translated and adapted many works into Turkish, prepared the first books introducing many branches of modern Western sciences, and taught them in the school.

American traveler J. de Kay, who visited the school in 1831-32, writes, "Upon asking for the principal, we were directed to a door through which (after stumbling over a huge pile of slippers) we were ushered into a spacious matted chamber. . . . There were some fifty to sixty young men in the room . . . all seated in various positions on the floor, and had their papers before them, copying literally after the dictation of the lecturer."[18]

The new educational system under his chief instructorship is narrated in the daily newspaper *Takvim-i Vakayi,* dated 22 October 1833. The daily teaching program of the senior class consisted of five lessons, one being fieldwork. Thirty-six would-be engineers of the senior class came to the ˙ibrary every morning and sat in their chairs. They were divided into groups of three, and each group would be on duty on different days to draw the diagrams, make calculations, and write the text on the blackboard. Though the lessons differed, the technique of teaching was the same. First, the teacher gave a lecture; the students on duty wrote it on the blackboard, while the other students copied it on their individual slateboards. The teacher then evaluated their work; calculations were made, question-and-answer exercises were practiced, and the students rested in their rooms until the second lesson.

In the first lesson, a text from Bézout's French mathematics book was translated and the subject of "hydraulics" from the same book was discussed. In the second lesson, mechanics (*ilm-i cerr-i eşkal*) from Ishak Efendi's *Mecmua-i Ulûm-i Riyâziye* was taught. Following the noon prayers, readings were made from the *Isagoci* book during the lesson on logic. During the fourth lesson, which consisted of the Hoca's *Usul'üs-Siyaga,* ten students and a draftsman were taken on fieldwork to practice the day's lessons. Each class had its own teacher.[19]

In 1834, while still the chief instructor, Ishak Efendi was sent abroad to repair the Holy places (Ebniye-i Mübareke) in Medina.[20] He died in February 1836, on his way to Istanbul from Medina.[21] A stone in his memory was placed in the cemetery near the Mühendishane in Hasköy, with the inscription, "Divan-ı Hümayun Sabık Serhalifesi ve Mühendishane-i Berri-i Hümayun Başhocası el Hacc Hafız Ishak Efendi." Thus his official duty, which had started in Medina in the company of Hüseyin Rıfkı Tamani, ended on his return from the same city in 1836.

Ishak Efendi was a hardworking man and when he was not teaching at the Mühendishane he read, wrote, or translated books. Numerous works published in a rather short span of eight years are proof of his hard work. He also had a special knowledge of student psychology. Among his students were famous scientists and scholars: chemist Derviş Paşa, Müşir Emin Paşa, Mirliva Esad Paşa, all of whom contributed to the development of modern Western sciences in the Ottoman state.

In J. de Kay's words, "Ishak Efendi was a much respected man among Turks,"[22] and the *Takvim-i Vakayi* qualified him as "the second Kâtip Çelebi of the period."[23] He was admired and respected in his circle for his ability and intelligence. His strong personality impressed those around him but also made enemies of some of his rivals.

IV

He performed the pilgrimage during his first visit to Mecca in 1816. It is understood that he memorized the Holy Koran between the years 1827 and 1830 and thereafter was called Başhoca El Hacc El Hafız Ishak Efendi. He was known to have enjoyed the *nargile* (water pipe) at all times.

Because of his weakness for fame and money, he never could gain the full trust and appreciation of the sultan and his statesmen. There are some obscure points in his life which have not yet come to light. Some original information about his character and daily life is provided by those who knew him well and from whom M. Esad Efendi, the writer of *Mirat-ı Mühendishane,* had collected his reports, as well as by J. de Kay, who met him at the Mühendishane.

Different posts and positions held by Ishak Efendi in two decades created several opportunities for Sultan Mahmud II to become aware of him. Within this period, the sultan's interest in him is seen in the imperial decrees and also in the documents relating to him: these documents also reflect the importance the sultan gave to Western sciences.

Sultan Mahmud II most probably first came to know about Ishak Efendi in the year 1817, while Ishak was assisting chief instructor Hüseyin Rıfkı Tamani in Medina. When Hüseyin Rıfkı Tamani died that year, Kasım Ağa of the Harameyn recommended Ishak Efendi as his replacement. In this instance the governor of Egypt, Mehmet Ali Paşa, asked for the sultan's instructions, saying "mumaileyhin fenn-i mez burda maharet ve mişvari erbabından tahkik olunup, müteveffa-yı mumaileyhin yerine mi ikamesi iktiza eder." But since the sultan did not know Ishak Efendi well, he preferred to assign another engineer, Ahmed Bey of Istanbul, to replace Hüseyin Rıfkı. Ishak Efendi continued to work as the assistant.[24] There was a reluctance on the part of the sultan, which can also be seen in Ishak Efendi's later appointments. Five years later, on the eve of Ishak Efendi's designation as dragoman, the sultan ordered him to go through an examination to prove his ability. The sultan's reservation and prudence towards him, which was also evident when he was removed from the post of dragoman, manifested itself again during his appointment to chief instructor at the Mühendishane in 1830.[25]

After Ishak Efendi became Başhoca, he wrote a petition to the sultan requesting a medal of honor and even attached his own design for the medal. The sultan's uncertainty about Ishak Efendi's work and his refusal of this request on the grounds that such a medal was not suitable for a man of Ishak Efendi's position are further examples of this mistrust.[26] Consequently, another medal suitable for Ishak Efendi's position was considered. Later, seeing the first volume of *Mecmua-i Ulûm-i Riyâziye,* the sultan awarded him 1,000 rub'iye gold pieces.[27] After twenty years, the sultan's appreciation of Ishak Efendi had reached this moderate level.

IV

162

ISHAK EFENDI'S WORKS

The introduction of modern sciences and the educational reforms called for the translation of science books. Ishak Efendi's translation of basic educational books of modern science is his main contribution to the development of Ottoman science and education. As far as we were able to establish, during the eight years between 1826 and 1834, he published a total of ten books in thirteen volumes. These books which were translated or adapted from European sources in such a short time had the backing of the sultan and the state officials.

Upon the completion of the first book of the four-volume work, namely *Mecmua-i Ulûm-i Riyâziye,* he was awarded 250 adlî pieces of gold (1,000 rub'iye) by Sultan Mahmud II. Priority was given to the book for its publication in the State Printing House (Matbaa-i Amire).[28]

The terminology used in these books need special attention. Though the works were based on foreign sources, he was careful to use Ottoman terms. Arabic expressions that were difficult to use were replaced by their equivalents in the Western languages. This was an important contribution to Ottoman scientific terminology. Following are samples of chemistry terms proposed by Ishak Efendi: *Hava-i Memat* for Memphitis (Nitrogen), *Esas-ı Muhdesetü'l-milh* for salifiable base, *Hava-i Nesimi* for atmospheric air, and *Hava-i Hayat* for respirable air. Some of the terms coined by him were still used in Turkey in 1930 and one can find them even today in some Arabic works. Many writers assume that Ishak Efendi adopted the terms *müvellidü'l-humuza* and *müvellidü'l-ma* to replace 'oxygen' and 'hydrogen'. But it is Hekimbaşı Mustafa Behçet Efendi who first used these terms, which were to remain for many years in Turkish and Syrian chemistry terminology.

Most of Ishak Efendi's works were prepared during his chief instructorship. His two books on military art and science, *Rekz ve Nasbü'l Hiyam*[29] (1826) and *Tuhfetü'l Ümera fi Hıfz-ı Kıla*[30] (1827) were printed while he was the dragoman of the Imperial Court. The first book that he published while he was the chief instructor of the Mühendishane is *Medhal fi'l Coğrafya* (1831). This is a summary of the geography section of the book *Heyet,* written by his teacher Hüseyin Rıfkı Tamani.[31]

His major work which brought him fame is the four-volume *Mecmua-i Ulûm-i Riyâziye* (Istanbul, 1831-1843; reprinted at Bulaq in 1841-1845), based on European science books, mainly French. It is a compilation of Western scientific works and his own writings, printed by the order of Sultan Mahmud II. Considering the Ottoman scientific literature printed until 1831, the importance of this book is in its being the first attempt to compile the Turkish texts of natural and mathematical sciences – such as mathematics, physics, chemistry, astronomy, biology, botany, zoology, and mineralogy – in a compendium.[32]

In the Introduction to the first volume, Ishak Efendi wrote, "warfare depends on modern sciences," and he emphasized the necessity of translations from Western sources while at the same time taking care not to antagonize the scientists of the old school. When Kaymakam Paşa (Deputy of the Grand Vizier) presented the first volume of *Mecmua-i Ulûm-i Riyâziye* to Sultan Mahmud II, he explained that this work was compiled from European sources to facilitate the learning and teaching of "military sciences." This expression reflects the outlook of Ottoman statesmen regarding modern sciences.[33]

It is significant that the last volume of this work includes the first treatise on modern chemistry printed in Turkey. In this twenty-nine-page text, Ishak Efendi introduced chemistry as a new science (*Ilm-i cedid*) and pointed to the necessity of this science for the war industry. From his description of chemistry, we may deduct that he considered it only as a science of analysis and synthesis. His understanding of chemistry is in conformity with that of Lavoisier and with the general understanding which dominated in the West at the end of the eighteenth century.

Ishak Efendi mentioned chemistry in different instances as *sanat* (art), *ilim* (science), or *fen* (technique), probably because chemistry was called "science" in the Western sources and sometimes "art," as it was referred to in the Middle Ages. In later Turkish books, the word *sanat* was replaced by *fen* and *ilim*. In our previous studies we indicated that this treatise was in conformity with Lavoisier chemistry but we were not able to cite a definite source. Recent research has established that its source is definitely Lavoisier's *Elements of Chemistry*.[34] Thus Lavoisier's viewpoints and experiments were first introduced to the Ottoman state through this text.

In this treatise Ishak Efendi referred largely to the first part of *Elements*, namely "Of the Formation and Decomposition of Aeriform Fluids, of the Combustion of Simple Bodies, and the Combination of Acids with Bases, and the Formation of Neutral Salts." On the other hand, he made efforts to find Ottoman equivalents for Lavoisier's nomenclature. By giving the number of metals as eighteen instead of Lavoisier's count of seventeen, he shows his awareness of the chemistry literature after Lavoisier.[35] *Mecmua-i Ulûm-i Riyâziye*, which contains this text on chemistry, was taught for a long time at the Mühendishane-i Berri-i Hümayun and at Mekteb-i Harbiye and was instrumental in the wide usage of the new chemistry terms in Turkey.

Ishak Efendi devoted 235 pages to astronomy in the fourth volume of *Mecmua-i Ulûm-i Riyâziye*. The information that he gives about modern Western astronomy is most detailed when compared to Turkish and Arabic literature of science found in his time. The introduction of the theories of modern astronomy to Ottoman science goes as far back as the middle of the seventeenth century. Copernicus was first mentioned in Ottoman scientific

IV

164

literature in 1660, in the book *Secenceliü'l-Eflâk* by Tezkereci Köse Ibrahim from Zigetvar. Later on, Ebu Bekir b. Behram el-Dimaşki in his translation of *Atlas Major* and Belgrade dragoman Abdülmennan Efendi in his translation of *Geographia Generalis* briefly touched on this subject.

In our study on the introduction of modern astronomy into the Islamic world, we gave an overview of the process of introduction of Copernicus' new astronomy concepts into Ottoman science and the developments between the years 1660 and 1850, covering a period of about two centuries.[36] According to our findings, the most detailed early Turkish text on the subject of modern astronomy is the twenty-three-page supplement to Katip Çelebi's *Cihannüma,* written by Ibrahim Müteferrika. The information on Copernicus' astronomy given by Ibrahim Hakkı of Erzurum in his *Marifetname* is based on Müteferrika's work and it is one of the most widely known Turkish texts on this subject. Both Müteferrika and Ibrahim Hakkı discussed the matter in general terms, mainly to make the reader aware of this new concept of astronomy, although each had different motives in mind.

Later writings on this subject are repetitions of Ibrahim Müteferrika's text. On the other hand, the chief instructors of Mühendishane-i Berri-i Hümayun, namely Hüseyin Rıfkı Tamani and Seyyid Ali Bey, were both faithful to old concepts of astronomy in their notes and published works. At the Mühendishane, astronomy lessons in accordance with the new theories started with chief instructor Ishak Efendi's work *Mecmua-i Ulûm-i Riyâziye.*

In the astronomy section of this work, Ishak Efendi gave basic information about the subject and explained the major cosmic systems (Ptolemy, Tycho Brahe, and Copernicus) with diagrams. His explanations of the Copernican system were the most elaborate and most technical ones found in the Ottoman works written so far. He explained the development of this theory in a few words and, without comparing it with the older systems, stated that many astronomical events could be explained on the assumption that the earth is moving. Though he expressed that the view accepting the Sun as the center of the universe was in conformity with the laws of physics and astronomical observations, he left a margin of prudence and stated, "there might be a possibility of error."

Sections on mathematics and mechanics are based on Etienne Bézout's (1730-1783) *Cours de Mathématiques*[37] with additions to the text from various sources. In preparing the sections on astronomy, botany, biology, chemistry, physics, mineralogy, he depended on different technical educational books used in Europe. The contents of these sections are comparable to those published in Europe at the beginning of the nineteenth century and show that the teaching of modern sciences in the Mühendishane was more or less similar to Western scientific education.

Usul'üs-Siyaga is another book by Ishak Efendi dealing with the casting of cannons and was possibly printed between 1831 and 1833. In the Introduction, he writes, "this book is an abridged translation of European science books," and that he hopes it might prove beneficial to Tophane-i Amire (Imperial Arsenal of Ordnance and Artillery).[38] Two other books were also translated and adapted from European sources. Usul-i Istihkâmat (1834) was about military engineering and mainly fortification.[39] The second one, Aksü'l-Meraya fi Ahz-il-Zevaya (1835) dealt with the calculation of angles, directions, distances, altitudes, and the instruments used for this purpose.[40]

Though some sources accept that he is the author of the treatises entitled Küre (Sphere) and Hikmet (Philosophy), the prints or manuscripts of these treatises were not found; it is more possible that he does not have individual works under these names. On the other hand, it was found that the treatise Alat-i Kimyeviye attributed to Ishak Efendi was written by his student Bostanizade Hac Mustafa Bey.[41] Encyclopedist Semseddin Sami, who introduced Ishak Efendi as "the most famous scientist of the later Ottoman period," states that he translated Maison Rustique under the title of Hane-i Zürra.[42] It was later established that this work did not belong to him. In addition to his printed works, Ishak Efendi had three manuscripts. The first was named Kavaid-i Ressamiye,[43] which he wrote during his term of chief instructorship. In the Introduction, he says that he had "invented a method to facilitate understanding these techniques." His second manuscript is the Risale-i Ceyb,[44] the subject matter of which is related to his book Aksü'l-Meraya fi Ahz-il-Zevaya.

His third manuscript deals with sea mines. It is mentioned in some sources as Deniz Lağımı Risalesi. According to our research, its original title is El-Risâlet el-Barkiyye fi'l-Alât el-Ra'diyye.[45] Ishak Efendi translated this work from the French version of Torpedo War and Submarine Explosions written by Robert Fulton (1765-1815), an American mechanical engineer, discoverer of steamships and builder of the first paddle-wheel submarine. This work, which deals with maritime mines, contains two parts. The first part is an explanation of the figures in the book while the second part is devoted to some observations and explanations about these engines.[46]

Ishak Efendi spent his life in the service of the Ottoman state and was instrumental in modernizing the Tercüme Kalemi (Translation Office) and Mühendishane-i Berri-i Hümayun, two important institutions of the state. He was one of the most important personalities who fostered the introduction of modern Western sciences into the Ottoman State. More information about his life will come to light by the discovery of new archival documents. On the other hand, extensive research should be conducted about his works. His contribution to the derivation of new scientific terminology and his influence in establishing new scientific expressions should be assessed.

IV

166

Thus, Ishak Efendi's place in Ottoman scientific and cultural history would easily be established.

Ishak Efendi's influence in the transfer of Western sciences was not limited to the Mühendishane-i Berri-i Hümayun, but it also spread to the other modern military and civilian schools. This effect was also felt in Egypt as we can easily see from the 1841-45 reprint of the *Mecmua-i Ulûm-i Riyâziye* published in Bulaq, Egypt.

We may briefly say that Ishak Efendi was an Ottoman engineer and scholar who started his education at the Medrese and continued at the Mühendishane. The translation movement in which he effectively participated and the educational reorganization he started at the Mühendishane were his main contributions to the development and introduction of modern sciences into the Ottoman state. In a way, he paved the way for a new generation of scientists who followed his footsteps and whose number grew during the Tanzimat period.

Notes

* Head of the History of Science Department, Istanbul University, Faculty of Letters and Director General of Organisation of the Islamic Conference, Research Centre for Islamic History, Art and Culture (IRCICA) Istanbul, and founder of the Turkish Society for History of Science. After this paper was delivered, an article by the author entitled "Mühendishane-i Berri-i Hümayun Başhocası Ishak Efendi, Hayatı ve Çalışmaları Hakkında Arşiv Belgelerine DayalT bir Değerlendirme Denemesi" was published in *Belleten* 53 (August-December, 1989); Türk Tarih Kurumu, and his book devoted to Ishak Efendi's life and works. *Başhoca Ishak Efendi, Türkiye'de Modern Bilmini Üncüsü*, Kültür Bakanlığı Kaynak Eserler Dizisi No.36 (Ankara, 1989), 146 pp.

1. Following are the works of writers who claimed that Ishak Efendi was the son of a Jew who converted to Islam. Mehmet Esad, *Mirat-i Mühendishane-i Berri-i Hümâyûn* (Istanbul, 1312), 34-39; for other sources based on Mehmet Esad, see: Salih Zeki Bey, *Kamus-i Riyaziyat*, 2 (printed section) (Istanbul, 1924). Among those who claimed that it was Ishak Efendi and not the father who was a convert are: Ahmed Cevdet Paşa, *Tarih-i Cevdet*, (Istanbul, 1309), 12:105; Şemseddin Samı, *Kamus-i alâm* (Istanbul, 1306), 2: 899-900; Mehmet Süreyya, *Sicill-i Osmani* (Istanbul, 1308), 1: 328.

2. Bursalı Mehmet Tahir, *Osmanli Müellifleri* (Istanbul, 1327), 2: 254-55; Mehmet Fuad Köprülü, "Hoca Ishak Efendi, *Cumhuriyet Gazetesi*, 1380 (10 March 1928): 1 and 2.

3. Esad, *Mirat-i Mühendishane*, 34-49, 60.

4. Avram Galante, *Histoire des Juifs de Turquie* (Istanbul: ISIS, n.d.), 5: 318-21, "Un Mathématicien Juif, Hodja Ishak Efendi," *Haménora* 2, no. 12 (8th year: December, 1930): 358-60.

5. Faik Reşid Unat, "Başhoca Ishak Efendi," *Belleten* (Ankara) 28, no. 109 (January 1964): 89-145.

6. Salih Zeki Bey, *Kamus-i Riyaziyat.*

7. Record of promotions dated 14 Şevval 1221 (25 December 1806) given by the headmaster of the Mühendishane; "Başbakanlık Osmanlı Devlet Arşivi" (BOA), Istanbul, Cevdet Maarif, no. 567.

8. BOA, Cevdet Maarif, nos. 6388, 3926.

9. M. Franco, *Essai sur l'Histoire des Israélites de l'Empire Ottoman Depuis les Origines Jusqu'à Nos Jours* (Paris, 1897), 141.

BAŞHOCA ISHAK EFENDI   167

10. M. Esad, 36-37.

11. BOA, Cevdet Maarif, no. 567, 4964, 6388, 3926.

12. Hüseyin Rıfkı Efendi, (d.1817) was originally from the Crimea. He was the chief instructor of Mühendishane-i Berri-i Hümayun during the years 1806-1817. He was one of the pioneers instrumental in the introduction and spread of modern Western science in the Ottoman state. In addition to his various official duties, he translated many science books into Turkish. Many of his manuscripts and printed translations were used in the Mühendishane as textbooks. Among his works were *Usul-i Hendese* (1797); *Mecmuatü'l- Mühendisin; Imtihan'ül- Mühendisin* (1802), *Telhisü'l - Eşkal fi Marifeti Terfi'ül Eskal fi Fenni'l- Lağım (1794), Usul-i İnşa-i Tarik, Irtifa Risalesi, Müsellesat-i Müsteviye.*

13. Hatt-i Hümâyûn (H.H.), dated 1239 (1823-4), no.16749.

14. H.H., no.16749; Carter V.Findley, *Bureaucratic Reform in the Ottoman Empire. The Sublime Porte* 1789-1922 (Princeton, 1930), 133 states that Yahya Naci Efendi is of Bulgarian origin while Mehmet Esad refers to him as "Rum'ul-asıl" meaning origin. M. Esad, 33. 33.

15. Ahmet Cevdet, 12: 105.

16. Imperial Decree H.H., no. 43333 based on a brief report dated 1244, which was presented to the sultan. Compare the dates of travel to and from the Balkans. Petition of Ishak Efendi dated Muharrem 1245. Cevdet Hariciye no. 3481 and Petition of Halil Esrar Efendi dated 29 Muharrem 1245, Cevdet Hariciye, no. 2222; Ahmed Lütfi Efendi stated that Ishak Efendi was sent to this post because of Reisül Küttab Pertev Paşa's enmity, *Tarih-i Lütfi,* 2: 143.

17. Imperial Decree H.H., no. 28635 based on a brief report dated 1246H, decision dated Evail-i Receb 1246H, Cevdet Maarif, no. 5588.

18. J. D. Kay, *Sketches of Turkey, 1831-1832* (New York, 1833), Part 15, 138-44.

19. *Takvim-i Vakayi,* no. 69, 7 Cemaziyelâhir 1249 (22 October 1833), 34, lines 1-2.

20. Cevdet Evkaf, no. 21277, 19635, 10777; Cevdet Maarif, no. 1832.

21. In the book *Mirat-i Mühendishane* by Mehmet Esad Efendi, and other works based on this book, it is stated that Ishak Efendi died in Suez on his return from Medina. In a new document that we found, there is no definite indication of his death taking place in Suez; it mentions, however, that he died "on his return to Alexandria" while trying to reach Istanbul. Undated Imperial Decree (1253/1836), H.H., no. 27254.

22. Kay, *Sketches of Turkey,* 138.

23. *Takvim-i Vakayi,* no. 33, 7 Rebiülahir 1248 (3 September 1832), 3, line 1.

24. Note from Mehmet Ali Paşa, Governor of Egypt, dated 23 Rebiülevvel 1232 (10 February 1817), Cevdet Maarif, no. 3458, and appendix.

25. Cevdet Maarif, no. 5588.

26. H.H., No. 18293.

27. Imperial Decree based on a brief report sent by the Superintendent of the Grand Vizier, H.H., no. 93074.

28. Ibid.

29. Ishak Efendi, *Rekz-i ve Nasb-i Hiyam,* Dersaadet 1242 (1826).

30. Ishak Efendi, *Tuhfetü'l- ümerâ fi Hıfz-ı Kıla,* Dersaadet 1243 (1827), 132.

31. *Medhal fi'l Coğrafya,* ed. Ishak Efendi, Dersaadet 1247 (1831), 77, pl. 8. A detailed account on this work is given in the astronomy chapter of this article.

32. Ishak Efendi, Mecmua-ı Ulûm-i Riyâziye, 4 vols., Matbaa-i Amire 1247-1250 (1831-1834). Ekmeleddin Ihsanoğlu, Feza Günergun, "Mecmua-ı Ulûm- Riyâziye," Philosophy, Logic and History of Science Symposium, 19-21 December 1986, Ankara.

33. Imperial Decree, probably dated 1246 (1831), based on a brief report submitted. H.H., no. 33074.

34. Antoine Lavoisier, *Elements of Chemistry,* by Robert Kerr, Edinburgh 1790 (Facsimile, New York, 1965), 176.

35. After the publication of the list containing the seventeen metallic elements of Lavoisier, the first metallic element found in 1797 was Vauquelin's "chromium" isolated from lead chromate, i.e. the chrome metal. This is probably the same metal mentioned by Ishak Efendi as the eighteenth one, an evidence of his awareness of the works of the French Chemist L. H. Vauquelin (1763-1829).

36. Ekmeleddin Ihsanoğlu, "Introduction of Modern Astronomy to the Islamic World (1660-1860)." International Symposium in Modern Science and the Muslim World, First Meeting on Science and Technology Transfer from the West to the Muslim World from the Renaissance to the Beginning of the XXth Century, 2-4 September 1987 Istanbul, to be printed.

37. Etienne Bézout, *Cours de Mathématiques à l'usage du Corps de l'Artillerie,* 1er tome (Paris, VIII, 1798).

38. Ishak Efendi, *Usul'üs-siyaga, Matbaa-i Amire 1247-1250 (1831-1834); Ekmeleddin Ihsanoğlu, Açıklamalı Türk Kimya Eserleri Bibliyografyası (Basmalar 1830-1923) ve Modern Kimya Biliminin Türkiye Cumhuriyet'nin Kuruluşna Kadar olan Durumu ve Gelimesi* (Istanbul, 1985), 53.

39. Ishak Efendi, *Usul-i Istihkâmat,* Matbaa-i Amire, 1250 (1834).

40. Ishak Efendi, *Aksul-Merâyâ fi Ahziz-zevâyâ,* Matbaa-i Amire 1250 (1834). A manuscript copy, under Number 25, is to be found in the Library of the Research Center for History of Science and Technology, Istanbul Technical University.

41. Ekmeleddin Ihsanoğlu, *Açiklamali Türk Kimya Eserleri Bibliografyası, 29-40.*

42. Semseddin Sami, *Kamus-i Alâm* (Istanbul, 1306), 2: 899-900.

43. *Kavaid-i Ressamiye,* Istanbul University Central Library, Turkish Manuscripts (T.Y.), 6829 Rika, 168 folios, Library of the Research Center for History of Science and Technology, Istanbul Technical University, no. 24, Rika, 173 + 16 folios.

44. *Risale-i Ceyb,* Istanbul University Central Library, T.Y. 714; Kandilli Rasathanesi, no. 354/2, Nesih, 14b-27b; Bagdad, Dar Saddam Li'l-Mahtutat, no. 6429.

45. *El-Risâlet el-Barkiyye fi'l-Alât el Ra'diyye,* Kandilli Observatory Library, no.168/2, Rika, folios 21a-48b. The diagrams in the last section of this copy are missing.

46. Robert Fulton, *Torpedo War and Submarine Explosions,*(New York, 1840). The French version translated by Ishak Efendi is *Le Torpedo, ou Moyen de Faire Sauter en Mer les Navires Ennemis.*

V

SOME CRITICAL NOTES ON THE INTRODUCTION OF MODERN SCIENCES TO THE OTTOMAN STATE AND THE RELATION BETWEEN SCIENCE AND RELIGION UP TO THE END OF THE NINETEENTH CENTURY

The introduction of the modern sciences into the Ottoman State and society came about by three different means. The first of these was translations from western languages, either directly or via third languages. The second was the reports prepared by Ottoman ambassadors in Europe. As these reports were solely for the information of high-ranking statesmen, their sphere of influence was very limited. In comparison to these two means, the third undoubtedly constituted a far more influential factor in the spread of knowledge of the modern sciences among the Ottomans. This was the new educational institutions founded by the State.

The primary objective of the continuing endeavour to modernise the army, which began in the reign of Ahmed III and continued during the reign of Mahmud I and Mustafa III, was to increase its strength and to produce soldiers versed in the new knowledge and techniques of Europe. With this objective in mind new institutions and schools were opened, known by such diverse names as *Humbarahâne*, *Lağımcı Ocağı* and *Hendese-hâne*. But these institutions were short-lived and inconsistent. They taught only arithmetic, geometry and geography but no physics or chemistry[1].

Systematic technical education in Turkey began first at the Imperial Naval Engineering College (*Mühendishane-i Bahr-î Hümâyûn*), established in 1773[2].

After this date new fairly stable educational institutions were started. During the course of the research no evidence was found that chemistry,

---

1. Titles of the primary works on the history of general and technical education in Turkey, which will often be referred to in this study, are abbreviated as follows:

T.M.T. — Osman Ergin, *Türk Maarif Tarihi*, Istanbul, 1977

Y.M.O. — C. Uluçay and E. Kartekin, *Yüksek Mühendis Okulu*, Istanbul, 1958

M.M.B.H. — Mehmed Esad (Niğdeli), *Mir'at-ı Mühendishâne-i Berri-i Hümâyûn*, Istanbul, Karabet Matbaası, 1312.

2. Feyzi Kurtoğlu, *Deniz Mektepleri Tarihçesi*, Istanbul, 1941.

V

physics or similar sciences were included in the curriculum of the Imperial Naval Engineering College until the establishment of the Infantry Engineering College (*Mühendishane-i Berrî-i Hümâyun*) in 1795.

The decree dated 1210 (1795) and issued by Selim III described exactly what this school was expected to provide and which subjects had priority[3].

The same decree commanded that the engineering schools provide a thorough knowledge of geometry, arithmetic and geography, and teach the theory and application of the military sciences. It also accorded superior status to the students of these schools, over both medrese students and other military classes[4].

The curriculum of the four years of study remained strictly within the limits laid down by the *firman*. Among the subjects taught were the following: languages (Arabic, French), mathematics (arithmetic, geometry, plane trigonometry, algebra, calculus), mechanics, conical cross sections, geography, military engineering, astronomy, ballistics, military history, and the training of soldiers. Physics, chemistry and other natural sciences were not in the curriculum[5].

This list reveals that contemporary concepts of modern scientific education were based on a highly pragmatic and practical approach to the philosophy of education, which did not recognise a need for study of the natural sciences and of physics and chemistry.

In 1801 (1216) an amendment to the decree was issued under which both theoretical and practical education was confined to the subjects described above[6].

From its establishment until the 1830s there is no evidence that the Engineering School taught chemistry, physics or any natural sciences. These subjects were introduced by İshak Efendi, who was appointed school principal in 1830.

İshak Efendi's four-volume work on the natural sciences and technical subjects, entitled *Mecmua-i Umum-i Riyaziye,* is a comprehensive representation. The first and second volumes were published in 1831 (1247 H), the third in 1833 (1249 H), and the fourth in 1834 (1250 H) in Istanbul.

3₁ *Mühendishane-i Sultâninin Tesis ve Küşâdını âmir Sultan Selim Han Sâlis Fermanı,* Istanbul, 1328 (This work contains the facsimile of the original *firman* as well as its printed copy. Often referred to in our study, it will be mentioned in abbreviated from as *S.S.F.* It must be noted here, however, that there are certain discrepancies between the original text of the decree and the printed text).

4. *S.S.F.* (The original text p.1)

5. *S.S.F.* (The original text p.2)

6. *Y.M.O.* pp.494—497

The article published in the *Takvim-i Vekâyi* newspaper in 1833/1249 H. tells us that in the same year İshak Efendi taught the third volume of a French book entitled *Ulum-i Riyaziye* (Mathematical sciences) by "Bezu"[7] in the mornings[8]. The fact that hydraulics was among the subjects taught and that İshak Efendi covered this subject himself in the 3rd volume of his own four-volume work may offer a clue as to the sources used by İshak Efendi. This new theory appears to be more viable than the evidence put forward by M. Esad in his book *Miraat* to the effect that İshak Efendi's work is a translation from Latin—a language which in the period in question was no longer the academic language of Europe.

After teaching hydraulics in the mornings, İshak Efendi's second lesson was mechanics, which he taught from his own book, published in Turkish in the same year (1834).

The section in the fourth volume of İshak Efendi's book on post-Lavoisier modern chemistry, was the first treatise on the subject to appear in Turkey. The subject was first taught in Turkey in 1834.

The research has also revealed that technical training in arms manufacture and gunnery, both closely related to chemistry, began before the teaching of chemistry.

Since in two earlier studies İshak Efendi's treatise on chemistry has been analysed from the viewpoint of the history of chemistry and history of science in Turkey[9-10], it will be more beneficial here to look at some other aspects of his contribution to science in Turkey.

İshak Efendi's book was for a long time the textbook of the Engineer-

7. This French autor's name is spelled in various forms; sometimes as Bezu and at other times as Pezo or Bezout. It is more possible that he was Etienne Bezout (1730-1783).

8. An article entitled *Fünûn* in issue No. 69 (1249 = 1833) of the Ottoman official gazette *Takvim-i Vekâyi*, defining "the actual state of teaching and instruction at the Infantry Engineering College...", extensively describes the educational system and the daily programmes of the college, particularly during the presidency of İshak Efendi. A summary of this article was first quoted in *M.M.B.H.* (pp.58-60) but with no reference as to the date or number. Later *M.M.B.H.* itself became a source of reference for many other studies and researches. A close comparison of the original article and its quoted summaries reveals however that the latter contain several mistakes. Examples:

(i) İshak Efendi's work *Usul-üs siyağa* to which reference is made by O. Ergin (*T.M.T.* vol. 2, p. 331) is mistakenly called *İsağocı*, a work on logic.

(ii) The 3rd volume of the French author Pezo or Bezout's work in French is mixed up with 3rd volume of İshak Efendi's work (*T.M.T.* vol. 2, pp.330-331).

9. E. İhsanoğlu: "Türkiyede Basılan ilk Kimya Eseri, İlim Tarihi Açısından bir Değerlendirme", *TÜBİTAK, VIIth Scientific Congress*, (1980), pp.113-123,

10. E. İhsanoğlu, "The First Treatise on Modern Chemistry Printed in Turkey", *1st International Congress of Turkish-Islamic History of Science and Technology* (14-18.September 1981), vol. I, pp.123-133.

V

238

ing School and its influence abroad has been long-lasting, particularly in Egypt.

With the establishment of the War College in 1834 it was decided to teach the sciences, but the implementation of this decision was delayed until 1847. Among the subjects in the science curriculum was "*İlmi-hall* and *terkib-i ecsam*", terms used by İshak Efendi for modern chemistry in his work *Ulum-i Riyaziye*, demonstrating that chemistry was included in the curriculum of the War College.

Of relevance here is *Usul-i Kimya* by the chemist Derviş Paşa, published in 1848[11]. Derviş Paşa was a scientist and statesman who graduated from the Engineering School and was one of the students who were sent to Europe in 1250 or 1251 (1834/1835) to study with the objective of training employees for the Cannon foundry and affiliated gunpowder factory, cartridge factory, foundry and other departments, and well-qualified teachers for the newly-opened Imperial War College. Derviş Paşa was subsequently appointed physics and chemistry teacher at the Medical School and chemistry teacher at the War College. Derviş Paşa's book was the first chemistry textbook published in Turkish, and gives a detailed account of modern inorganic chemistry. It might be said that he laid the bases of chemistry in Turkey. Although Derviş Paşa translated the technical terminology as far as possible, he leaves some in their original form where no suitable equivalent is available[12].

He also defended the view that the use of Latin characters as symbols was more appropriate than substituting them by Arabic characters.

Modern chemistry was first taught in brief and theoretical form at the Engineering School in 1834, and it was not until 1850 that a laboratory was set up.

However, at the Engineering School, cannon casting, a subject which requires detailed chemical and metallurgical knowledge was taught soon after its establishment in 1795, during the reign of Selim III (1795-1808). It has also been established that İshak Efendi taught this subject from 1830-1833 and also published a book an the same subject.

Thus, we can deduce that almost from the first day the Engineering School began instructions about arms manufacturing as specified in its founding decree. Since the school gave priority to the objective of serving the

11. Derviş Paşa, *Usul-i Kimya*, Istanbul, 1268 (1834).

12. For further details on the history of chemistry, see: E. İhsanoğlu; *Açıklamalı Türk Kimya Eserleri Bibliyografyası (Basmalar / 1830-1923) ve Modern Kimya Biliminin Türkiye Cumhuriyetin Kuruluşuna kadar olan Durumu ve Gelişmesi*, Istanbul, 1985.

war industry and aimed to reach this objective by the shortest route, it conveyed only very limited practical information about metallurgical science based on chemistry, and for a long time (1795-1834) chemistry was not taught.

When the decree of 1210 is examined, we see that apart from its educational role, the engineering school played an important role in state service and development. This role was the technical supervision of arms manufacturing at the Hasköy Foundry and the Cannon Foundry, as laid down in a firman.

As the above illustration reveals, the Ottoman state was concerned with the practical aspect of modern western science rather than its fundamental principles. Without appreciating the interdependence between pure and applied sciences, it embarked on a programme of establishing a war industry based on the application of scientific knowledge. The preface to İshak Efendi's four-volume work gives an account of why the sciences were essential for war. In the preface to his book, the chemist Derviş Paşa, who was a generation younger than İshak Efendi, also explains why chemistry was necessary for military officers and staff, after giving an extensive and detailed account of the benefits of chemistry. For subsequent generations this objective gradually lost its importance, superseded by civil objectives, such as medicine, agriculture, and industry.

When it reached the time of Ziya Gökalp in 1918, *Fen*, the supposed Turkish equivalent of ''science'' was still meaning ''techniques'' and according to him modernism meant being capable of manufacturing and using all the instruments that were manufactured and utilized by developed nations [13].

After İshak Efendi's first translation, the translation of modern scientific works and terminology into Turkish was continued by subsequent generations. In this sphere a divergence of attitude arose between Istanbul and Cairo, both important Ottoman centres of learning. In Istanbul, the trend was to use words of Arabic origin to replace the terms. Whereas in Cairo the original terms were retained, occasionally with minor spelling alterations. For example, in Istanbul oxygen was called *müvellidülhumuza*, while in Cairo it was called *oksijen*. This custom continued to the end of the nineteenth century. Ottoman translators of modern science held the view that ''just as Latin and Greek were employed in the creation of new

13. Ziya Gökalp, *Türkleşmek, İslamlaşmak, Muasırlaşmak*, Istanbul, 1918, p. 9.

V

terms in European languages, Arabic could be a source of new terms in Turkish"[14].

The most systematic and wide-scoped survey of the derivation and/or creation of new Ottoman terms from Arabic and Persian to express modern concepts is the *Lügât-i Tıbbiye*, a French-Turkish dictionary published in 1873 by the Ottoman Medical Association. Later this dictionary became a major source for Arab universities attempting to "Arabicise" their education after the First World War.

In spite of Derviş Paşa's defense of the use of Latin rather than Arabic characters for symbolisation, later authors and translators generally chose to use Arabic for this purpose, or a combination of both. This practice continued until the end of the nineteenth century. In particular after the second constitutional period (1908) the use of Latin characters became increasingly widespread.

The most conspicuous shortcoming in the concepts and implementation of Ottoman science is the lack of research. No research was carried out at the Engineering School during the first century of its existence.

Some graduates of the Engineering School and Medical School who went to Europe for further studies, carried out researches, results of which were published in the countries where they studied. One example is the thesis of Georges Della Suda, also known as the Chemist Faik Paşa, which was published in Paris in 1855[15]. The first doctoral thesis on chemistry to be published in Turkish in 1919 was a translation of the thesis written originally in German by Mustafa Azmi Bey and published after receiving his doctorate from the Munich Technical College[16].

At least five doctoral theses were written by Ottoman scholars from 1855-1919, some of which were published in Europe.

The first doctoral thesis on chemistry written in Turkey was submitted to the Science Faculty of Istanbul University in 1940[17], approximately 100 years after chemistry became part of the curriculum at the Engineering School, and 150 years after its establishment.

The establishment in 1861 of the Ottoman Scientific Society, the publication of the first scientific periodical called *Mecmua-ı Fünûn*, and the first endeavour to establish an Ottoman university (1863) were all aspects

14. Dika Bey: *Kimya-yi uzvî-i Tıbbî*, Charles-Adolphe Wurtz. Translator: Dika ve Limonidis, Istanbul, 1883. Preface.

15. G. Della Suda, *Thèse sur l'ammonium*, Paris, 1855, p.56.

16. M. Azmi, *Di Fenil Amin'in Usul-i Elektriki ile Oksidasyon ve Bromlandırılması*, 1919, p.126 + 4.

17. E. İnönü, *1923-1966 Dönemi Türkiye Kimya Araştırmaları Bibliyografyası ve Bazı Gözlemleri* Istanbul, 1982, p.13.

of the attempt to acquaint the masses with modern science. The first course of lectures at the university was in physics. The first lecture, during which electrical experiments were carried out on the podium, was attended by 300 students and attracted such interest, that within a short time the audience increased to 400, the maximum capacity of the lecture hall, and those who arrived late were turned away[18].

It is seen from the first article of the Statute of the Ottoman Scientific Society, determining its objectives, that the aim of the society is to publish books and translations in modern sciences and to enlighten the people[19]. This reveals that a second group of Ottoman scholars taking up modern sciences pioneered by Münif Paşa has not been able to accord its due importance to research.

With the second attempt to establish a university in 1870, twelve public lectures were arranged, all of which were about physical and natural sciences thus demonstrating the importance which Ottoman intellectuals attached to the spread of modern sciences[20]. However, what influence these activities had on intellectuals and the general public is unknown. And it is not possible to contend that these had any effect different from previous endeavours on the formation of scientific mentality among Turkish intellectuals.

As far as the publications are concerned, in general the material published in Turkish until the end of the nineteenth century in the field of modern sciences consists of translations from French. Broadly speaking the written material is, of a general nature and with a few exceptions for each discipline, no high quality publication exists.

One of the major aspects of any study aiming to understand the pattern of development of these sciences in the Ottoman State, must be the analysis of the relationship between these new sciences and the religion, which was the common background of Ottoman intellectuals raised in a society where the main educational system was that offered by the *medreses*. We must analyse this relationship as it existed at the time when Ottoman intellectuals first came into contact with these sciences.

The impression given by the only general study of the history of Turkish science ever written by A. Adıvar, is that in the eighteenth century, the mentality which dominated in society was one of opposition to all sciences other than religious sciences[21].

18. M.A. Aynî, *Dâr'ül-Fünûn Tarihi*, Istanbul, 1937, pp.13-14.
19. *Cemiyet-i İlmiye-i Osmaniye, Mecmua-i Fünûn*, 1st Year, No: 1. Muharrem 1297 (1862).
20. M.A. Aynî, *op. cit.*, p. 20.
21. A.A. Adıvar, *Osmanlı Türklerinde İlim*, Revised 4th ed.; compilors: A. Kazancıgil and S. Tekeli, İstanbul 1982, pp.159,176.

V

The author is also of the opinion that the term *"Fen"* rather than *"İl-im"* used for these modern sciences by the *medrese*s is an evidence of the disdain in which they were held[22].

Subsequent studies describe the role played by the *medrese*s in specific terms, and in the context of clearly-defined subject areas. According to S.C. Antel, the *medrese*s had deteriorated into strongholds of ignorance, fanaticism and corruption, which obstructed all forms of progress. Antel states bluntly that the *medrese*s "methodically and obdurately strove to prevent European science and technology gaining acceptance in the country"[23].

E.Z. Karal, in his study of education during the reign of Mahmud II, describes the attitude of the *medrese*s to the educational reforms in the following terms: "Not only were they not in a position to assist, but were actually hostile to western thought"[24].

S.C. Antel[23] and S. Irmak[25], both conclude that the hostility is attributable to religious fanaticism. In his study of the history of modern Turkish thought, H.Z. Ülken expresses the same widely-held opinion, and gives a specific example to illustrate this: "This fanatic mentality obstructed the progress of modern medical education, which had been imposed by the state."[26]

S. Irmak offers a different perspective on the issue with the following theory which gets to the essence of the problem by refferring to the relationship between the new sciences and religion on the conceptual plane. "This was a time when contradiction between the interpretation of the Quran and scientific fact began to materialise"[27]. He tells us that "this fanatic attitude was the official view held by the *medrese*s and the Office of Sheikhulislam, as well as being the official doctrine of the state." And Irmak takes his theory to a greater extreme with the claim that "it was courageous endeavours of heroic pioneers in the first half of the nineteenth century, very often at the risk of lossing their lives, which enabled positive science institutions to be established in the Ottoman Empire"[28].

The summaries and quotes given above illustrate the theory that the

22. A.A. Adıvar, opt. cit,. p.222.
23. S.C. Antel, *Tanzimat Maarifi*, (Tanzimat, Istanbul 1940), p.441.
24. E.Z. Karal, *Osmanlı Tarihi*, vol. 4, T.T.K., 1983, p.158.
25. S. Irmak, "Pozitif Bilim Kuruluşları ve Dâr'ül-fünûn' dan Üniversiteye Geçiş", *Cumhuriyetin 50. yılında Istanbul Üniversitesi*, Istanbul 1973. p.100.
26. H.Z. Ülken, *Türkiyede Çağdaş Düşünce Tarihi*, Istanbul, 1979, p.27.
27. S. Irmak, *op. cit.*, p.100.
28. S. Irmak, *op. cit.*, p.101.

V

medreses, or theological colleges, resisted the acceptance of modern science in Turkey; that the fanatic bigotry of the *ulema* caused them to reject modern science; and that a kind of controversy between religion and science existed in Ottoman society. In this study this theory will be subjected to close scrutiny. But before embarking on this it should be pointed out that A. Adıvar, in his book *Tarih Boyunca İlim ve Din* in which he analyses the conflict between science and religion, A. Adıvar states the need for a separate study of the relationship between these two subjects and the situation of science in the Orient from the eighteenth to the nineteenth centuries[29]. From this, we may deduce that the view expressed by Adıvar in his earlier work, *Osmanlı Türklerinde İlim*, is not based on a thorough study, as he himself admits. Moreover, the authors of the statements quoted above have undertaken no such research themselves, as a perusal of their works clearly reveals.

The analysis will consist of clarifying the following points by studying the original sources: On the conceptual plane, what was the attitude of Muslim Ottoman intellectuals to the positive sciences and its consequences?

What was their reaction to those aspects related most closely to religion?

What was the role played by the *ulema* from the late eighteenth century onwards?

The Imperial Naval Engineering School established in 1773, the Imperial Medical School established in 1838 and the War College established in 1834, were the first Ottoman educational institutions where modern technical education assumed permanency. The opening ceremonies of all these institutions were attended by top-ranking *ulema*, who led the prayers. Moreoever, graduates of the medreses taught religion, oriental languages, and general culture. Besides, the first teachers of mathematics and medicine, the major subjects of these institutions, were people who had been educated according to traditional concepts.

At the Imperial Naval Engineering College mathematics was taught by the famous mathematician and *müderris* (professor) Gelenbevî İsmail Efendi (1730-1791) who later became a *kadı* (judge)[30]. Gelenbevî had been educated under the traditional system, and not only wrote books on arithmetic, geometry, logarithm, and trigonometry, but also on literature, *fıkıh* (Muslim canonical jurisprudence), theology, logic, philosophy, and other classical subjects of *medrese* scholarship, most of which were published[31].

29. A.A. Adıvar, *Tarih Boyunca İlim ve Din*, Istanbul, 1944, vol.1, p.113.
30. Salih Zeki, *Kamu-ı Riyâzıyât*, 1308, vol. 1, pp.318-321.
31. Y.A. Sarkis, *Mu'cem'ül-Matba'ât al-Arabiye ve'l Mu'raba*, Cairo, 1928, pp. 1165-1166, 1365.

V

The two people who figured most prominently in the introduction of modern medical science into the Ottoman State were the historian and physician Şanizade Ataullah Efendi (1796-1826) and Chief Physician to the Sultan, Hekimbaşı Mustafa Behçet Efendi (1774-1843), both of whom had received a *medrese* education. Their translations and adaptations of European medical books made available the new medical knowledge which was developing in the West. They exerted great efforts to introduce modern medicine to the State, and their published works pioneered in this field[32]. Chief Physician Behçet Efendi also played a major role in the establishment of the Imperial College of Medecine in 1838, the institution which was responsible for establishing modern medical education in the Ottoman State. Behçet Efendi's brother Abdülhak Molla, who contributed to the establishment of his school, was also a physician educated at the *medrese*[33].

A detailed study of the early years of these forerunners of systematic and well-established modern scientific education reveals that most of the original teaching staff were graduates of the *medreses*. Yet they did not hesitate to take up posts in these newly-founded institutions. This shows that the *ulema* were not opposed to the new educational institutions. Nevertheless, it is undeniable that a divergence of attitude later emerged between the intellectuals who graduated from these new educational institutions and those who were graduates of the *medreses*. This divergence undoubtedly resulted in the situation described by many researchers as "dualism" in Ottoman thought, particularly in the sphere of science. An important question here is whether this dualism, or conflict between the modern and traditional, was actually the conceptual conflict at issue. In other words, was it the result of a conflict between the concept of science based on Islam and Islamic scientific heritage upheld by the *ulema* and represented by the *medrese* teachers, and the one that was upheld by the newstyle intellectuals trained in contemporary science? To find the answer to this question it is necessary to ascertain the attitudes and views of the Ottoman scholars educated under the influence of religious education, on the subjects related to religion. Astronomy is the subject which is generally most sensitive to discrepancies between religion and science. Modern astronomy is based on the Copernican theory, which introduced a new concept of the universe, superseding the theory of Ptolemy that the earth was the centre of the universe, as espoused by some Muslim scholars.

32. A. Kazancıgil, "Istanbul Üniversitesi", in *İstanbul Üniversitesi Kuruluşu, Tarihçe, Teşkilat ve Öğretim üyeleri*, vol. I, 1453-1981, Istanbul 1983, p.73.

33. A.S. Ünver, "Osmanlı Tababeti ve Tanzimat Hakkında Yeni Notlar", *Tanzimat*, Istanbul, 1940 pp.936-937.

Kâtip Çelebi (1609-1657), one of the formost thinkers and scholars of the first half of the seventeenth century, was the author of *Cihannûma*, a geographical treatise which he began writing in 1648, and which was based not only on classical oriental sources but also on contemporary western sources. Kâtip Çelebi puts forward several pieces of evidence which lend validity to the theory that the earth is round and sets out to prove that this fact does not conflict in any way with the Islamic religion. He quotes from the *Tehafût* of Imam Gazâli to support this view. Kâtip Çelebi makes no direct reference in this work to the Copernican theory, which by then was well-known in Europe. Adıvar interprets this omission as possibly stemming from "the double fanaticism of both Islam and Christianity"[34].

On the other hand, H. Z. Ülken explains this omission as the consequence of "the opposition to Copernican theories among some clerics in Europe". On this subject Ülken says the following:

> This is not at all surprising and should not hastily be attributed to intellectual reactionism. In the seventeenth century these new ideas were the subject of controversy. Belief in the Copernican system was not as widespread or well-established as had been assumed. Consequently, although Kâtip Çelebi was aware of the new concept of the universe, it was natural that he should regard them with suspicion. This conservatism in Kâtip Çelebi's geography was the result of the opposition to these ideas, which were still relatively new in Europe, by his assistant Sheikh Mehmed Ihlasi, a former French priest and convert to Islam.[35]

In 1685 not fourty years after Kâtip Çelebi wrote the *Cihannûma*, Ebu Bekir bin Bahram al-Demaski was the first, according to Adıvar's findings, to make a brief reference to the Copernican system. Al-Demaski describes the Ptolemaic theory that the sun and all the stars revolve around the earth as "more probable and easy to understand."[36] It therefore becomes apparent that the interpretation of Copernicus has no connection with religious concepts.

Later on, in 1732 İbrahim Müteferrika published Kâtip Çelebi's *Cihannûma*, adding an appendix in which he refers to various theories of the universe, including those of Tycho Brahe and Copernicus. But when describing the Copernican theory, he cautiously remarks that it is not necessarily true, and that "it is not accepted by the Muslim *ulema*"[37]

34. A.A. Adıvar; *Osmanlı Türklerinde İlim*, ..., pp.144-145.
35. H.Z. Ülken, "Kâtip Çelebi ve Fikir Hayatımız" *Kâtip Çelebi, Hayatı ve Eserleri Hakkında İncelemeler*, Ankara, 1957, pp.178-179.
36. A.A. Adıvar, *op. cit.*, p.155.
37. A.A. Adıvar, *op. cit.*, p.171.

This observation by Müteferrika requires further investigation. He himself did not elucidate the connection between this cautious attitude and the Islamic religion. He probably opted for a prudent approach because as a convert from Christianity to Islam he could not predict the reception which these ideas would have.

After Müteferrika, İbrahim Hakkı of Erzurum (1703-1780) was the next to refer to this subject. He was a scholar educated along totally classical lines at *medrese*s far from Istanbul. He was also a member of a *sufi* sect. In his *Marifetnâme* written in 1760 he gives his views of Copernican astronomy, which he calls *Heyet-i Cedide*, in the perspective of Islam[38].

İbrahim Hakkı tells us that this new concept of astronomy had become renowned over the past few years, that its principles were simple and brief, that it had won many adherents, and that those who endeavoured to spread belief in this new theory, ''were stoned with blame and recrimination by the laymen'', by which he meant that the uneducated failed to comprehend this new theory.

İbrahim Hakkı goes on to describe the relationship between this issue and religion, and expresses clearly the stance towards science taken by Islam, which is that so long as one does not claim that the universe came into being of its own accord and accept the existence of an omnipotent creator, then any explanation of the system of the universe put forward by scientific method will be compatible with religion. He also says that ''trust and belief'' in these matters is not a matter of religion or a dictate of faith.

When these statements by İbrahim Hakkı are analysed in conjunction with those of Kâtip Çelebi as summarised above, the broadmindedness of those Ottoman scholars educated traditionally in Istanbul or at the furthest corners of Anatolia, whether at the *medrese*s or not, on the subjects of astronomy and cosmology, will enable us to make a re-evaluation of Islamic attitudes.

In 1834, 74 years after İbrahim Hakkı wrote his *Marifetname*, the Principal of the Imperial Engineering School, Ishak Efendi, deals with the subject of astronomy in the second essay in the fourth volume of his book *Mecmua-i Ulum Riyaziye* in quite a different way. Ishak Efendi explains the systems of Ptolemy, Tycho Brahe and Copernicus at length and in full technical detail, drawing comparisons between them:

Although the Copernican theory may be mistaken, it is a point worth

38. İbrahim Hakkı, *Marifetnâme*, Istanbul, 1330, pp. 144-151.

considering because the aformentioned theory of the movement of
the celestial bodies is a plausible explanation ... and according to
this hypothesis ... it appears appropriate and in keeping with the laws
of nature and astronomical investigation that the sun be the centre
of the planets.

Although İshak Efendi considered the Copernican system capable of
explaining the movements of the heavenly bodies, he says:"that is quite
possibly mistaken"[39].

İshak Efendi's minor reservation on the Copernican theory is similar
to his attitude to the elements from the point of view of the relation between traditional and modern science. In an earlier published study on İshak
Efendi's book on modern chemistry, I dealt with the subject from the aspect
of science history in general, and the history of chemistry in particular.
On the subject of the elements İshak Efendi says:

> Some earlier scholars held that there were only four elements, but
> modern scholars of chemical science claim that there are more than
> four, and that water, air, and earth, previously considered to be
> elements by old scholars, actually consist of more than one element.

Thus İshak Efendi makes a circumspect and tactful transition from
traditional to modern scientific concepts. This judicious approach is reminiscent of that displayed by Müteferrika, as referred above. It seems likely
that what prompted such caution in both men was the fact that both were
converts to Islam. However, this reserve on the part of İshak Efendi, who
lived a century after Müteferrika, cannot be attributed to fear of "rejection by the Islamic *ulema*," because the explicitness of the *Marifetname* of
İbrahim Hakkı of Erzurum demonstrates that such fear hardly existed.

In the Islamic world in general and in eighteenth century Ottoman
Turkey in particular, where the clarification of the relationship between
religion and science is concerned, İbrahim Hakkı quotes the following *hadith*
of the Prophet, "you are more knowledgeable on worldly matters" and
expresses the view that "explanations of material facts" in non-religious
matters should not be judged according to religious precepts.

Adıvar's claim that attempts to introduce the modern sciences into
Turkey were largely made by scholars who had not received a *medrese*
education[40] is therefore invalid. E. Kuran's viewpoint is also along the
same lines. He claims that "only a few scholars and thinkers, such as Kâtip
Çelebi, İbrahim Müteferrika, and Gelenbevî İsmail Efendi, who were not
educated at the *medrese*s" had any knowledge of the positive sciences of the

39. İshak, "El-hâc El-hâfız", in *Mecmua-i Ulûm-i Riyâziye* Istanbul, 1834.
40. A.A. Adıvar, *op. cit.*, p.176.

248

western world[41]. Müteferrika, who came to the Ottoman State as a young educated man, should be dealt separately from the others as a result of his conversion. But where Kâtip Çelebi and Gelenbevî İsmail are concerned, they had certainly studied the subjects and read the books taught at the *medrese*s, or graduated from the *medrese*s. Gelenbevî İsmail Efendi, in particular, cannot be an outsider to the *medrese* system, since he held the ranks of *müderris* and *kadi*. Furthermore an accurate evaluation of İbrahim Hakkı's contributions in this sphere clearly indicates that the attitude of the *medrese*s towards the modern sciences has been misunderstood from the very beginning.

In Turkish books written by pioneers of the positive sciences in the first half of the nineteenth century, one cannot see any mention of a conflict between religion and science. For example, in the works of İshak Efendi described in detail above, and the earliest Turkish works on diverse aspects of modern science written by the chemist Derviş Pasha, Hami Pasha, Aziz Bey of the Crimea, and others, there is no mention of any religious-scientific controversy, but on the contrary, one continuously comes across expressions to the effect that in the golden age of Islamic civilisation, religion was the motivating factor in the progress of science. Moreover, they claim that Islam actually encouraged the study of science. It would be very difficult to say that the Ottoman pioneers of positive science even felt any anxiety about a possible conflict of this kind. It is difficult to suppose the existence of anyone who defended the old sciences against physics, chemistry, astronomy, and other branches of modern science.

Meanwhile, the existence of dualism between old and new is undeniable, but it would be erroneous to seek the reasons for it on the plane of thought alone.

The privileged status of the members of the new educational institutions which met the changing needs of the Ottoman society in the late eighteenth and early nineteenth centuries, undoubtedly contributed to this dualism. For example, when we examine the decree concerning the establishment of the Engineering College by Selim III in 1210, the privileges accorded by the state to the teachers and students at this new institution are remarkable. Under this decree the engineering students were not only granted the rights accorded to *medrese* students, but being students of science, they had additional privileges not granted to other military ranks.

On the subject of this dualism Adıvar says:

41. E. Kuran, "Müsbet Bilimlerin Türkiye'ye Girişi (1797-1839)". *Turkish History Congress*, Ankara, 25-29 September 1970. Papers. vol. II, Ankara, 1973, pp.671-675.

While on one hand, it was desired to introduce mathematics and
positive sciences into the country in order to effect the modernisation and reform of the army; on the other hand, the *ulema* of the
medreses continued with their former works and translations of the
traditional sciences, thus creating dualism in the world of sciences[42].

This statement is hard to understand. Adıvar implies that the introduction of modern science somehow necessitates the disappearance of of traditional scholars overnight. Moreover, he does not provide any historical
evidence supporting his assumption of the existence of a school of thought
desirous of maintaining the old sciences. Most probably this dualism was
caused by the fact that scholars educated along traditional lines continued
to write books in the old vein due to the cultural environment in which
they had been raised, rather than a mental attitude of conscious rejection
or opposition to the modern sciences. To interpret this phenomenon as ir-
reconcilability between the two schools of thought and to attribute the kill-
ing of Seyid Mustafa Efendi and Raif Efendi at the incident of 1807 to
this, is an exaggerated attempt to construe a political incident as a clash
of belief which did not exist in the society.

Following the proclamation of the *Tanzimat*, educational institutions
spread and progressed at an increased rate, and under their influence a
new concept of science arose which naturally began to supplant the old concept. But the translation and publication of theoretical works which served
to systematise this new concept of science lagged behind. The first Turkish
translation in this sphere was of a work by the Italian logician Pasquale
Galuppi, and entitled *Miftah-ul-fünün* (1860). The work is a general exposi-
tion of method, analysis, and composition; and these subjects are dealt with
quite differently from the traditional treatises on logic, and the book also
includes a new scientific classification. This transitional period is illustrated
in the reconciliatory attitude with which both old and new are present. Ali
Sedad Bey (1857-1900), who made a significant contribution to spread and
acceptance of modern science and logic in Turkey, compiled three books
on logic. The first and most wide-scoped of these is *Mizam-ul-ukul fi'l mantık ve'l-usul* (1855). In N. Öner's study of this book, it is concluded rightly
that although it was more progressive than the aforementioned *Miftah-ûl
Fünün*, it is nevertheless the result of an attempt to reconcile old and
new[43]. In this context, I wish to refer briefly to a study on which I am cur-
rently engaged on Ali Sedad's little-known work, *Kavaid al Tahvuulât fi
Harekât al Zerrât* (1883) which as far as I have been able to establish has

42. A.A. Adıvar, *op. cit.*, p.214.
43. N. Öner, *Tanzimattan Sonra Türkiye'de İlim ve Mantık Anlayışı*, Ankara, 1967, p.49-61.

not been treated at all by Turkish researchers of the history of science and thought.

Atomism, thermodynamics, Darwin's theory of evolution and aspects of modern science and philosophy are treated by Ali Sedad Bey in some detail and with a very original style and method of approach. In his preface to the book he explains his basic objective as follows:

> It is my objective to ensure, disregarding the technical details, that the basic principles of these sciences be spread in our country, and to prove, with the aid of some philosphical methods, that these sciences which are exploited by the enemies of Islam, actually lend validity to Islamic ideas[44].

With "enemies of Islam", he is probably referring to certain orientalists and a handful of new-style Turkish intellectuals whom they influenced, who declared their hostility to Islamic science and philosophy. This was a trend which reached a peak in the writings of Ernest Renan at a later date.

In connection with the relationship between Islam and modern science, I wish to comment on *Es'ile-i Hikemiye* by İshak Efendi of Harput (?-1892), published in 1862. İshak Efendi of Harput received an entirely traditional education in the *medrese*s, and unlike Ali Sedad had relatively little knowledge of western science, philosophy, and logic. His book consists of his replies to diverse questions asked by two Europeans whom he met on a ship while travelling back to Izmir from Damascus. It is a vivid source of information about the attitude of the *ulema* raised in the traditional cultural environment towards the new sciences and inventions. In one of the questions, the claim is made that "some of the principles of modern astronomy conflict with religion". In his lengthy reply to this question, İshak Efendi of Harput takes a stance reminiscent of that of İbrahim Hakkı of Erzurum, declaring that these subjects have no connection with the basic principles of religion and that only the ignorant common people, under the influence of former customs, oppose to new concepts[45]. To a question about the view that "the Islamic concept of fate is responsible for the backwardness of Muslims in technology and science", he responds with a question of his own: "Why have the Russians, and the Christians, Jews, Druse, and Nusairis living in the Islamic world lagged behind the Europeans?"

In the preface of his Turkish translation of Dreper's work entitled *Science and Religion*, Ahmed Midhat Efendi studied the relationship between religion and science in more extensive terms than any other scholar. He tried to

---

44. Ali Sedad, *Kavâid al-Tahavvulât fi Harekât el-Zerrât*. Istanbul, 1883, pp.8-9.
45. İshak Efendi (Harputî), Istanbul, 2nd imp., 1301, pp.104-107.

demonstrate the conflict between Christianity and science, which never affected Islam:

> If we define the status and condition of Islam in the light of the
> sciences, it will be seen that the subjects of contention between Chris-
> tianity and the sciences are, on the contrary, subjects of approval
> and accord between Islam and science[46].

From the mid-seventeenth century onwards and particularly in the
last quarter of the eighteenth century, the attitude of Ottoman intellectuals
towards the modern sciences introduced by the new educational institutions, and their view of the relationship between religion and sciences, are
explained by these words of Ali Suavi (1839-1878): "The expression of old
and new science is meaningless. The tradition of clinging to the old and
opposing the new is a Christian one. There is no place in Islam for such
a tradition"[47].

We may conclude that the Ottoman scholars of traditional Islamic
cultural background were not against the newly introduced sciences and
in general, they did not conceive any conflict between science and religion.
On the contrary, they were generally open minded and tried to accomodate
the modern techniques.

Finally, it can also be pointed out that the idea of conflict between
science and religion originated after the emergence of new types of Ottoman
intellectuals who believed in positivism and biological materialism and
whose attitude towards religion was negative; later, this idea took new intellectual, social and political dimensions.

46. Ahmed Mithat, *İslam ve Ulûm*, Istanbul, 1313, vol. 1, pp. 11-12.
47. H.Z. Ülken, *Türkiyede Çağdaş Düşünce Tarihi*, p. 87.

VI

THE INITIAL STAGE OF THE HISTORIOGRAPHY OF OTTOMAN MEDRESES (1916-1965)

The Era of Discovery and Construction

The study of the history of Ottoman medreses, spanning a period of approximately six centuries — from 1331 until 1924 — commenced in the Second Constitutional Era and continues today. Even though medreses were established throughout the whole of the Ottoman Empire, the areas of Anatolia and Rumeli quite naturally formed the initial and most prolific geographic areas for constructing these centres of Islamic learning. In this article we will investigate from an historiographical point of view studies produced in the period stretching from the Second Constitutional Era to the publication of İ. H. Uzunçarşılı's *Osmanlı Devletinin İlmiye Teşkilâtı* in 1965, a work which has become regarded as a turning point in the field. We will scrutinise this period, which we have termed the era "discovery and construction" of the historiography of the Ottoman medrese, with regard to the political and ideological ideas that affected society as a whole. Additionally, we will investigate the sources of the, in our view, subjective approach to the field of Ottoman history and particularly address the manifestations of this approach in the works dealing with the history of medreses.[1] At the same time we will

---

* I would to express my gratitude to my esteemed colleagues Prof. Dr. Mehmet İpşirli, Dr. Hidayet Nuhoğlu and Dr. İsmail Kara for reading the first draught of this article and supplying me with constructive criticism, at the same time it would also be a pleasure for me to extend my appreciation to my associates working at IRCICA, Hümeyra Zerdeci and Süreyya Subaşı, for their titeless efforts in helping me during painstaking research undertaken and to Mr. Can Erinton for his assistance in translation.

[1] We will consider articles and books relating to the topic at hand which are thought to be of importance. The main reason for proposing the date 1916 as the starting-point is to be found in the fact that at the time the first indepent study exclusively dealing with the history of the Ottoman medrese appeared. However information on Ottoman medreses as well as on the so-called İlmiye class can also be encountered in a variety of historical sources. We refer to the following invaluable study for a critical bibliography and the examination of certain as-

42

try to demonstrate, by means of various examples, how a conceptual model of the Ottoman medrese was devised on the basis of unfounded comparisons between the Islamic institution of the medrese and European universities. We will try to demonstrate how a number of superficial views, held by the authors under consideration, have led to unfounded conceptions that have influenced academic research in a negative way. The mentioned authors had acquired these superficial views from their misinterpretation of academic studies executed by a number of orientalists. Subsequently, we will discuss how these factors have been shaped by certain political and ideological agendas in a period stretching across half a century.

In this article we will analyse significant publications, comprising books and articles on the topic under investigation, published in the mentioned time-frame, and subsequently subject them to a scrutiny that would juxtapose their contents with the era's intellectual atmosphere and thus point to the origins and development of certain concepts and literary forms of expression used to convey historical material affecting the profession of history-writing.[2] In other words, these works will be investigated from a histori-

---

pects of the topic: Hans Georg Majer, *Vorstudien zur Geschichte der Ilmiye im Osmanischen Reich*, München, 1978. It is also the case that numerous information on the topics relationg to medreses, ulema and particularly in connection with the so-called ıslah-ı med ris reform policy can be found in publications aside of the contributions of Turkish as well as foreign historians dealing with Ottoman history. A list of such non-academic contributions can be in the following work: Abdullah Ceylan, *Sırat-ı Müstakim ve Sebilürreşad Mecmuaları Fihristi*, Ankara, 1991. In connection with works comparing the educational systems of the medrese with modern educational applications, it is possible to refer to the following study as a reference work: Muallim Cevdet, *Mektep ve Medrese*, ed.Erdoğan Erüz, İstanbul, 1978. In our study we have only consulted such works in the sense that they provided source material for independent studies and writings that appeared after 1916.

[2] This study of the era, which we have dealt with as the first stage of the historiography of Ottoman medreses, has become a publication of documents that could constitute a positive influence upon historical research undertaken in this field, and that could thus be regarded as an important source of sound information. The first one of these sources is the Arabic vakfiye of Fatih Mehmed II published in 1935 by the Deutsche orientalische Gesellschaft in İstanbul. Whereas, the Turkish text of the vakfiye was published in 1938 by the general directorate of Pious Foundations [Vakıflar Genel Müdürlüğü]. In this publication the original versions and Turkish transcriptions of vakfiyes relating to the Eyüp Külliyesi can be found as well, next to the ones relating to the Fatih Külliyesi. Prof. M. Tayyib Gökbilgin's *XV.-XVI. Asırda Edirne ve Paşa Livası*, published in 1952, contains a systematic evaluation of the Ottoman archival and vakıf registries relating the medreses founded in Edirne and thus offers numerous documents and useful information on this topic. In this important work, the vakfiyes promulgated by Bayezid II, in connection with his külliye and the medreses found within this külliye, were published and evaluated for the first time. The third important basic document published in the same period is the Süleymaniye Vakfiyesi, furnished by Kemal Edip Kürkçüoğlu. This vakfiye was published in the year 1962 by the

ographical point of view, particularly with regard to their critical rendering of source material, and their selection of information from primary sources. These selective pieces of information will then be subjected to a critical reading which will attempt to determine their validity in the face of historical reality. The concept of historiography in this instance relates to the gathering of historical evidence and to the edition of historical sources. In addition the term historiography will be employed in the sense that the influence of ideas and notions relating to developments outside of academic composition of historical narrative will be investigated. It is a point worth remembering that the notion of historiography does not correspond to mere history-writing. One should never lose sight of this distinction. As such, the term "historiography" has recently become identified with the notion of the "history of history writing" comprising all of its ramifications. The term should thus not be understood as simply comprising the history of the history writings but rather as a branch of intellectual history or even as a sub-branch of the sociology of science. In this article the concept of historiography has been employed in this way. The aim of our article is to purvey a new approach to the study of the medrese, as one of the basic elements of the history of Ottoman civilisation, by means of applying the described methodology to studies produced in the initial period of the historical study of the Ottoman medrese.[3]

The history of Ottoman medreses began with the foundation of a medrese in İznik by the second Ottoman Sultan Orhan Gazi in the year 1331. The end of the institution, on the other hand, was brought about by the proclamation of the *Tevhid-i Tedrisat Kanunu* [Law on the Unification of Education] on 3 March 1924. This Law, reorganising the educational system, was promulgated after foundation of the Turkish Republic, and as a result medreses were abolished. Ottoman medreses thus stretched out over a period of six centuries and expanded over a wide geographical field. The reform movements that started in the eighteenth-century Ottoman Empire, and that became particularly widespread throughout the nineteenth century, also affected the field of eduaction in a substantial manner. These reforms, on the one hand, gave rise to changes in the whole of the educational system and in the actual make-up of institutions, and on the other, led to outright neglect of

---

general directorate of Pious Foundations [Vakıflar Genel Müdürlüğü] and furnished by an introducctory piece by Kemal Edip Kürkçüoğlu. As these works constitute the editions of source material, they have not been subject to a critical anlysis in this context.

[3] In relation to historiography, cfr. J.H.Hexter, "Historiography - The Rhetoric of History" in *International Encyclopedia of the Social Sciences*, VI, p. 368; and in relation to historical criticism, cfr. Leon Halkı, *Tarih Tenkidinin Unsurları*, ed. B.Yediyıldız, Ankara, 1989.

44

medreses. As such the reform movements resulted in removing the topic of medreses from public attention.[4] But in contrast during the Second Constitutional Era official interest again became focused on the topic of medreses, that had been neglected since modernising movement of the *Tanzimat*.[5] Academic interest in the history of medreses started with the tangible reform and modernisation of medreses, that would bring them in line with the general educational system. These modernising moves were initiated with the proclamation of a reform movement by means of directives issued in 1910 and 1914. With the foundation of the Republic substantial changes, particularly in the field of education, were introduced and quickly realised. As a result, shortly after the Republic's proclamation medreses were abolished. A number of academic studies relating to the history of medreses were executed, giving prominence to the reasons underlying this policy of closure. As a result, theses concerning medreses that had been developed in the Constitutional era, were reviewed in the early years of the Republic and while safeguarding a number of its critical elements new ideas were added. Instead of the notion of "developing medreseses by means of reform", the idea that "as medreses were in no position to be reasonably reformed" gave way to the "inevitable abolition" of the institution of the medrese. The period mentioned as the temporal framework of our article thus really has to be regarded the era of the discovery and the construction of the history of medreses, as the most important educational, cultural and scientific establishments in Ottoman history.

The studies on the history of the Ottoman medrese really began along two different paths. The first path was constituted by writings aiming at investigating medreses within the framework of Ottoman cultural history and the so-called *İlmiye* [the ulemâ class] organisation, in an academic manner and based upon primary source material. The latter studies, on the other hand, were writings giving rise to the composition of an "image", in tandem with at the first reform and later abolition policies promulgated. As such these studies still affect contemporary attitudes towards the academic study of the history of Ottoman medreses. In these latter-mentioned studies on the history of medrese during the Ottoman heyday, comparisons with European universities were made, and subsequently this comparative approach was

---

[4] For a general evaluation of the studies on the history of Ottoman medreses, cfr. Ekmeleddin İhsanoğlu, "Osmanlı Eğitim ve Bilim Kurumları", in *Osmanlı Devleti ve Medeniyeti Tarihi II*, İstanbul, 1998, pp. 223-250.
[5] For the reforms undertaken in the late-Ottoman period, cfr. Yaşar Sarıkaya, *Medreseler ve Modernleşme*, İstanbul, 1997; the text of the "Med ris-i İlmiye Nizamnamesi", dated 1920, cfr. *Düstur*, tertib-i s ni, İstanbul, 1330, II, pp. 127-138; the text of the "Islah-ı Medaris Nizamnamesi", cfr. *İlmiye Salnamesi*, İstanbul, 1334, pp. 652-662.

also applied to later periods in the history of Ottoman medreses. This in turn gave rise to a different understanding of the institution of medreses. At the outset of these comparisons it is noted that in Ottoman medreses, just as in the examples of European universities, many "faculties" could be found, and, it is then indicated that here many different academic disciplines were taught. The ensueing "image" of the Ottoman medrese has subsequently been evaluated according to accustomed brilliant-tarnished or progressive-regressive couplings oftentimes encountered in the context of Ottoman history, as will be illustrated by means of relevant examples further on. This ensuing "image" of the Ottoman medrese has been turned into a causal link by means of adding unfounded historical element so as to arrive at the notion that Ottoman medreses should be regarded as the basis of the Istanbul University. In our opinion, it is necessary to base studies on Ottoman medreses on a trustworthy historical foundation. These studies should depend upon researching how this "image" was arrived at, and subsequently upon the proper history of the medrese in Islam and on an accururate appraisal of the development of European universities and whether or not any relationship between the two ever existed. In this piece certain aspects of the historiography of the Ottoman medrese will be investigated according to the outlined methodology so as to demonstrate how this incorrect image has been constructed, while at the same also attempting to illuminate a number of issues mentioned earlier.

The year 1916 has to be regarded as the starting-point of the study of the history of the Ottoman medrese. On this date two very different texts on the topic at hand were published. The first one, written by the professor in Ottoman history at the *İstanbul Dârülfünun* [İstanbul University] Ârif Bey, was published in the *Edebiyat Fakültesi Mecmuası*;[6] the latter, on the other hand, was written by the teacher [*muallim*] Emin Bey, who had long served as a literature teacher in military schools, and was issued in the *İlmiye Sâlnamesi*, an appendix of the *Meşihat* [office of the Şeyhülislâm]'s official periodical the *Meşihat-ı celile-i İslâmiye'nin Ceride-i resmiyesi*.[7] Whereas the article published in the *Dârülfünun Mecmuası*, which was a university publication, was aimed at researching the history of the Ottoman medrese as a foundation supplying religious education which should be regarded as the natural context of the institution; the article published by the *Meşihat* [office of the

---

[6] Ârif Bey, "Devlet-i Osmaniyye'nin Teessüs ve Takarrürü Devrinde İlim ve Ulem ", in *Dârülfünun Edebiyat Fakültesi Mecmuası*, nr. 2, (May 1332), pp. 137-144.
[7] Emin Bey, "Tarihçe-i Tar k-i Tedris", in İlmiye Salnamesi, İstanbul, 1334, pp.642-51. For another edition of a great part of this text, cfr. Osman Nuri Ergin, *Mecelle-i Umûr-i Belediyye*, I, İstanbul, 1995, pp. 258-63.

46

Şeyhülislâm], a religious institution par excellence, on the other hand attempted to compare medreses to universities. These two characterisations show the extent to which both writings appear to be contradictory in nature and how both writings approached the subject from a different orientation.

Ârif Bey's work, investigating the issues of the learning and the ulemâ, has to be regarded as the first serious research enterprise in the field. In the work, the author alotted space to such works as the *Aşıkpaşazâde Tarihi*, Neşr 's *Cihannüma*, Taşköprüzâde's *Şekâyık*, as well as, to various translations and appendices of the mentioned texts, while also including references to Ottoman chronicles and other historical sources. In his research Ârif Bey employed the printed version of Mecd Efend's translation of the above-mentioned *Şekâyık*, as well as manuscript versions of the other works available in various İstanbul libraries. Furthermore, the author also included an excerpt from the *Haremeyn Mukataa Defteri*. In this article the appearance of the Ottoman *ilmiye* class and the early stages of its further development have been dealt with relying upon sound historical source material. For instance, Ârif Bey critically evaluated the material relating to the early stages of the *ilmiye* organisation found in the *Şekâyık*, pointing out that Taşköprüzâde had adapted this material from Âşıkpaşazâde 's work, and that the material presented by Âşıkpaşazâde in turn had been strongly politically motivated:

> The first historian to supply information about the Ottomans ulemâ was Âşıkpaşa-zâde, recording the names of a number of individuals and supplying information on the biographies of some in sections of his work. These records are more of a political and administrative than scholarly nature.[8]

Ârif Bey explains the scarcity of material relating to the activities of scholars and the general cultural atmosphere [*ilmiye*] in the early stages of the Ottoman State in Taşköprüzâde's work, written approximately two and a half centuries after the State's foundation, in the following manner:

> In all likelihood, the Şekâyık-ı Nu'mâniye's author [Taşköprüzâde] regarded the historical works on the ulemâ of the early period, and, particularly Âşıkpaşazâde's history as [sound] authorities and [subsequently] recorded the legendary accounts contained in them without any further investigation.[9]

Ârif Bey indicated the necessity of having to consult sources, outside the ones mentioned earlier when conducting research on the subject of the scholarly life and ulemâ of the early Ottoman period. Furthermore he pointed to two important works found in a general collection of biographies of jurists [*fakih*] of the Hanef school of law [*mezheb*] to which Ottoman scholars were

---

[8] Ârif Bey (May 1332), p. 137.
[9] Ârif Bey (May 1332), p. 139.

affiliated to. The first work was *Ketâib a'lâm el'Akhyâr min Fuqahâi mez-heb el-Nu'mân el-Mukhtar*, written by the author Muhammed bin Süleyman, known as Kefev (died 1582), and the latter, Taqiyeddin b. 'Abdulqadir (died 1601)'s *el-Tabakhât el-Seniyye fi Terâcim el-Hanefiyye*.

Another important issue dealt with by Ârif Bey's article was the issue of the first Ottoman medrese. While not paying attention to the definite statement made in the work *Şekâyık*, he indicated in a speculative manner that prior to the İznik medrese founded by Orhan Bey in 1331, that the same ruler had in fact founded an earlier medrese in İzmit in the third year of his reign. In Ottoman sources the conquest of İzmit (738/1337) has been variously dated in the period 727-731/1326-1331, and Ârif Bey thus gave credence to one of these legendary accounts, however his proposal has not been dealt with in later works, and even in Uzunçarşılı's *İlmiye Teşkilatı* this issue is not even considered.[10] Apart from his incorrect evaluation of the date of the earliest medrese, Ârif Bey's article, in spite of not being acknowledged sufficiently in later studies, has to be seen as an important opening move for the historiographical study of Ottoman medreses and the *ilmiye* organisation. Particularly in view of the fact that the article relied upon primary source material, and that the author possessed a certain analytical and critical methodology, as well as a scholarly style and willingness to approach the subject from a broad perspective.

One can witness that, apart from Ârif Bey's study, the writings relating to the history of medreses published subsequently, presented a new "mesage" and "image" to the public at large, a public not well-versed in the history and development of Ottoman medreses. This message was formulated in accordance with the policies of the *İttihad ve Terakkî Fırkası* [Party of Union and Progress], in an attempt to emphasise the religious authority of the Ottoman Sultan as the Caliph of the Islamic world, and, at the same time, to strengthen the party's influence over Muslims living within and without the Ottoman lands. Simultaneously, the Turkist ["Türklük"] element present in said policies became apparent in the fact that the roots of Ottoman medreses's "glory" period were stressed to come from "Türkistan", or to use the then popular phrase sublime Turkistan ["Türkistan-ı ulyâ"]. The central ideas of the message were that, on the one hand, these Ottoman medreses in their glory period as educational centres offering tuition in the Islamic sci-

---

[10] For a discussion on the different dates supplied for the conquest of İzmit in Ottoman sources, cfr. İ. H. Danişmend, *İzahlı Osmanlı Tarihi Kronolojisi*, I, pp. 22-23; and in connection of the issue of the first medrese, cfr. Mustafa Bilge, *İlk Osmanlı Medreseleri*, İstanbul, 1984, pp. 11-12 and İ. H. Uzunçarşılı, *Osmanlı Devletinin İlmiye Teşkilâtı*, 2nd ed., Ankara, 1984, 1-3.

ences represented an exemplary standard for the whole of the world of Islam, and, on the other, that, using then current progressive slogans, that the so-called "müspet ilimler" [positive sciences], comprising the exact sciences, medicine and engineering, in fact sciences necessary for the further development of state and society, had been taught at these medreses as well. Thus one of the themes around which this message revolved was that in Ottoman medreses an environment sympathetic toward "liberal ideas" had been present and that consequently philosphy lectures had also been part of the curricula of the said medreses. These ideas were developed as a result of the necessity of renewing and modernising Ottoman medreses at the time. By stressing that keeping medreses in line with contemporary developments had long been neglected, the message implicitly blamed Sultan Abdülhamid II for this neglect. Traces of this image, created in the context of the reform movements of the Second Constitutional Era, can still be detected today. In view of the fact that some of these idea are still widely accepted today, this new medrese image would appear to be an important issue to deal with at some length.

THE CONSTRUCTION OF A "MEDRESE MODEL" AND ITS SOURCES

The main source of this "modern" image of the medrese can be found in the article entitled "Tarihçe-i Tar k-i Tedris", written by the "Muallim-i muhterem Emin Beyefendi" published in the *İlmiye Salnamesi in* the year 1916. In one of our earlier articles we have already mentioned Emin Bey's article, particularly in connection with the views presented on the so-called law and other curricula supposedly promulgated by Ali Kuşçu and Molla Hüsrev at the Fatih Medreses, and the impact of this notion upon the historiography of the Ottoman medrese. In this instance, however, we will consider Emin Bey's article in a broader sense, and while trying to trace the sources of his views regarding Ottoman medreses, we will try to demostrate that the terminological confusion was a result of the desire to apply a Western system of education to the Ottoman-Islamic system. Emin Bey opens his text by stating that the medreses to be found in the 'ruler of all the lands' [*sultanü'l-akaalim*] Istanbul had in the course of the Ottoman Sultanate 'created miracles' in augmenting the faith of Islam, safeguarding cultural values and ascertaining knowledge. But, he added, these once-lofty institutions in due time had fallen into the hands of uncivilised individuals that had caused the medreses to lose their essential characteristics, and as he put it "zıya'a uğratıldığı" [had been reduced to naught]. This piece, written in a style quite florid for its day, was full of words of praise for the services rendered by the *ulemâ* and medreses, dwelt at length on the importance of the medreses founded by Sultan Mehmed II, known in Turkish as Fatih, and Süleyman the Magnificent, and even presented a whole set of information on these institutions without mentioning any sources however. The article instead sufficed in enumerating the names of famous scholars, active throughout Ottoman history, and heaping verbose praise on them. One of the most interesting aspects of the article is the fact that it draws attention to the lofty positions taken up in state and society by members of the ulemâ in the past, and Fatih's reign is employed as a point of reference in this instance, relating how the Sultan dressed in the attire reserved for members of the ulemâ class as an indication of the extent to which he identified himself with them. As will be shown in detail below, Emin Bey related a great deal of information from the history written by Ahmed Cevdet Paşa, without including any reference though, and in doing so he effected a number of mistakes. And, as such, Emin Bey presented a number of unfounded claims as historical fact which were to influence the historiographical treatment of medreses. On the other hand, it would seem necessary to locate the root of these statements in the chaotic circumstances upsetting the intellectual climate of the Second Constitutional Era. The true sources upon which Emin Bey based himself were M. Şemseddin [Günaltay]'s writings on Islamic civilisation and Namık Kemal's works on Fatih, as will be demonstrated below.

For instance, in talking about Fatih's educational regulations (*tertibat-ı tedrisiye*), Emin Bey mentioned the following — however without indicating any source or historical substantiation:

> If we were to transpose the different educational stages found in them [Ottoman medreseses] to our own time [Second Constitution], the Sahn-ı Semân medreses would correspond to [our] darülfünun [university], the Tetimme medreses, that had received the name musıla, would correspond to [our] idad [senior high school] schools, [whereas] the İbtidâ-ı dâhil medreses would corrspond to [our] rüşdiye [high school], [and] the İbtidâ-ı hâriç, on the other hand, to [our] mekâtib-i ibtidâiye [primary schools]. A student, who had completed his education a the musıla would be admitted to a department in the Sahn-ı Seman medreses to carry on his education on a higher level in that department, since the medreses of the Sahn-ı Seman were established on the basis of specialisation... .[11]

Emin Bey pronounced these comparisons based upon the classification of primary school, secondary school, high school and university, of the Western style modern educational system that had entered the Ottoman world during the nineteenth century. By the Second Constitutional Era this system had become firmly established. As a result he reached very bold conclusions that did not reflect the realities of the education offered at medreses. In his text

---

[11] Emin Bey 1334, p. 645.

50

he elaborated in the following way:

> In the homeland [Ottoman Dominions] different types of instruction in arts and
> sciences were offered at these medreses, and all judges, medical doctors and engi-
> neers were graduates of these institutions.[12]

Emin Bey ventured even further than that, and in the section of his article dealing with the Süleymaniye medreses, he implied that during the institutions' glory days education of a standard commensurate with "European universities" was offered there:

> The Süleymaniye medreses, which functioned as diffusion centres of the age's
> learning for the Ottomans, [and] maybe even for the whole of mankind, in the era
> of [Sultan] Süleyman, when the Empire was expanding in a glorious fashion and in
> order for doctors and engineers to be employed in every corner [of the Empire] a
> number of medreses were constructed for the purpose of instruction in the medical
> and mathematical sciences.[13]

Among the sources Emin Bey employed to glorify Fatih's era was also a book written by Nâmık Kemal. Within the context of the reform and modernisation efforts of the Tanzimat era Nâmık Kemal's aim had been to eradicate Ottoman feelings of backwardness in the face of the West's superiority. Additionally, Nâmık Kemal also claimed that Islamic and Ottoman contributions could be found at the root of the contemporary European civilisation. He did this by means of presenting heroic personalities and the glories of the Ottoman and wider Islamic past in literary hyperbole. Behind the favourable outlook of Muslim intellectuals of the late era, including the figure of Nâmık Kemal, lay the desire to stress the gravity of the current situation by using history as an ideological argument — quite apart from mere nostalgia. In this connection, Nâmık Kemal put forward the following appraisal of the cultural atmosphere pervading the reign of Sultan Mehmed II, better known as Fatih in Turkish literature, in the relevant of his famous biographical series entitled *Evrâk-ı Perişan*:

> Fatih's progressive ideas induced him to choose the road towards innovation, also
> with regard to the topic of education, causing him to abolish the [traditional] model
> of course regulation in the medreses he had founded assigning certain medreses to
> the intruction of ulûm-i şer'iye [canonical sciences], such as jurisprudence [fıkıh]
> and Prophetical traditions [hadis], and others to rational sciences, such as medicine
> and geometry. [Today] one of the main reasons behind the [West's] civilisational
> superiority we witness lays the [rational] distribution of labour — which only be-
> came widely practised during the last century in Europe — and if upon Fatih's in-
> stitution [of his mentioned reforms] a continuity and a general acceptance would
> have ensued, Istanbul would have undoubtedly remained the focal point of culture,

---

[12] Emin Bey 1334, p. 645.
[13] Emin Bey 1334, p. 646.

as it had been in Fatih's day, and would have [thus] endured as a centre dispensing civilisation.[14]

As can be deduced from reading these lines, in praising Fatih and his era Nâmık Kemal wanted to express the view that 'the source of the superiority noticeable in Europe's contemporary civilisation can be found with us, so much so that with regard to specialisation in education, the fact that a separate educational foundation was assigned to each branch of learning had in fact been transferred from Islamic civilisation to Europe'. However, in advancing this thesis space was given to a number of unfounded claims, comparable to the example just mentioned. These unfounded claims have, in time, been accepted and advanced as historical realities by generations of writers, with the "muallim-i muhterem" Emin Bey taking the lead.

This picture of medreses drawn by Nâmık Kemal and Emin Bey is in direct contrast with the traditions of medreses and education in Islam. Such ideas are also contradicted by the statements we have access to in numerous vakfiyes [deed of trust of a pious foundation] supplying basic information on the nature of education in medreses. Besides, it is a well-known fact that in the history of education in Islam medreses were established for the sole purpose of instructing religious sciences, particularly Fıkıh [jurisprudence]. The development of specialisation such as had taken place in European universities never occurred within medreses. Apart from such medreses known as Dârülhadis and Dârülkurrâ, which respectively specialised in Prophetic traditions [hadis] and Quranic studies, it appears rather obvious that the only educational institution displaying a parallelism with the kind of professional specialisation in Europe were medical medreses. In spite of the obvious contradiction, the image that had thus been constructed became solidified in the years to come. The model of the Ottoman medrese constructed according to this image, insinuating specialisation, was particularly reiterated in future studies of the instruction of engineers ["mühendis"]. These unfounded comparisons did not take account of the historical development of European universities, particularly keeping in mind that the classical model of the university kept in use until the late eighteenth century was structured along four basic disciplines or faculties [Liberal Arts], theology, law, and medicine.[15] The instruction of engineers was in fact undertaken in the form of a master-apprentice tradition, as had been the case with a great many professions until the advent of the industrial revolution. The education of this type of instruction was not undertaken on a university level until well into the middle of the

---

[14] Namık Kemal, Evrak-ı Perişan, İstanbul, 1289, p. 236.
[15] For a history of European universities, cfr. A History of the University in Europe, I-II, ed. Walter Ruegg, Cambridge, 1992-1996.

nineteenth century.[16] In other words, the underlying idea of the appearance of "European universities" had been developed without taking account of the historical realities of learning in Europe. Whereas the separation of institutions of higher education according to different professions did not come about until the middle of the nineteenth century as a result of certain economical and technological developments, it appears rather obvious that to expect such an event to have taken place in a mid-sixteenth century Ottoman context to be nothing but the result of a misunderstanding and intermingling of conceptions of Ottoman glory and of the ascendent glory of Europe — as has been the case in many other instances related to cultural and scientific matters in Ottoman history.

Emin Bey based a good deal of his fundamental ideas and information upon Ahmed Cevdet Paşa (1823-1895)'s famous *Tarih-i Cevdet*. He particularly relied upon the sections of this famous book dealing with the decay of the *ilmiye* class, however without remaining faithful to the original and without including a reference. In the first edition of his *Tarih*, Cevdet Paşa put forward a number of general observations containing a peculiar analytical value. These valuable observations were spread out across several volumes and appeared separate from the historical narrative being developed. These scattered fragments were re-organised under a separate heading entitled "Devlet Kanunlarının Nasıl Bozulduğuna Dair" (On the Reasons of the Dissolution of the State's Legal Structure) in the work's new edition of 1891 (1309). The main thrust of the section is explained as follows:

> To know how changes occured in the State's ancient laws and traditions ... to explain how decay occured periodically ... to be of help in understanding the new regulations ... Every state's basic obligations can be divided into two, the first is the establishment of justice and the protection of subjects' rights, the latter, on the other hand, is to to safeguard its borders from foreign aggression.[17]

In order to achieve this goal Cevdet Paşa examined the two basic institutions of the Ottoman State, the *askeriye* [the military-administrative organisation] and the *ilmiye* [the legal-educational organisation]. By means of his wide historical knowledge and rich cultural background he was able to scrutinise those two establishments thoroughly. Cevdet Paşa attributes the waning of the Ottomans' military prowess to the disintegration of the so-called "timar" system, linking it to the dissolution of the justice system as a result of the

---

[16] In England where the industrial revolution fist occured, "courses in engineering" were introduced in the course of the 1840s. The first example was to be found at the University of Glasgow, whereas the fist academic degrees were conferred at the University of Oxford in 1865. M. Sanderson, *The Universities in the Nineteenth Century*, London, 1975.

[17] Ahmed Cevdet Paşa, *Tarih-i Cevdet*, I, İstanbul, 1309, p. 88.

corruption of the "ilmiye" class and to the disintegration of the ilmiye's main foundation, the medreses.

> ... as we explained higher with the continuation of the decline of the possessors of timar and zeamet holdings they reached the stage of utter annihilation (muzmahil) ... there had remained no force in their hands to impose discipline and order amongst their corps. As explained higher, with the dissolution (inhilal) of the mülkiye and askeriye organisations, in turn the profession of the İlmiye (tar k-i ilmiye) also disintegrated including its rules and regulations.[18]

In the sections dealing with the *ilmiye* class Cevdet Paşa puts forward general information concerning the place of medreses and stresses the importance of the Fatih and Süleymaniye medreses. After having considered the fact that the rank of *müderris* [professor at a medrese] could lead to higher positions, such as *molla* [judge of canonical law] or *kazasker* and was therefore reliant upon a great degree of knowledge and merit on the part of the candidate, he mentions that the novitiate system ["mülâzemet] constituted the introductory phase of the careers of education and judiciary. If a candidate wanted to attain a position in the novitiate system, he was held to undertake a thorough study [iştigâl"] as an advanced student ["dânişmend"] in medreses. Cevdet Paşa then made a reference to Koçi Bey's *risale*:

> After the year 1,000 the rules and regulations of the İlmiye profession commenced to disintegrate and personal matters became intermingled with business affairs and indulgence towards [candidates] unworthy was shown so as to confer high office upon them ... the position of judge ["mevleviyet"] became venal ... as the novitiate system ["mülazemet"] was the entry into a career in education ["tar k-i tedr s"] as well as to a career in the judiciary ["tar k-i kazâ"], these two became upset ["muhtel"] as well ....[19]

The second source mentioned by Cevdet Paşa in his writings on the decay of the *İlmiye* profession is the memorandum [*lâyiha*] written by the high-ranking Rumeli official Tatarcık Abdullah Efendi. Rather than describing the history of Ottoman medreses, Cevdet Paşa's aim was to analyse the decline ["inhilâl"] of the *İlmiye* class while at the same time presenting certain general information about the Fatih and Süleymaniye medreses. In bringing these matters to his readership's attention he employed a "modern" terminology. For instance, while talking about eight medreses [*Tetimme*] constructed in tandem with the eight medreses [*Sahn-ı Seman*] of the *Fatih Külliye* [complex], constituting a high level of learning, which had been meant to function as a preparatory education, he referred to them as "idad ". Whereas the term was used in Cevdet Paşa's time to refer to a secondary level of public educa-

---

[18] *Tarih-i Cevdet*, p. 108.
[19] *Tarih-i Cevdet*, p. 112.

tion, the basic lexical meaning of the phrase denotes "preparatory". At a later stage however, particularly during the Second Constitutional Era, the term was employed outside its lexical meaning in connection with other comparative descriptions so as to insinuate a kind of equivalent status ["muadelet"] between the Ottoman medrese system and modern educational systems.

In another part of his text talking about the glory days of Ottoman medreses, Cevdet Paşa expressed the contention that all men of scientific and professional learning had been educated at these medreses.

> ...and because men of science and culture ["erbâb-ı ilm ve maarif"] including the likes of engineers and physicians [were] students ["talebe-ulûm"] educated at these medreses, [which thus] formed the centre of the public instruction of the day ["terbiye-i umûmiye"] ... [20]

From these general statements and studies on medreses appearing later, particularly during the Second Constitutional Era, such claims, with no basis in historical reality, pertaining to the existence of medreses specialised in science ["Fen"] and engineering ["Mühendislik"], which had never existed within the Ottoman medrese system or the medrese system in Islam upon which they relied, were constructed. Cevdet Paşa investigated historical events with a view to causal relationships relating to the development and change of society and its institutions based upon a sociological methodology influenced heavily by the work of Ibn Khaldun. He put forward his analyses of the changes affecting institutions as being interconnected, as if comparable to the interrelated wheels of a clock. Later writers distorted these general pieces of information and views concerning the history of Ottoman medreses and the *ilmiye* class by means of employing them outside of their original context. A point requiring our fullest attention in this connection is the exact significance of the word "mühendis" [engineer]. In classical texts the term is employed as denoting a geometer, but later on in the modern age it started to be used as the equivalent of the term "engineer", as used in various European languages. One could arguably put forward that with the institutionalisation of the engineering training in nineteenth-century Europe and the simultaneous establishment of so-called *mühendishanes* [centres for the instruction of engineers] in the Ottoman Empire, the old and new meanings of the term *mühendis* seem to have been confused in the literature on Ottoman medreses.

Foremost amongst the sources, supplying Emin Bey with intellectual background material on the topic of Ottoman civilisation in general and Ottoman medreses in particular, stands M. Şemsettin [Günaltay] (1883-

---

[20] *Tarih-i Cevdet*, p. 110

1961). The Turkist-Islamist author Şemsettin [Günaltay]'s writings on religious, philosophical and literary topics influenced a wide readership during
the Second Constitutional Era and later periods as well. M. Şemsettin Bey,
who had access to the writings on the history of Islamic civilisation of orientalists, through his knowledge of French, rather superficially investigated a
plethora of subjects in his works, based upon the basic guiding principle of
the dichotomy progress-regress. He dealt with Islamic, and particularly Ottoman medreses, within the context of a dichotomy progress-regress, under the
heading of "müterakk " [progressive] and "müteredd " [degenerate] medreses.

> Progressive ["(m)üterakk "] medreses have instructed a great many important per-
> sonalities in the context of virtuous geniuses to the homeland ["memlekete"] ...
> [These] (p)rogressive ["müterakk "] medreses have in time been founded upon ra-
> tional principles, [and] in response to a great need ... .[21]

In the year 1913 Şemsettin [Günaltay] published his famous work *Zulmetten
Nura*, in a style reminiscent of his many passionate and enthusiastic articles
issued in various periodicals at earlier dates. In this book he put forward a
number of important claims, without including references however, that
strongly influenced later historians of the Ottoman medrese. These claims
were appropriated *verbatim* by Emin Bey in his piece that appeared in the
İlmiye Salnamesi and later an important number of those dealing with the
"History of the Ottoman Medrese" unquestioningly accepted these views as
well. For example, in relating Fatih's well-known penchant for the sciences
and its practitioners in a highly florid style, it is claimed that the Sultan had a
committee set up to establish the Ottoman Medrese System, however without relying upon any kind of historical proof:

> Fatih ... and his ulemâ ... inviting them to the capital [İstanbul] to set up a scientific
> committee to assign them with the regularisation of medreses.[22]

In his second piece entitled "Müteredd Medreseler" Şemsettin [Günaltay]
developed this notion of a committee further, and mentions the names of Ali
Kuşçu and Molla Hüsrev

---

[21] M. Şemseddin [Günaltay], *Zulmetten Nur'a*, İstanbul, 1331 [1913], pp. 170-171. According
to Seyfettin Özege this work had been published in 1915 (*Eski Harflerle Basılmış Türkçe
Eserleri Kataloğu*, V, 2130). Since Özege accepted that after the proclamation of the Second
Constitution (1908) publications bore dates in accordance with the *Rumî* calender, this
would indeed indicate the year 1915. But my coleague Dr. İsmail Kara countered that as a
result of his research he had come to the conclusion that the given date of 1331 had not been
in accordance with the *Rumî* calendar, but instead according to the Hegira calender, in which
case the date would coincide with 1913. In our article we have accepted the latter view.
[22] M. Şemseddin [Günaltay] 1331, p. 172.

56

Fatih's medrese regulations, laid down by Ali Kuşçu, who had been summoned
and invited from sublime Turkistan ["Türkistan-ı ulyâ"], together with Molla Hüsrev and comparable geniuses [of the day], lost some of its initial vigour in time and
were able to be upheld until the eleventh century of the Hegira.[23]

According to this assertion the committee thus assembled divided the medreses into four components. These were the *İbtidâ-ı hâriç*, *İbtidâ-ı dâhil*, *Musıla* and *Sahn* medreses. However, if one were to look at the *Fatih Kanunna-
mesi*, providing information on the ranks of the ulemâ during Fatih's reign,
and, the detailed material supplied in Âli's *Künh el-Ahbar*[24] one can easily
conclude that the facts supplied by Şemsettin [Günaltay] do not appear to be
consistent with historical reality. It would appear to be helpful to compare
the above-quoted paragraph with the following piece from Emin Bey's *İl-
miye Sâlnamesi* article:

Fatih ... succeeded in regulating the educational system of his day ["tertib-i tedrisat
ve nizamaları"] by means of the strenuous efforts of two geniuses like Ali Kuşçu,
whom he had summoned and invited from "Türkistan-ı ulyâ", and ... Molla Hüsrev
... and a great number of scientists present at the time ... .[25]

According to this claim the regulations ["Nizam"] that had been set up by
the mentioned committee ["Heyet"] laid out in broad strokes how a diligent
student could enter the medrese system at the lowest level and end up at the
last and highest stage, the *Sahn*. But, in our view, such a contention is nothing but the result of a literary topos that has distorted the historiography of
the Ottoman medrese.

The eight medreses (Sahne) that can today be found in the vicinity of the Fatih
Mosque were constructed corresponding to an educational system [comprising] a
Dârülfünun [university], the Tetimme medreses (Musıla), in other words the İdad
[high school], and the medreses built on the periphery, the (İbtidâ-ı dâhil) or the
Rüşd [secondary] and the (İbtidâ-ı hâriç) or the İbtidâi [primary schools].[26]

Emin Bey repeated M. Şemseddin's words in his article and presented them
in an even more florid way to his reader. This contrived educational model
devised to correspond to the nineteenth- and twentieth-century modern-style
Ottoman educational system that had been imported from the West, from
France to be exact, does not conform to the traditions of the Islamic medrese
and, as such, does not correspond to historical reality. On the other hand,
Şemsettin [Günaltay], while on the one hand stating that Islamic civilisation
and the institution of Ottoman medreses in their heyday had constituted an

---

[23] M. Şemseddin [Günaltay] 1331, p. 180.
[24] E. İhsanoğlu (1998), pp. 238-239.
[25] Emin Bey 1334, p. 643.
[26] M. Şemseddin [Günaltay] 1331, p. 172.

example for Europe, claimed that they should also function as a guiding light for comtemporary Ottomans. In order to supply this thesis with additional arguments, he also mentioned certain scientific degrees that had entered the European university tradition from the Islamic world (*Bachelier, Licenciă* and *Docteur*). However, in the model of European universities and Ottoman medreses that he had developed, Şemsettin [Günaltay] glossed over the fact that the the legal status and the educational systems of both institutions, as well as, the fact that the socio-cultural and historical conditions leading to their respective development had been quite different. The attempt to find a parallel between European style universities — as disciplined and specialised institutions of higher education and instruction such as the Istanbul Dârülfünun to which Şemsettin [Günaltay] had been attached since 1914 as professor [*müderris*] — and Ottoman medreses does not appear to be anything but a blending of two different traditions and a combination of the terminology used in two different civilisational entities, without however having a base in sound realities. One of the most striking elements of this model is the fact that Şemsettin [Günaltay] presented every single medrese of the complex of the eight *Sahn* medreses as a university in its own right:

> ... students ... that had completed their education ... [at an] İdad [high school] could gain specialised training in their chosen branch of science at one of the eight (Dârülfünuns) [universities] which had received the name (Sahn) ... .[27]

Information, which would substantiate the existence of different educational programmes mentioned cannot be found either in the traditions of the Islamic medrese, nor in the Fatih Vakfiyesi nor even in the biographies of the ulemâ that had studied at the *Sahn* medreses. Therefore, it seems obvious that this kind of claims was made in the context of the heated debates and controversies current at the time of writing.

One can detect that M. Şemsettin profited from the works of a number of European orientalists. And he enthusiastically presented some of their findings to his readers. Şemsettin [Günaltay] aimed at enlightening the Turkish readership on historical topics of Islamic culture and science unknown to them, by means of presenting in a florid style the findings he had made in the French books he had read. We will briefly make reference to the findings of some orientalists which, in our opinion, might have exercised an indirect influence upon Şemsettin [Günaltay].

The first one is Daniel Haneberg, whose *Abhandlung über das Schul- und Lehrwesen der Mohammedaner im Mittelalter* dealing with the history of education in Islam, was published 1850 in Munich. In the book the author

---

[27] M. Şemseddin [Günaltay] 1331, pp. 172-3.

contended that the graduation system found in European universities, in other words the *licenciate*, had been appropriated from the *icazet* system of Islamic educational traditions: "Ich vermute, dass unser Licenciat von dieser Mohammedanischen Einrichtung hestammt".[28] The second one is the Spanish orientalist Julian Ribera, who, towards the end of the nineteenth century, claimed that the educational system prevailing in world of Islam had strongly influenced the universities of Romance Europe. He particularly put this view forward in his lectures on Islamic education given during the academic year 1893-94 at the university of Saragoza. Ribera had developed this viewpoint as a result of research conducted on a number of *phenomena* — such as the conferring of diplomas and dignities — absent in the Greek, Roman or Christian West in eras predating the appearance of Islamic civilisation. In the West this study carried important discussions in its wake about the influence of the medrese system upon the development of European universities — as a result of the fact that there had been no previous comparable institution in the western civilisation.[29]

One of Şemseddin Bey's basic sources used to ascerain specific information on the history of Islamic civilisation was the French orientalist Gustave le Bon 's *La civilisation des Arabes*, published in 1884. Le Bon stated in this work that for the duration of a number of centuries European universities developed the scientific and philosophic legacy inherited from Muslim institutions:

> Next to schools supplying basic education in big centres, such as Baghdad, Cairo, Toledo and Cordoba, laboratories, observatories and rich libraries, in other words universities in possession of all the materials needed for the realisation of scientific research could be found. Only in Spain seventy popular libraries were present. ... They were in the position of students able to consult Greek works that constituted earlier sources of knowledge, but before long they understood the value of experience and observation over mere book wisdom, ... Generally, the replacement of the authority of teachers by experience and observation which forms the basis of the modern scientific method is attributed to Bacon; but today one has finally to accept that this change could only have taken place under the auspices of the Arabs. But this view has been acknowledged by all scientists who have read their works, and particularly by Humboldt. In relating how the highest scientific level relies on one's[the scientist's] ability to create events at will, in other words upon the ability to conduct experiments, this famous observer adds: 'The Arabs attained this highest level quite unknown to the ancients'. ... The Arabs conducted experiments and for a long time remained the only ones to comprehend the importance of this methodology. ... At the same time they distributed this knowledge by means of universities and books. Therefore the influence they exercised upon Europe is tre-

---

[28] In connection with the studies of Daniel Haneberg, cr. G. Makdisi, *The Rise of Colleges: Institutions of Learning in Islam and the West*, Edinburgh, 1981, p.275.

[29] For Ribera's research in connection with this topic, cfr. G.Makdisi 1981, pp.294-6.

mendous. ... Until the advent of the modern age our universities based their education upon translations of Arabic books. ... .[30]

These studies explaining the influence of the world of Islam upon the Christian world, were met enthusiastically by Muslim intellectuals, and quite naturally led to a number of misapprehensions. The biggest misapprehension
was equalising Islamic medreses and European universities. In other words,
the fact that higher education in Islam and higher education in the West were
regarded as identical in academic nature and status. In addition to the Ottoman Turkish intellectuals cited, one could also refer to the Egyptian historian
Ahmed Zeki Paşa (died 1934) and later Arab historians of education who repeated similar mistakes. Thus one could easily interpret this phenomenon as
a common reflex reaction of Muslim intellectuals.

On the other hand, as proven by European mediaevalist historians, universities should be understood as the product of a certain form of social organisation peculiar to Western Europe. In thirteenth-century Europe all kinds
of so-called *corporations* were being developed, and universities have thus
to be understood as institutions of higher education advanced in this context.
In societies outside Western Europe institutions of higher education developed in different socio-cultural structures and thus assumed different shapes.
In Islam the medrese emerged as a *vakıf* [waqf] institution. In general, one
can put forward that the world of Islam and correspondingly the Ottoman
Empire only became acquainted with the institution of universities in the
nineteenth century, and started to appropriate them from Europe at that
stage. The Ottomans understood this new institution of higher education as
separate from medreses, and clearly wanted to differentiate it from them and
the learning they offered (*ulûm*). This new institution, where the new sciences (*fünûn*) were taught, was thus given the name "Dârülfünun".[31] George
Makdisi, the author of the basic reference work on the history of the Islamic
medrese, pointed out that education in Islam was not organised according to
a university system but according to a college (*medrese*) system. Islam's significant contribution is thus to be found in the principle of higher education
and in a number of basic elements relating to methodology, which were developed in the medrese system, and transferred to the West and later incor-
porated into the West's own educational system. The great contribution of
the Romance West, on the other hand, was its organisation of the instruction

---

[30] Gustabe Le Bon, *La civilisation des Arabes*, Paris, 1884, pp.341-4.

[31] E.İhsanoğlu, "The Genesis of the "Darulfünun" An Overview of Attempts to Establish the
First Ottoman University", in D.Panzac (ed.), *Histoire âconomique et sociale de l'Empire
ottoman et de la Turquie (1326-1960). Actes du sixième congrès international tenu à Aix-en-
Provence du 1er au 4 juillet 1992*, (Paris, 1995), pp. 827-842.

VI

60

of knowledge, containing a large element of Islamic-Arabic origin whose existence cannot be denied. In the Romance West this knowledge was further
developed and elaborated in the college system so as to lead to the emergence of the institution of the university.[32]

THE POPULARISATION OF THE MODEL

In a short while Emin Bey's article assumed the form of a basic source on
the subject discussed. The first one to appropriate the article's stance with
great appreciation was Osman Nuri Ergin. In his work, entitled *Mecelle-i
Umûr-ı Belediyye*, the author presented certain information on the topic the
kadı [judge in canon law] organisation, while dealing with the services rendered by and the organisation of municipalities in Islamic countries. After
having considered the *kadı* organisation in the Ottoman State, he moves to
the topic of medreses, and, quotes M. Emin Bey's 'extremely well-informed'
['gayet vâkıfâne'] article *verbatim*.[33] In his text O. Nuri Engin points towards the benefits of medreses and mentions the following:

> Consequently, if within a nation ["millette"] only a kadı and no müfti were to be
> present, only a legal court but no medrese, then public order ["asayiş"] would
> seemingly ["zahir "] be in place, however, without the intrinsic [bâtın "] moral vir-
> tues [necessary].[34]

Ergin offers certain proposals in connection with the regulation of courts and
medreses, and states that if medreses and other institutions providing religious education, falling at present within the authority of the *Evkaf Nezareti*
[Ministry for Pious Foundations], were to be linked to the *Meşihat* [Office of
the Şeyhülislâm] these institutions would be secured a positive development
and subsequently says that "in this way religion would be elevated ["teâli"],
and the state strengthened". O. N. Ergin, a staunch critic of the "Aile Kanun-
namesi" [Family Code], promulgated at the time, states that the unification
of *şeriat* and secular courts, in other words 'Tevhid-i mehâkim' [Unification
of Courts], would be wrong and would only give recourse to changing the
appointment of certain conventional *şer'i* judges and to introducing a couple
of judges unfamiliar with the *şeriat* legal code into the judicial system. After
having expressed this opinion, Ergin violently defends the continued presence of religious education, and thus implicitly of medreses, and elaborates

[32] G.Makdisi 1981, p. 293.
[33] Osman Nuri Ergin 1995, I, pp. 257-263.
[34] Osman Nuri Ergin 1995, I, p. 271.

his firm judgement that separating ['tecrid'] the state from religion would lead to important social upheavals.[35]

Further than simply acquainting a new readership with M. Emin Bey's image of Ottoman medreses, O. N. Ergin violently defended the continued existence of the medrese as an educational institution in his book published in 1922. This piece carries an important place, in view of the establishment of the Republic the following year and the subsequent new political and ideological ideas that were to become dominant in society giving rise to a new pronunciation of the concept of the "medrese" and influencing O. N. Ergin to distance himself from the ideas he had promoted in this text. Particularly, if we keep in mind that Ergin published a work entitled *Türkiye Maarif Tarihi* in 1939, in which he defended these "new" ideas in direct contrast with his views expressed earlier. While vigorously defending the continued existence of medreses in 1922 he rejected the idea of a "Tevhid-i mehâkim", later Ergin became a virulent defender of the "Tevhid-i Tedrisat Kanunu" [Law on the Unification of Education], as we shall see at a later stage. He thus stated that medreses had reached a phosilised state by that time.

Ahmed Hikmet Müftüoğlu (1870-1927) constitutes an early and prominent example of somebody who in the first years of the Republic dealt with the history of medreses in a populist way, in the manner of Şemseddin Günaltay and Emin Bey. In other words, Müftüoğlu attempted to present views congruent with Republican theses developed at the time. He displayed the framework of his ideas in very bold strokes in two articles dealing with the sources of seventeenth-century Turkish culture and began his writing in the following way:

> The first half of the eleventh century of the Hegira (seventeenth century A. D.) marks the end of the era of order and conquest, and the beginning of the times of dissolution and stagnation. The second half, then, was the time when the rabble became prominent and [thus marked] the moment of the decline of intellectual and governmental power.[36]

In this piece Ottoman educational institutions were dealt with under the heading *Medrese ve Tekke*, and medresses and tekkes were presented as two teachers, one *abûs* [sulky faced] and the other *beşûş* [smiley face]:

> In their heyday medreses were the sources of the intellectual classes... thinking that those graduating from medreses ["icazet alanlar"] only specialised in religious sciences is absurd ["abes"]. As medreses were the dârülfünuns [universities] of the olden days, all kinds of useful sciences were instructed there. It is possible to di-

---

[35] Osman Nuri Ergin 1995, I, pp, 284-285.

[36] Ahmed Hikmet [Müftüoğlu], "Onbirinci Asr-ı Hicr de Türk Menabi-i İrfanı - I", in *Mihrab*, nr. 19-20, (İstanbul, 1340), p. 623.

vide the teachings provided by medreses in the seventeenth century into four faculties: 1 -Religious sciences and religious philosophy (Kelâm) 2 - Law, Fıkıh [Ca-
nonical Jurisprudence] 3 - Literature (Arabic literature) 4 - Sciences (Medicine,
Mathematics, Astronomy).[37]

Ahmed Hikmet Bey elaborated upon a great number of issues found in Emin
Bey's earlier article. He put the issues in his own style and did not include a
reference to Emin Bey or the *İlmiye Salnamesi*. In the same way, he dealt
with Kâtip Çelebi's ideas on the teaching of philosophy, discussed below.
Let us quote Ahmed Hikmet Bey's relatively simple sentences, expressing
ideas for which Emin Bey had not disclosed a source for either:

> While setting up his medreses Fatih included the condition that such books as Ha-
> şiye-i Tecr d and Şerh-i Mevâkıf were to be read at the Sahn-ı Seman, these courses
> were prohibited in the twelfth century, upon the charge that they constituted phi-
> losophy ['felsefiyattır'] and these works were then substituted by a number of
> books as Hidâye and Ekmel.[38] [These and other books mentioned can be found in
> the appendix].

As shall be discussed later, some information contained in these phrases,
which had been written by Kâtip Çelebi in response to the Kadızadeli movement, were presented outside their original context and thus distorted historical reality.

The most striking part of Ahmed Hikmet Bey's piece was the attention
given to the Kadızâdeli movement and the contention that it represented the
"greatest proof of the backwardness ["irtica"] of medreses". Ahmed Hikmet
Bey, who adorned his writing with a number of poems and historical anec-
dotes, conducted a debate between *tekkes* and *medreses*. Medreses were shut
down as a result of the promulgation of the *Tevhid-i Tedrisat Kanunu* (3
March 1340/1924). He took up a position against medreses but in favour of
tekkes and ended his piece in the following way:

> As a result of the conflict with regard to the instruction and training offered in me-
> dreses and tekkes, an emotional and rational consensus was not achieved among
> the contemporary population, [whereas] the social body associated with the in-
> struction of the tekke [was] elegant, humble, generous, [and] gentle; the set of peo-
> ple brought up in medreses [was in contrast] rude, argumentative, tight, [and] ma-
> terialistic and this [contrast] has continued up to this day.[39]

Ahmed Hikmet Bey wrote down the following final negative judgement on
medreses as being in direct contrast with tekkes and zaviyes, which he per-

---

[37] Ahmed Hikmet [Müftüoğlu], "Onbirinci Asr-ı Hicr de Türk Menabi-i İrfanı - II", in *Mihrab*,
nr. 21-22, (İstanbul, 1340), p. 715.

[38] Ahmed Hikmet [Müftüoğlu], "Onbirinci Asr-ı Hicr de Türk Menabi-i İrfanı - II", p. 723.

[39] Ahmed Hikmet [Müftüoğlu], "Onbirinci Asr-ı Hicr de Türk Menabi-i İrfanı - I", p. 633.

ceived as positive. Ironically, these latter-mentioned institutions were to share a similar fate the following year (1925):

> Rather than being a source of culture medreses had become a hearth of malice In-
> stead of science and knowledge, ignorance and extremism started to be instructed
> there ... an obstacle to the development of ideas and knowledge, they had taken the
> shape of a calamitous danger. In this way they attracted the loathing of the people,
> and as a result of [the] neglect these institutions [experienced], they became
> crowded by vagabonds and fanatics and thus started to collapse by themselves. [40]

The views that were broadcast in A. H. Müftüoğlu's writings, appearing in the first years of the Republic and carrying clear traces of the winds of change sweeping through these days, undoubtedly also affected academic circles.

We have recognised the higher-mentioned article written by Arif Bey, during the epoch of the Dârülfünun, as the first "academic" study on the history of Ottoman medreses. And following the reform movement of 1933, which was to transform the Dârülfünun into the İstanbul University, a number of studies were carried out as well at the that university's Faculty of Letters. Particularly its two famous "ordinaryüs" professors Mehmet Şerefeddin Yaltkaya and İsmail Hakkı Uzunçarşılı were active in this respect. The latter's long and arduous efforts wich were to find completion in the publication of his famous book in 1965, constituting the basic reference work on this topic and on the understanding and research of the whole of the *İlmiye* organisation, will be dealt with in a separate article. However, the research paper prepared by the third-year student of the Institute of History [*Tarih Enstitüsü*] Remziye Beksaç, under Uzunçarşılı's supervision and with the assistance of Yaltkaya, on the topic of the "Sahn-ı Semân Medreseleri" completed in 1936 has to be regarded as the first academic thesis on this topic.[41] As it was impossible to locate a second thesis prepared by R. Beksaç on the topic of the Süleymaniye Medreses, also prepared under Uzunçarşılı's supervision, we will in this instance only deal with her study of Fatih.[42] Beksaç consulted a great number of primary source material, which constitute the real sources of the history of the Ottoman medrese. First among these are the Ottoman chronicles, either in printed form or in manuscript versions to be

---

[40] Ahmed Hikmet [Müftüoğlu], "Onbirinci Asr-ı Hicr de Türk Menabi-i İrfanı - II", p. 724.

[41] Remziye Beksaç, *Sahnu Seman Medreseleri*, İstanbul University Faculty of Letter, unpublished History Thesis, 1936, 16 pp.

[42] İsmail Hakkı Uzunçarşılı reiterates this thesis in the bibliography of his *Osmanlı Devletinin İlmiye Teşkilâtı* (p. 289). In Prof. Dr. Şehabettin Tekindağ's *Cumhuriyetin 50. Yılında İstanbul Üniversitesi*, published on the occasion of the 50th anniversary of the Republic's foundation, the complete reference to this thesis is given as follows: Remziye Beksaç, *Süleymaniye Medreseleri*, Thesis, Edebiyat Fakültesi Tarih Semineri Kütüphanesi, no.118.

VI

64

found in certain İstanbul libraries, the translation of the work *Şekâyık* made by Mecd Efendi, and the Arabic text of the Fatih Vakfiyesi, prepared for publication by Tahsin Öz in 1935.

This piece of research gives concise yet sound information based on reliable sources on the medreses constructed prior to Fatih's reign. After mentioning the medreses which had been founded by Fatih employing extant Byzantine buildings, R. Beksaç explains in a detailed manner the different aspects of the medreses found within the *Fatih Külliyesi* [complex of buildings and institutions surrounding the Fatih Mosque]. Next to the text of the *Vakfiye*, Beksaç employed relevant passages of Âli's *Kühn el-Ahbâr* as a basic source. However, Beksaç added a new element to the historiography of the Ottoman medrese based on information supplied by Kâtip Çelebi. After explaining how she was able to use M. Ş. Yaltkaya's lecture notes she discloses the following information:

> As yet I have not encountered a programme relating to the courses taught at the Medâris-i Seman. One can however be certain that the courses taught here belonged to the [so-called] ulûm-ı âliye [high sciences] (kelâm, tefsir, hadis, fıkıh, ferâiz, usûl-ı fıkıh). However, beyond these mentioned courses, others belonging to the [so-called] ulûm-ı tahsiliye, such as hendese [geometry], hey'et [astronomy] and ulûm-ı Arabiye [arabic] were not neglected either. As a matter of fact, in the Vakfiye Fatih Sultan Mehmed had included two lessons to be taught at these medreses. The first one was 'Şerh-i Mevâkif'' and the latter 'Haşiye-i Tecrit'. As the object of these medreses was first of all to instruct judges able to run the country in a just and fair fashion ["adaletiyle"], these [students] were obliged to study (Fıkıh) and to learn the Arabic language so as to be able to study this topic from [the original] Arabic texts. Consequently, if we would not have been told by books, informing us of the past, that subjects such as tefsir, hadis and fıkıh were taught there, we would still have been able to find out as a result of [acknowledging] the goal of these medreses. Courses difficult to track were philosophy and (kelâm), which were combined. My teacher Şerefeddin Bey discovered while researching that the vakfiye mentioned by the scholar Kâtip Çelebi, who is deserving of extraodrinary trustworthyness, is not identical with the vakfiye kept in the Evkaf Müdüriyeti [Directorate of Pious Foundations]. In this vakfiye the courses instituted by Fatih in his vakfiye as recorded by [Kâtip Çelebi] are not present, [nor] does the language [employed] bear resemblance. In the printed [version] vakfiye these [courses] are absent as well.[43]

In our view the following passage from Kâtip Çelebi's *Mîzânü'l-Hakk* gave rise to the issue under discussion: "Sultan Mehmed Han had the Medâris-i Samâniye constructed and had written down in [its] vakfiye that it should be managed and provide instruction in accordance with tradition (kanun) and that the courses of '*Hâşiye-i Tecrid*' and '*Şerh-i Mevâkıf*' would be taught

---

[43] Remziye Beksaç, *Sahnu Seman Medreseleri*, pp. 10-11.

there [as well]".[44] As can be seen quite plainly no course or even book-title is mentioned in the Fatih Vakfiyesi. Furthermore, in the same way as the book 'Hâşiye-i Tecrid' is a work on ilm-i kelâm employed in lower ranks of the medrese system (rank of twenty or twenty-five), the volume 'Şerh-i Mevâkıf' is a work used in medreses pertaining to the rank of forty, as Kâtip Çelebi himself indicated in his Cihannüma. In other words, it should be quite clear now that Kâtip Çelebi had erroneous judgements on the instruction offered at medreses, which deceived later researchers and subsequently gave rise to the pronunciation of new errors on the instruction offered at the Semâniye medrese, which belonged to a rank of fifty.

Apart from Kâtip Çelebi's erroneous writings, all writings on the Fatih medreses rely on the contents of two sources. According to these two basic sources there is no piece of information, proof or even conjecture to be found relating to a course programme or set of educational regulations associated with the Fatih Küliyesi.[45] The first source is the Fatih Vakfiyesi. This vakfiye relates the general educational framework of the medreses as well as the various provisions and privileges they offered. The second source, on the other hand, is Taşköprüzâde's work Şekâyık, offering a great wealth of information on the era's ulemâ. The only texts not relying upon this latter source are naturally the chronicle written by Fatih's contemporary Tursun Bey and other early accounts of Fatih, as their authors proceeded the time of Taşköprüzade, the author of the Şekâyık.

Yaltkaya and his pupil did not consider the fact that Kâtip Çelebi could have made a mistake nor did they take account of the possibility that he had been developing an argument in his conflict with members of the contemporary Kadızali movement, a movement well-known for harbouring animosity towards philosophy. Instead they considered that provisions — even though they appeared contradictory to the principles of the legal framework for Pious Foundations (vakfiye) — could have been added to the text of the vakfiye at a later date:

Consequently it seems necessary that these vakfiyes must have been written at a later date. As I have not been able to benefit from this vakfiye which is accessible today, my teacher [Yaltkaya] states that (I was able to verify the courses taught at these medreses by means of sifting through the memoir Şekâyık-ı Numâniye and its appendices and [numerous] icazetnames [diplomas] containing scholarly traditions). (Fatih's successors abolished these courses, saying "these courses are phi-

---

[44] K tip Çelebi, Mizânül-Hakk fi İhtiyari'l-Ahakk, ed.O.Şaik Gökyay, İstanbul, 1980, p. 21.
[45] For a discussion of similar topics, cfr. Ekmeleddin İhsanoğlu, "Fatih Külliyesi Medreseleri Ne Değildi! Tarih Yazıcılığı Bakımından Tenkit ve Değerlendirme Denemesi, in İstanbul Armağanı I, İstanbul, 1995, pp. 105-133; and Ekmeleddin İhsanoğlu, Büyük Cihad'dan Frenk Fodulluğuna, İstanbul, 1996, pp. 39-84.

losophy" ["bu dersler felsefiyattır"] and seeing it fit to replace them with courses from (Fıkıh) over (Hidaye) to (Ekmel) [and] truly in the view of many the occupation with (Kelâmiyat) caused one to become impious. The ancient Greek ([branches of] science, [such as] metaphysics, physics, chemistry and geology comprising all interpretations and being of a nature to [further] develop individualistic and more realistic ideas were only prohibited by the ignorant Sultans, believing the words of bigoted and hypocritical ulemâ, who succeeded Fatih's reign. As a matter of fact, Fatih attempted to secure the continuity of this scientific toleration in his vakfiye, as we mentioned higher, he included the tuition of two famous books of kelâm, the Haşiye-i Tecrit and the Şerh-i Mevakıf. Another work that was taught was the text (Metâliilenzar) that had become renowned under the title (Şerh-i metalı), written by Esfehanlı (Şemseddin Mahmut bin Abdurrahman), commenting on the work (Tevaliilenvar) that had been composed by the famous müfessir [Quranic exegetist] (Kadı Beyzavi — died 685).[46]

Beksaç indulged in the pronouncement of numerous interpretations, relying upon the information supplied to her by her teacher Uzunçarşılı on the so-called "Kanunname-i Talebe-i Ulûm" which he found in the University Library. Whereas Uzunçarşılı in his later works dated the work to the early sixteenth century, his student placed the kanunname in the periods of Fatih and Bayezit. In spite of the fact that the text of the kanunname itself makes no mention of the Fatih medreses, as a result of a piece of information communicated to her by her teacher she explains that the kanunname contains regulations pertaining to the *tetimme* medreses. In an article written by M. Ş. Yaltkaya, which will be dealt with at a later stage, this kanunname, which makes no reference to the Fatih medreses, is mentioned. Many subsequent authors repeated this reference to Yaltkaya's article. Süheyl Ünver even transformed this text into the actual "course regulations" of the Fatih medreses, as shall be explained in detail below.[47]

---

[46] Remziye Beksaç, *Süleymaniye Medreseleri*, Thesis, p.11.

[47] Remzye Beksaç stated that this note given to her by her supervisor indicated that the mentioned kanunname was to be found under the following heading: İstanbul Üniversitesi Halis Efendi Kitapları no: 207. Yaltkaya, on the other hand, claimed that the work was part of the same collection undert the catalogue number 206. In Uzunçarşılı's book the work is again mentioned as being under catalogue number 206. Süheyl Ünver, on the other hand, claimed that "[t]he thesis mentioned in these quoted lines has not been found in the named library and under the described catalogue number. Professor Yaltakaya mentioned that he had received this note from Professor İsmail Hakkı Uzunçarşılı: "[w]e are still looking for [the kanunname]. We have not been able to locate it in different libraries amongst books on political issues. We regard these lines to have sprung from course regulations composed in Fatih's day". After having finished the quotation, Ünver says that "... if these lines indeed relate to course regulations composed by Ali Kuşcu and Molla Hüsrev in Fatih's reign, then their value is inestimable", in this way indicating his reservations on the topic. In 1948, two years after his study on the Fatih Külliyesi, Ünver published a work on the figure of Ali Kuşçu, the brilliant mathematician and astronomer active in Fatih's reign and who had

Remziye Beksaç, in repeating the issue of comparing the Sahn-ı Seman medreses to a university, summarised the then prevalent view on the topic of medreses in her conclusion section:

> The Fatih medreses were safeguarded in all of their glory until the era of Sultan Süleyman the Magnificent. The medreses constructed in the vicinity of the Süley-maniye mosque did not endanger the outstanding position of the Sahn medreses that were as a result assigned to deliver instruction in the high sciences ["ulûm-i âliye"]. The Süleymaniye Medreses, on the other hand, became the medreses where the rational sciences ["ulûm-i akliye"] were taught. At a later time, depra-vations occured in these hearths of high learning and culture ["irfan ocakları"], the Sultans neglected their expenses, bigots were raised, the position of professors ["müderrislikler"] were givin to unworthy individuals, diplomas ["icazetnameler"] became venal, lecturers sent replacements to teach their courses. Science was continuously degraded, the tolerance that had been prevalent in Fatih's day was replaced by an extremist fanaticism. We can see how the era of Fatih, which was a time redolent of scientific discussion, [and] liberal thinking ["hürriyet-i efkâr"] lapsed in the reign of his son Bayezid II. An example of the influence of fanaticism [at the time] is the execution of the mathematician Mevlâna Lutfullah, a pupil of Ali Kuşçu who had even been a professor ["müderris"] at the Semaniye, who was accused of heresy. Koçi Bey, who presented a risale [treatise] to Sultan Murad IV, which in an open and plain manner brought this corruption of the sciences to the attention of the sovereign, and [who] expressed the opinion that one of the reasons [underlying] the first flowering of the sciences had been the fact that such scientific abuses [as witnessed at the time] had been absent. If this high medreses organisation, on a par with our first universities, had not succumbed to disintegra-tion, [and had instead] annually witnessed a steady progress, then our contempo-rary modern university would have been founded in the seventeenth century ... But these organisations continued to exist with all of their material and spiritual defects up to a very recent period ... On 18 September 1330/1914 it was attempted to re-form these [organisations] under the heading of Dârülhilafe but [instead] they were taken down an even more ruinous road. We have only been saved from these me-dreses through the proclamation of the (REPUBLIC).[48]

The Faculty of Letters "ordinaryüs" professor Mehmed Şerefeddin Yaltkaya (1879-1947) subsequently presented the historical quotations and arguments, used by Remziye Beksaç in her assignment, in his contribution to the book

---

introduced the school of Samarqand into the Ottoman lands, in which he presented some information on the figure supplied by Ayvansaraylı Hüseyin Efendi. He then set out to combine this information with the material published by Yaltkaya: [i]f these medrese regulations ["medreseler nizamnamesi"] have indeed been written by Molla Hüsrev with Ali Kuşçu's approval, then their importance knows no bounds. The fact that its formulation was made in the mode of expression then common is a further argument in favour of its authenticity". In other words, in the first quotation this Kanûn-ı Talebe-i Ulûm" was presented as a course programme for the Fatih medreses, in the second instance, however, they have become simply regulations concerning medreses. Additionally, this programme of regulatory notice has been repeatedly ascibed to Ali kuşçu and Molla Hüsrev.

[48] Remziye Beksaç, *Sahnu Seman Medreseleri*, pp. 15-16.

68

Tanzimat, published fours years later by the Ministry for National Education commemorating the centennial anniversary of the Tanzimat.[49] The article entitled "Tanzimat'tan Evvel ve Sonra Medreseler" is a piece oftentimes used by scholars writing on this topic and a source continually present in bibliographies on medreses. It is certain that Yaltkaya's background conferred a peculiar status on to his writing. He had been brought up against a religious background, and even received a diploma [icazet] in religious sciences next to being educated in a modern way at the Rüşdiye and the Dârülmuallimin, in addition to being engaged in lecturing [müderrislik] at the medreses of İstanbul, prior to being appointed at the İlâhiyat [Theology] and Edebiyat [Letters] Faculties.

Yaltkaya considered the history of medreses against a good period/bad period bifurcation. He mentioned Cevdet Paşa's higher-mentioned views on the collapse of the ilmiye profession outside of their context and without including a reference either. While referring to Koçi Bey's memorandum [risale], that had served as a source to Cevdet Paşa, Yaltkaya quoted this work through Cevdet Paşa rather than consulting the printed version of the mentioned text. In his risale Koçi Bey pointed to the year 1003/1594 as a turning point for the fortunes of the ilmiye class. Koçi Bey illustrated this point by means of various examples.[50] In this connection, however, Cevdet Paşa rounds off the date and states that "Koçi Bey [stated] ... by the year 1000" or "after the year 1000 the ilmiye path started to deteriorate".[51]

Yaltkaya communicates the information Cevdet Paşa quoted from Koçi Bey without mentioning its source and proceeds to present Koçi Bey's views, in combination with the information provided by Kâtip Çelebi. Yaltkaya communicates the two views found in Kâtip Çelebi's *Nizâmü'l-Hakk* — subsequently, this information was frequently repeated and in all writings dealing with medreses they were unconditionally accepted as historical realities up till recently. The first view is that the works relating to the teaching of İlm-i Kelâm — in other words works relating to the philosophy of Islam — *Haşiye-i Tecrid* and *Şerh-i Mevâkıf* were supposedly mentioned in Fatih's Sahn medreses vakfiye as being under obligation of being taught there. Together with the fact that it is not very difficult to prove that such a claim was never made in Fatih's vakfiye, it remains true that these words, which Yalt-

---

[49] M.Şerefettin Yaltkaya, "Tanzimattan Evvel ve Sonra Medreseler", in *Tanzimat I*, İstanbul, 1940, pp. 463-467.

[50] Koçi Bey's memorandum has been printed at later dates. For the section relevant to the ulem , cfr. *Risale-i Koçi Bey*, London, Mösyö Vats Tabhanesi, 1277, pp.9-12; *Koçi Bey Risalesi*, İstanbul, 1303, pp. 16-46.

[51] *Tarih-i Cevdet*, pp. 109, 112.

kaya borrowed from such an authority as Kâtip Çelebi, have up to our own day become one of the most popular "Leitmotifs" on the topic of Ottoman medreses.

One could have expected Yaltkaya's article to deal with this topic in a manner different from Remziye Beksaç's treatment in her research paper. Because, if Yaltkaya would have researched whether or not such a provision or expression was indeed to be found in the Fatih Vakfiyesi then this error regarding medreses would not have been repeated for the duration of fifty years. The fact that the actual text of the vakfiye had not been subjected to a serious reading led the way to accepting Kâtip Çelebi's words as absolute truth. But it seems more reasonable to suspect Kâtip Çelebi had either incorrectly remembered the information supplied in Âli's *Künh el-Ahbar* or that he had simply communicated the said information in a wrong way. The unconditional acceptance of Kâtip Çelebi's critique, without verifying it against historical sources and realities, while at the same time isolating it from the context of the Kadızadeli movement has led to a certain confusion. Kâtip Çelebi's advocacy of the teaching of İlm-i kelâm and his complaints against its exclusion from the teaching offered in medreses should be confined to the context of the Kadızadeli movement's hostility towards philosophy, while taking account of this movement's influence on Ottoman cultural life. These factors have added to the ready adoption of the aforementioned medrese image. This image is amenable to a bifurcated model of the opposition between müterrak /progressive–müteredd /regressive poles. Yaltkaya himself supplied one of the most manifest proofs that the science of kelâm and its works as *Haşiye-i Tecrid* and *Şerh-i Mevakıf* had not been discontinued from being taught following the year 1000. Because he had himself received a diploma [icazet] for instruction in this field from his medrese teachers, and, he had himself taught the *history of Kelâm* at the Faculty of Theology [İlâhiyat] and courses on Islamic religious philosophy at the Faculty of Letters [Edebiyat]. In other words, Yaltkaya, who had been instructed after the year 1000 of the Hegira, in his article did not succeed in reflecting correctly the medrese tradition. He was also not able to evaluate properly either the *Fatih Vakfiyesi*, that at the time had been accessible in printed version, in Arabic (1935) as well as in Turkish (1938), nor the Fatih Medreses nor the general topic of Ottoman medreses, and as such, he opened the way for the misleading views put forward by later generations of scholars.

Another common trait of the progress-regress bifurcation of Yaltkaya as well as Emin Bey is the fact both authors, after having listed the names of numerous famous scholars active in the so-called backward or dark era, proclaim the fact that each individual case cited should be regarded as an excep-

VI

tion. Another error, which Yaltkaya propagated from Emin Bey and Ahmet Hikmet Müftüoğlu, is the claim that in the *Süleymaniye Külliyesi* "a medrese reserved for the mathematical sciences" was to be found as well.

One of the contradictions drawing attention in Yaltkaya's article is his attitude towards the "Kanunname-i Talebe-i Ulûm", which had been uncovered by Uzunçarşılı. While on the one hand asserting that "medreses had been left to their own fate and not subjected to any kind of reform in the long years between the era Sultan Süleyman the Magnificent and the proclamation of the Tanzimat", he interprets the text of the Kanunname that he quoted as providing "information on the courses and books instructed at the ancient medreses and the progress made by students". Whereas in fact this kanunname had been devised as a way of rectifying the crumbling educational system of medreses.

These views concerning medreses, first advanced by Yaltkaya in his courses at the Faculty of Letters and later in the notes he gave to his student Remziye Beksaç, would receive many echoes in subsequent research executed. The fact that a figure who had been a medrese professor [müderris] in the late-Ottoman period, an university "ordinaryüs" professor in the early Republican era, and eventually even a President of the Department for Religous Affairs as well as the holder of a diploma [icazet] in the sciences of kelâm, advanced such opinions does confer a prestigious status to the views expressed.

In the same period Osman Nuri Ergin (1883-1961) published his five-volume *Türkiye Maarif Tarihi*. This massive undertaking contained in an unmistakable and detailed manner the intellectual and political climate of the early Republic. In this work, published between 1939 and 1943, the history of Turkish education is divided into three stages. The first stage received the label *Araplaşma ve skolastik tedris devri (1453-1918)* [The Era of Arabification and scholastic Instruction], the second *Garplaşma ve yenilik devri (1773-1923)* [The Era of Westenisation and Innovation], whereas the third period started in 1923 with the proclamatoin of the Republic and was thus termed *Türkleşme ve milliyetçilik devri* [The Era of Turkification and Nationalism]. Ergin, whose original occupation had been *vilayet mektupçu* [chief secretary of provincial administration], had written a history of Ottoman medreses in a plain style, sometimes even promulgating haphazard expressions. As such, he composed his work based upon all kinds of sources, either original or not, second-hand scholarly and journalistic news items, and a number of authentic documents, so as to give rise to a work falling short of the scholarly requirements of history and in contrast with the methodology of historiography as refered to at the beginning of our article. In this book

Ergin thus displayed a "methodological" framework, that was clearly lacking in systematical consistency. He explained himself in the following way:

> In literary history, I remember the following anecdote generally attributed to Abdülhak Hamik: What is the point of always writing the things we see? Let us instead try to write the way we see fit ! As such, I cannot claim not have benefitted from the works listed above. Moreover, I can safely claim that I did not write this book according to what I saw, but that instead I wrote it as I saw fit ["düşündüğüm gibi yazdım"].[52]

Ergin, who was apparently not able to distinguish between creative and academic writing of a historical nature, described the topic of medreses, which he had been a passionate and valiant defender of in 1922, as he "saw fit" — in accordance with the political and ideological conditions of his day. In other words, the issue at hand refers to the history of medreses, written differently, according to "different ideas", in the last years of the Ottoman existence and in the early Republican period.

After having pointed out that the first era of history of Turkish education, which he had termed the era of Arabification and scholasticism will not possess any value for the younger generations — apart from its historical importance — he explains his reasons for writing the sections on the ancient culture and its institutions in the following way:

> The reason that I devoted 60 pages to medreses, which have by now become totally fosilised ["müstehase"], was to record even to prove to what a degree these institutions had become degenerate ["müteredd "]. On account of their ancientness does one not confer value to fossils by means of collecting them in museums, by means of writing treatises on them, by means of writing books on them ? For this reason I have devoted a relatively excessive amount of space on them.
>
> If I would have come across a work dealing at length with these topics, I would not have devoted so much time on either medreses, or tekkes, nor on mosques as I have gathered in the fist volume. It is quite certain that those who will write on the history after me will not mention them at all or will suffice by mentioning them briefly.[53]

Employing the terminology current at the time, Ergin deals with Ottoman medreses under the headings of popular schools [halk mektepleri], professional and vocational medreses and professional and vocational schools, and in talking about the medreses to which he refers as popular schools [halk mektepleri] he divided them further into two sub-groups. In talking about new medreses the author referred to the Dârülhilâfetülâliye medrese and its educational programmes mentioned at the outset of our article, which had been established during the Second Constitutional Era in response to the so-

---

[52] Osman Nuri Engin, Türkiye Maarif Tarihi, I-2, İstanbul, 1977, p. XII.
[53] Osman Nuri Engin 1977, p. XIV.

called *ıslah-ı medâris* reform policy implemented at the time The information presented in this section was taken over from such contemporary
periodicals as *Beyânülhak, Sebîlülreşad* and *Sırât-ı Mustakimin*. With regard
to the topic of the older medreses next to Emin Bey's famous article, he also
mentioned Ahmed Hikmet Müftüoğlu's two pieces and the statements
written by Ali Suavi and other late-Ottoman intellectuals in numerous newspapers and periodicals on the subject of the medrese education one after the
other without much critical appreciation. With a view to clarifying Ergin's
style, it would seem very interesting to quote at length his interpretation of
the words of the Hungarian Ottomanist Fekete who had investigated the educational and cultural importance of medreses relying upon the *vakfiyes* and
registries of medreses built in Hungary during the Ottoman period:

> The ancient Turkish medreses, praised by a European [specialist] did not move in
> line with the times, gradually in Turkey as well in the West a whole set of innova-
> tions were introduced in the field of education and instruction, but whereas these
> old medreses in the West were turned into dârülfünuns [universities], ours proved
> unable to even uphold the previous standards and gradually decreased in value and
> importance. Our histories do not record one serious or essential enterprise in three
> [hundred] years to save the medreses from this downward spiral, to imbue them
> with a new spirit of revival; such thinkers as Kâtip Çelebi realised very well that
> this was a dead-end street; [and] even if works were written along this way official
> institutions nor the government realised the dire nature of the situation. The fact
> that Kâtip Çelebi was aware of the necessity to master a Western language, that he
> translated works from Western languages and did not suffice with the instruction
> offered in medreses [and thus] was able to broaden his personal field of know-
> ledge. The dire state of medrese curricula, the fact that the roads to education were
> insufficient could only be uderstood by man of such [exceptional] calibre.[54]

It would be possible to cite innumerable examples of Ergin's lack of consistency. In this context it should suffice to compare two sections dealing with
the above-mentioned issue of the *İdâdi* terminology:

> The tetimme medrese prepared students for entry into the sahn medreses and were
> thus a kind of idad s. The term İdad should not be understood as the equivalent of
> today's high school ["idad "="lise"], but rather as [a] preparatory [institution] and
> referred to by the older phrase mahreç.[55]

Ergin made a correct observation in this instance, but two pages later he
stated the following, indicating the extent of confusion in terminology and
denomination:

---

[54] Osman Nuri Engin 1977, pp. 101-102.
[55] Osman Nuri Engin 1977, p. 98.

Following the secondary level the tetimme medreses, or in other words the high schools ["liseler"] followed.[56]

In spite of the fact that Ergin proposed rather contradictory statements on the topic of medreses and modern education, a topic he thought himself to be well-versed in with regard to the two separate eras these institutions had gone through, it is nevertheless possible to detect that Ergin in one case appeared to have been a lot more reserved and even objective than the writers mentioned earlier. As mentioned above, there had been no separate medrese available at the Süleymaniye Külliyesi, offering exclusive instruction in mathematical and applied sciences and variously referred to as *Dârülhendese* or *Riyâziyât ve Tabiiyât Medresesi*. Ergin nevertheless mentioned that even İsmail Hakkı Uzunçarşılı shared this unfounded opinion put forward by Emin Bey in his prominent article. Ergin mentioned the following, referring to Uzunçarşılı's course notes:

In contrast with the statements made by Professor İsmail Hakkı, that have been repeated and corroborated by numerous other writers, at the Süleymaniye medreses no specialist medreses instructing mathematical and natural sciences had been present ... In the vakfiye there is no mention of specialist medreses other than the Dârüttıb [Medrese of Medicine] and Dârülhadis [Medrese of Prophetic Traditions].[57]

Ergin displayed remarkable attention on this issue, and after having quoted from Emin Bey's article, he added the following reservation:

Even if from reading [Emin Bey's] paragraph one is led to believe that a medrese called Dârülhendese [Medrese for Geometers and Engineers] was to be found in the framework of the Süleymaniye İmareti, the authors did not supply a source nor is it possible to come across the mention of such a professional or specialist medrese in the relevant vakfiye, and therefore until a corroborative document is to found one has to understand and treat these claims with the utmost reservation possible ["kayd-ı ihtiyat "].[58]

Apart from such correct historiographical evaluations, and a couple of examples can be encountered, this piece of writing, containing a great deal of unsubstantiated and adapted information, nevertheless appears bestrewn with erroneous information on the history of the Ottoman medrese, slandering Ottoman culture and scholarly life in the most negative way, and as such even reducing them to naught.

The four and half century history of medreses from the time of Fatih up to the reign of Abdülhamid II consists of this. What kind of benefit or service have these medreses rendered to the nation, to Turk-dom ["Türklüğe"] and to world of learn-

---

[56] Osman Nuri Engin 1977, p. 100.
[57] Osman Nuri Engin 1977, p. 100.
[58] Osman Nuri Engin 1977, p. 147.

74

ing? Which well-known scholars were educated by them ? Are there individuals amongst them worthy of international recognition ? There is no reason to study these topics at great length and we can suffice by stating that "they did not produce anybody of value" ["hiç kimseyi yetiştirmemiştir"].[59]

The three pages written on the Fatih Külliyesi by Dr. Adnan Adıvar in his *Osmanlı Türklerinde İlim* constitute the most careful and scholarly sound text — in spite of its rather narrow viewpoint — respecting historical realities and evaluating primary source material in a correct fashion, written at the time. Adnan Bey first prepared and published the book in French during his years of exile in the city of Paris (1939). After his return to Turkey, he prepared a revised Turkish version of the text (1943). In this concise piece of writing Adıvar presented the source material available on the Fatih medreses with great care and reservation and investigated the unpublished vakfiye texts as well. He compared the copy of the vakfiye kept in the Museum for Turkish-Islamic Arts in İstanbul with printed versions in Arabic and Turkish. Subsequently, he noted the differences and tried to prove and illustrate that the copy kept in the Museum for Turkish-Islamic Arts, which he had personally uncovered, was the oldest text available.

> The oldest information available on these medreses is an Arabic vakfiye, kept in scroll form. We have studied this document, kept in the Museum for Turkish-Islamic arts (No. 1872 vakfiye, No. 6354). The beginning and final part of the vakfiye have disappeared. According to the testimonies of manuscript specialists the writing resembles writing produced in Fatih's era. In the first section the constructions and territories donated to the mosque and medrese are considered, as such the text provides a wonderful topographical picture of contemporary İstanbul In our view this text constitutes the original and first Fatih vakfiyesi.[60]

Adnan Bey's words on Emin Bey's article in the *İlmiye Salnamesi* form the best illustration of the sensitivity and care he displayed with regard to the scholarly methodology employed in his text:

> According to a tradition that has been oftentimes repeated, it has been said that the educational programmes had been devised by the vezir Mahmut Paşa and Ali Kuşçu in Fatih's own life-time, [and] that at first the primary lessons were attended at the tetimme medreses, and that subsequently the transition to the real instruction at the Medâris-i Semâniye was effected and that the lecturers, in order to be promoted to higher-ranking medreses, had to undergo a competitive examination in the Sultan's presence. Additionally, in the İlmiye Sâlnamesi it has been written that courses in mathematics, astronomy and medicine were taught at these medreses, without any mention of source material.[61]

---

[59] Osman Nuri Engin 1977, p. 108.
[60] Adnan Adıvar, *Osmanlı Türklerinde İlim*, İstanbul, 1982, p. 44.
[61] Adnan Adıvar 1982, p. 46.

But Adnan Bey, without consulting available studies on the history of me-dreses or *bimaristan*s (*dârüşşifa*s) [hospitals] in Islam, claims that the provi-sions related to those two institutions mentioned in the Fatih Vakfiyesi are insufficient. But, on the other hand, the statements made in the vakfiye on these two institutions are nothing but the expression of a tradition which dates back to pre-Ottoman times. As has been put forward in studies dealing with the history of pre-Ottoman Seljuqid or medreses in general, the way in which courses were organised in the Fatih medreses was in accordance to a tradition, which had been kept in use for centuries. As we are able to deduct from the vakfiyes of various medreses, founded in different areas and in dif-ferent centuries, no mention was ever made of any "course programme". It is impossible to conceive that the issues mentioned in vakfiyes would bear any resemblance to curricula of high schools or universities founded at the end of the nineteenth century or at the beginning of the twentieth:

> It is regrettable ["yazık ki"] that the names of the courses instructed are not men-tioned, nor is there any indication as to whether or nor medical students at the dârüşşifa had to undergo practical training, in other words, it is not mentioned whether the hospitals carried the qualifications of a fully functional medical sur-gery.[62]

Adnan Bey's sadness, expressed in the phrase "[i]t is regrettable" ["yazık ki"], was intentional and the editors of the book's new extended version added a "course programme". These studies undertaken to prove the exis-tence of such a "programme" are nothing but a teleological exercise with no connection to historical reality. But as we have already dealt with this just-mentioned topic in a previous article[63], the issue falls outside the boundaries of this article and there is thus no reason to repeat it here.

With the approach of the year 1953 the anniversary's of Fatih's conquest of İstanbul entered Turkish public opinion, and as such the Fatih medreses were also part of this agenda. The well-known librarian of the period, Mu-zaffer Gökman, published a study under the title of *Fatih Medreseleri* in the year 1943. Gökman explained his goals in preparing this book as follows:

> To acquaint the younger generations [of Turkey], on the approach of the five hun-dredth anniversary of the conquest of İstanbul, with the Sahn-ı Seman medreses which were the Ottoman Turks, whose ideals had been ruined ["(k)ubbelerinde incir ağaçları bitmiş"], first university.

This monography, prepared on the basis of known historical sources such as the translation of the *Şekâyık* and the vakfiye of the Fatih medreses repre-

---

[62] Adnan Adıvar 1982, p. 46.
[63] cfr. Ekmeleddin İhsanoğlu, "Fatih Külliyesi Medreseleri Ne Değildi !".

76

sents a serious inquisitive study into the issue. In comparison with the earlier published studies, Gökman's monography presents a great deal of correct information. At the same time, the study was written in a style refuting the then popular hostile ideas concerning Ottoman heritage and medreses in particular. Gökman, who correctly employed the basic source-material, particularly made abundant use of Âli's *Künh-el Ahbâr*. At the same time, however, the author repeated the higher-mentioned errors on the topic of Ottoman medreses perpetuated by Emin Bey and other studies popular at the time. He also reiterated a series of mistakes that had been presented by Süheyl Ünver in his earlier works, as shall be demonstrated below. For example, he repeated Süheyl Ünver's claim that a Faculty of Medicine had been present among the Fatih medreses, even though this was not mentioned in any of the primary sources, used by Gökman himself.

Gökman put forward certain comparisons, without studying the cultural life of the West or its institutions, and without relying on studies that had been written on this subject. He shares in the sentimentality displayed by Turkish intellectuals with regard to this topic in the following interesting way:

> If we were to consider Ottoman cultural institutions as being placed on a line, we could notice how sometimes favourable and sometimes unfavourable movements were made. After having experienced an upward movement from the time of their first inception up to the era of Fatih, when this line safeguarded its vitality, during the reign of Beyazıt II [this line] slowly began to stoop[64] ... The educational scene of the 15th century was full of innovations able to draw the West's attention.[65]
>
> The second era of upward movement coincided with the reign of Süleyman the Magnificent. He had a medrese erected in his own name. He gave importance to the instruction of medicine and mathematics. During the first upward movement, in Fatih's day, the West was on a lower level, during the second upward move, in Süleyman the Magnificent's day, the West had reached our level.[66]

At the end of his writing, Gökman wonders how those who consider the instruction offered by Ottoman medreses to have been long and troublesome ["müz'ic"], who considered its education to have been "scholastic" and unpleasant will feel after having read his treatise. He thus ended his study on Ottoman medreses with the hope that one day they will be restored to their just position in the intellectual life of Turkey.

It is also necessary to consider two important aspects relating to the historiography of the Ottoman medrese in Süheyl Ünver's *İstanbul Üniveristesi tarihine başlangıç Fatih Külliyesi ve Zamanı İlim Hayatı*, published in 1946.

---

[64] Muzaffer Gökman, *Fatih Medreseleri*, İstanbul, 1943, p. 3.
[65] Muzaffer Gökman 1943, p. 45.
[66] Muzaffer Gökman 1943, pp. 3-4.

The fact that Ünver's book is a compilation rather than a work of composi-tion is accentuated by the characterisation "toplayan" [compiler] rather than "yazar" [author] on the cover of the book. The first important aspect of this work is the fact that it is a work of compilation. Ünver collected information and materials, such as Ottoman chronicles, archival documents, printed and manuscript works on the history of medreses that had hitherto been unknown and unused. The value of the work is greatly increased by the reproduction of some of this historical material, such as engravings, miniatures and photo-graphs, at the end of the book.

But the reason why it is necessary to carefully analyse this book is the model or picture of the Ottoman medrese put forward by Ünver. To compre-hend the true nature of this picture it is important to understand why this work was prepared and to see the mind frame surrounding the ideas devel-oped in the work:

> Since the year 1930 I have been involved in researching the beginnings of the Faculty of Medicine in Fatih's day and in this context I came across a great deal of documents of particular interest to the history of our sciences and our fine arts. To commence the history of the Univeristy of İstanbul with the conquest of İstanbul seems to be a correct starting-point.[67]

Ünver thus explained his goal in writing the history of the İstanbul Univer-sity's Faculty of Medicine. The President of the University, the medicine graduate Ord. Prof. Cemil Bilsel expressed this aim in the following way on the occasion of the Commemoration of the Graduates of Medicine on 14 March 1942:

> ... Because the conqueror of İstanbul founded a higher civilisation and a full-fledged university.[68]

As such, the authorities of the İstanbul University had thus expressed their desire to start the history of the university from the year 1453 onwards and Ünver, sticking to this teleological agenda, had carefully compiled historical materials which he subsequently presented in his book. But, there seems to have been an even more pressing reason for Ünver to investigate this subject. Ünver's father had been a graduate of the Fatih medreses. As a result he in-cluded the following dedication in his book:

> I present this work to the soul of my higly esteemed ["aziz"] father Tırnavalı Mustafa Enver, one of the last graduates of the Fatih Külliyesi and one of our first telegraphic operators.

---

[67] A.Süheyl Ünver, *İstanbul Üniversitesi Tarihine Başlangıç. Fatih, Külliyesi ve Zamanı İlim Hayatı*, İstanbul, 1946, p. IX.
[68] A.Süheyl Ünver 1946, p. 7.

78

A subliminal psychological motivation underlying Ünver's view-point of the topic at hand can be detected in this dedication. The author clearly talked about a graduate of the "Fatih Külliyesi", and not of the "Fatih Medresesi". But still graduation had been achieved throughout four centuries by means of receiving a diploma ["icazet"] from the hands of medrese professors. The term "külliye", which was a concept coined in the late-Ottoman period, assumed an equivalent status to the noun university in Ünver's terminology. Throughout the work the notion of equality was expressed in different ways, such as "Fatih Külliyesi (üniversitesi)" (p.X), "Fatih Üniversitesi" (p. 93). Moreover, Ünver elaborated upon the University President's above-quoted talk on Fatih as the founder of the "university", by even attributing University Presidency ["rektörlük"] to Fatih himself.

> ... In this the most important factor was undoubtedly that Fatih had held the position of University President ["Rektörlük mevkii"] for the duration of his life.[69]

In the first half of the book, Ünver interprets the rich materials he had collected in a misleading way so as to prove the idea of that the Fatih medrese = külliye = university, and in the second half, while dealing with the scientific and cultural life of Fatih's reign, he accentuated the concepts thus formed:

> A short look at the first president of the İstanbul Külliyesi and the head of the contemporary ulemâ corps Sultan Mehmed and the scientific activities of the scholars of the era.[70]

Ünver connects the date 1453 with the İstanbul University, founded in 1900 under the name of Dârülfünun. It is well-known that immediately upon conquering İstanbul Fatih turned a church and its attendant monasteries into medreses. Later he founded the Sahn medreses, that were to become known in his own name. (their construction was completed in 1470). These two types of medreses were supposedly conceived by Ünver as each other's continuations and in this way he presented the İstanbul University as the continuation of the Fatih *medreses* = *külliye* = university.

Ünver, without giving any historical or rational proof, proclaimed that all these medreses, which had been independently established in their own right, were related to the *Sahn* medreses. At the same time, he did not take account of the fact that there was no common ground between the newly established İstanbul University and the formerly established medreses. In a way, the causal chain proposed by Ünver is nothing but a collection of unconnected rings. Thus, the mentioned explanations are merely fanciful pronouncements. Ünver explained his fantastical composition in this way:

---

[69] A.Süheyl Ünver 1946, p. 120.
[70] A.Süheyl Ünver 1946, p. 157.

During the Byzantine era numerous religious instititutions had been present in İstanbul. A number of these were monasteries inhabited by priests. Two of these had been emptied. The first ones were the priests' cells in the Ayasofya, and the latter the cells in the Pantokrator Monastery in Zeyrek. In the days and months following the conquest, numerous scholars that had come here together with the Sultan, or that had arrived in the city at a later date, started to conduct courses here. With the conquest of İstanbul, instruction had started. Consequently, it would be incorrect to put the starting date with the completion of the buildings of the Fatih Külliyesi in 875 (1470).[71]

These first two centres of instruction became famous as the Ayasofya and the Zeyrek medreses. The İstanbul Külliyesi (University) had at first been founded in these temporary locations and this starting from the date of the conquest. Upon completion of the (Medâris-i Semâniye) Fatih had the institution and all of its lecturers moved from these temporary lodgings. But the medreses had only been completed 18 years after the conquest.[72]

On another page he repeated the same idea in this way:

Let us make it known, before we move on to the study of each of the individual buildings, the date 875 (1470) does not announce the foundation of the İstanbul University and its separate branches. The explanation we have given earlier show that we will have to make that date coincide with the conquest of the İstanbul. Teachers and students simply switched locations, as a matter of fact they continued with their earlier given assignments. The branches that had been opened at temporary locations were simply brought together. The professors possessed the highest scientific rank of professors at the Sahn Medreses ["Sahn müderrisi"] even while being active at the temporary locations. All the authorities upon whom we have relied emphasise this point.[73]

Ünver presented no proof for his claim of a causal link connecting the year 1453 with the year 1900. Nor did he indicate the way in which these independent educational institutions, each possessing its own vakfiye and its own administrative status, were supposed to have been interconnected so as to form a progressive line of educational promotion. He did not present any proof because his claims are fictitious. It would be wrong to assume that since the scholars of the age could have been promoted from one medrese to another, and that since students enrolled at a lower level medrese would have continued their education at a higher level medrese, these medreses were legally linked to each other, or to assume that the older medreses corresponded to the first stage of newer ones.

The educational activities which started in eight disused Byzantine buildings, converted to medreses immediately after the conquest, did not stop as Ünver had claimed with the establishment of the complex of medre-

---

[71] A.Süheyl Ünver 1946, p. 9.
[72] A.Süheyl Ünver 1946, p. 9.
[73] A.Süheyl Ünver 1946, p. 23.

80

ses around the Fatih Mosque, known as the Sahn Medreses. The Ayasofya and Zeyrek medreses, enumerated by Ünver, were independent medreses each possessing its own vakfiye that continued to be active after the foundation of the Sahn medreses. The educational activities of the Ayasofya medrese even continued up to the year 1924.[74] The educational activities of the Fatih Medreses also continued after the foundation of the Dârülfünun in 1900. The establishment of the fist Ottoman university goes back to the middle of the nineteenth century, when first attempts to establish an institution of higher education, independent of medreses, were made and this goal was finally achieved in the year 1900. The Dârülfünun was founded as an educational institution with no connection to the medrese system.[75] As a result of the reforms of 1933 certain professors that had been educated at medreses were dismissed and as a result the sole link between the İstanbul University and Ottoman medreses was broken.

The late Süheyl Ünver, in honouring his paternal love and memories, attempted to rid medreses of the bad reputation they had acquired over the years. At the same he wanted to raise the stature of Faculty of Medicine and the University, to which he was affiliated, among the wider universities of the world. As the 500th anniversary of the conquest of İstanbul was approaching, he attempted to heighten the memories of the Ottoman Empire and as a result presented a composite history of the Ottoman medrese, consisting of a large volume of incorrect information, such as has been described in the course of our article. In compiling important material, Ünver made a significant contribution for those who would later research the same topic. But based upon that information he composed an incorrect model of the Ottoman medrese, without any historical validity. Ünver indulged in illogical comparisons between medreses and modern university systems, the traces of which cannot be easily wiped away. In doing so Ünver contributed towards the survival of an erroneous model of the Ottoman medrese in academic circles.

A New Beginning in Academic Research

The year 1957 has to be regarded as an important starting-point in the historiography of the Ottoman medrese. In that year two publications, one in Turkey and the other in Europe, relevant to the issue were released. In spite of the fact that these two titles were prepared in accordance with the meth-

---

[74] Cahit Baltacı, *XV.-XVI. Asırlar Osmanlı Medreseleri*, İstanbul,1976, pp.274-280.
[75] E.İhsanoğlu, "The Genesis of the "Darulfünun". An Overview of Attempts to Establish the First Ottoman University", pp. 827-842.

ods observed in historiography and were on a higher academic level than the works discussed earlier, it still proved difficult for the authors to escape the influence of the consensus on the nature of medreses.

The first one was a study undertaken by Gibb and Bowen in their famous book "*Islamic Society and the West*", that considered the history of the Ottoman medrese in the framework of the history of education in the wider world of Islam. This broadly conceived work considered the influence of Western civilisation on the "Muslim culture" of the Near East. The authors considered the appearance of Ottoman medreses in the section dealing with education.[76] They attempted to connect Ottoman medreses with the development of medreses in the wider world of Islam, taking care however not to neglect their links with other Ottoman educational institutions.

The authors consulted D'Ohsson and Hammer in trying to delineate the general history of the Ottoman medrese, but also relied heavily upon the higher-mentioned two works by Osman Nuri Ergin (*Mecelle-i Umûr-ı Belediyye* and *Maarif Tarihi*). Even though it is not possible to encounter Emin Bey "Europeanised" model, one can still notice that the authors appropriated the general negative views on Ottoman culture and education, proffered by Turkish authors. Gibb and Bowen's use of the volume written by Adnan Adıvar, which has been referred to higher, and of a number of articles in the Turkish edition of the *İslâm Ansiklopedisi*, in progress at the time, clearly heightens the academic standard of the work. After having considered in detail the employment of medrese professors [*müderris*] in bureaucratic and judiciary institutions and the connections between their different duties, the eighteenth century, constituting the actual temporal framework of the work, was not really dealt with seriously. Only the general cultural life during the so-called Tulip Age and the innovations introduced by İbrahim Müteferrika were considered in the framework of the development of Ottoman modernisation. One cannot but notice that Gibb and Bowen shared the then commonly accepted views on Ottoman modernisation and that their analyses were made in accordance with these views. The unfounded and degrading accusations of an adventurer like the Baron de Tott on such topics as Ottoman science and culture were employed as persuasive arguments in favour of such views. According to these reports the ulema, stemming from a medrese background, were presented as violently opposed to the introduction of new ideas into the Ottoman Empire, and as a result the authors claimed that

---

[76] The first edition appeared in 1957, here the reprinted edition has been used: H.A.R.Gibb & Harold Bowen, *Islamic Society and the West: A Study of the Impact of Western Civilisation on Muslim Culture in the Near East*, I, 2nd section, London, 1965, pp. 139-164.

the contemporary Ottomans were all but "ill-informed" on then current affairs.[77]

This volume written by Gibb and Bowen stands out as the most reliable work written on the topic of the historiography of the Ottoman medrese written in a Western language until the middle of the twentieth century.

Towards the end of the period under discussion, the other important source constituting a new beginning in the academic study of the topic of the history of the Ottoman medrese is an encyclopaedia entry prepared on the basis of reliable historical sources. The *Encyclopaedia of Islam* was prepared by a group of orientalists, the first edition of which appeared in Leiden during the period 1913-1936. The Turkish Ministry for Education had a committee of members of the İstanbul Faculty of Letters prepare an enlarged Turkish version. The eighth volume of which appeared in 1957, containing J. Pedersen's entry on *Mescid*. Prof. Cavit Baysun wrote the part on Ottoman medreses under the sub-heading *Medreses of the Ottoman Period* ["Osmanlı Devri Medreseleri"] and contributed the most careful and methodical piece on the the history of the Ottoman medrese published up to that date.[78] In his entry Baysun described the history of Ottoman medreses based upon primary source-material, stretching over a period of six centuries from the foundation of the İznik medrese until the abolition of the institution in 1924, and consequently gave a summary of the development of the different stages attained by the members of the ilmiye class. In this piece of writing, still carrying an academic importance today, a certain influence of the higher-mentioned "erroneous" views can still be detected. Foremost amongst these is the claim concerning the Süleymaniye medreses.[79] Baysun mentions this claim while listing the names of the medreses contained within the Süleymaniye Külliyesi and in addition refers to a second unfounded claim made by Emin Bey and Ahmed Hikmet Müftüoğlu:

> According to this organisation, the instruction offered by Ottoman medreses was divided into two sections, the first being law, theology and literature which were [subjects] taught at the Sahn-ı Seman, and the other being mathematics and medicine, which could be completed at the Süleymaniye.[80]

And referring to the higher-mentioned exclamations made by Kâtip Çelebi, he proclaimed that the courses called *cüz'iyat*, comprising arithmetics, ge-

---

[77] H.A.R.Gibb & Harold Bowen 1965, p. 154.

[78] The part in which this section appeared, bearing the signature M.C.B., was published in 1957. *İA*, VIII, pp. 71-77.

[79] *İA*, VIII, p.73.

[80] *İA*, VIII, pp. 73-74.

ometry, astronomy and philosophy (hikmet) were abolished from the latter part of the sixteenth century:

> From the end of the sixteenth century onwards, these [courses] were removed as a result of being seen in contrast with religious dogmas.[81]

Encountering such claims in an encyclopaedia entry which had been written on the basis of primary sources and the writing of which had not been subject to any form political or ideological influence but had instead been composed on the basis of sound academic standards and methodology, clearly discloses the strong influence of a "model" that has been developed in time as a result of a variety of factors. In the field of the history of Ottoman science and culture numerous comparable problematic issue can be encountered. This clearly discloses the necessity of having to adhere to a stringent historiographical methodology.

The *image* of the Ottoman medrese that has been portrayed in the articles and books mentioned in this study was devised without any connection to historical reality. This image was constructed as a result of certain feelings of inadequacy felt in the face of Europe's superiority, and as a result of comparisons with European universities, containing numerous faculties that had entered the Ottoman and Islamic world in the course of the nineteenth century. These institutions had been developed in European societies reliant upon the base of so-called *corporations*, which had in turn been an outcome of the system of Roman Law. On the other hand, as the notion of Ottoman leadership became prominent in the political priorities of the Second Constitutional Era, and, as a result of a desire to foster allegiance to the office of the Caliphate in the Muslim world, reform in the field of religious education became a necessity leading to a policy of *ıslah-ı medâris* complementary to which the historical model of the medrese was developed. In tandem with the division of Ottoman history into two main thrusts of brilliant-dark or progressive-regressive, medreses of the earlier period were presented in a progressive light and those pertaining to a later period as indicative of regress and stagnation.

As explained above the image portrayed was based upon the acceptance of certain preconditions, particularly in connection with the history of the Ottoman medrese tradition, the development of educational institutions in Islam, and the appearance of European universities as well as comparative approaches with regard to this "image". The fact that this "image" was first propagated in the *Meşihat*'s own publication, the *İlmiye Salnamesi* as the official history of the medrese, has undoubtedly led to its general acceptance

---

[81] *İA*, VIII, p.75.

as a safe reference. As a result of the Tevhid-i Tedrisat [Unification of Education] law, which was promulgated immediately following the proclamation of the Republic, in combination with the abolition of medreses the "image", which had been developed in connection with the *ıslah-ı medâris* reform policy was elaborated upon and thus shown to provide a clear argumentation in favour of the Republic's new policy of abolishing the medrese institution. With the approach of the 500th anniversary of the conquest of İstanbul one attempted to rehabilitate this educational institution, and thus efforts were concentrated upon the example of the most brilliant of the "brilliant era", the Fatih medreses, and as a result this image that had been developed for half a century attained an even more complex state. The combination of several factors, such as literary expressions employed to bolster heroic historical theses, the fact that the medrese institution fell outside the new intellectual's sphere of knowledge combined with concepts and notions on modern education led to a confusion of terminology regarding the Ottoman medrese. This era saw the proliferation of knowledge not based upon historical sources and that was derived at without consulting primary source-material. This information was subsequently deformed in the hands of successive writers who did not abide by a judgement based upon the application of a critical methodology, but instead listed the material according to the requirements of political or ideological preferences. In the end one can safely claim that this time-frame has to be seen as the era of discovery and construction of the historiography of the Ottoman medrese. This confusion of notions and conceptions continued to be prevalent in İsmail Hakkı Uzunçarşılı's *İlmiye Teşkilatı*, constituting the first stage of the subsequent era, and in the works of those dealing with the subject writing after that.

While it is possible to conceive that comparable problems exist in other fields of academic study dealing with Ottoman institutions, it remains a fact that not enough work has been done on the topic of Ottoman medreses. In addition, as a result of the fact that pre-Ottoman medreses founded in Anatolia and medreses in Islam in general have to be considered as a civilisational unit could also explain in some ways why the history of Ottoman medreses has assumed such a confused apparition compared to the histories of other institutions.

APPENDIX: LIST OF BOOKS USED IN MEDRESE COURSES MENTIONED IN THE TEXT

Hâşiye-i Tecrid: The annotation [haşiye], written by el-Seyyid el-Şerif el-Cürcân (died 816/1413), on the şerh [commentary], composed by Şemsuddin Mahmud b. el-Isfahân (died 749/1348), on the work called *Tecrîd el-Kelâm* written by Nas ruddin el-Tûs (died 672/1274) on ilm-ı kelâm [theological philosophy]

Şerhü Mevâkıf: The şerh [commentary] on Azududdin el-İc (died 756/1355)'s work *el-Mevâkıf fî İlm el-Kelâm*, composed by el-Seyyid el-Şerif el-Cürcân (died 816/1413).

Tavâli' el-Envâr: The abridged version [muhtasara] of the work *Tavâli' el-Envâr min Matâli' el-Enzar* on kelâm, composed by el-Kadı'l-Beydâv (died 685/1286).

Metâli' el-Envâr fi'l-Mantık: The şerh [commentary], called *Levâmi' el-Esrar*, on the work of the kadı of Konya, Sirâcuddin el-Urmav (died 682/1283) on logic, composed by Kutbuddin el-Râz el-Tahtân (died 776/1364).

el-Hidâye: A şerh [commentary] on the work *Bidâyet el-Mübtedî* on the Hanef jurisprudence [fıkıh] written by Burhânuddin Ali b. Eb Bekr el-Merğ nân (died 593/1197), composed y the author himself.

Ekmel: The şerh [commentary] on the work *el-Hidâye*, composed Ekmeledd n el-Bâbert (died 786/1384) entitled *el-İnâye* [on Hanef jurisprudence].

THE GENESIS OF "DARULFÜNUN"

An Overview of Attempts to Establish
the First Ottoman University

From the beginning of the nineteenth century, the Ottomans turned their attention from the east to the west as regards science and changes occurred in the Ottoman thinking of science and education. Therefore, the Tanzimat period witnessed attempts to establish a new institution of higher education, besides the medrese. This institution was very seldom called "medrese-i cedide-i ilmiye" (the new medrese of learning) or "Darultahsil" (house of education). It was mostly called Darulfünun (house of the sciences), a name which was generally used in the meaning of university in later periods.

Apparently, the name Darulfünun must have been chosen for the Darulfünun to emphasize that this institution was different from the medreses of the time. In late Arab literature, the word "fen" meant a branch of science; its plural form "fünun" meant various branches of science. In Ottoman Turkish it meant sort, variety, category, state and branch of science. In the nineteenth century sciences based on observation and proof were called "fen". While the word "fünun" generally appeared in the compound phrase "ulûm ve fünûn" in the literature of that period, it was used alone separately from the word "ulûm" (sciences, mostly religious sciences) in the compound word of Darulfünun. Thus, apparently it was envisaged as an educational institution where mostly the instruction of the new kind of sciences of western origin, synonymous with the word "fen", would take place.

According to the findings of our research, the idea of the Darulfünun was first formulated in the Tanzimat period for the purpose of educating the people. It was conceived as an institution where all sciences would be taught. Since no regulation was prepared for the Darulfünun at its

VII

establishment phase, it appears that the statesmen and scholars of the
time did not have a clear idea of exactly what the Darulfünun would be.
Although the objectives and functions of the modern educational institutions established before the Tanzimat, namely the Mühendishaneler
(Engineering Schools), the Mekteb-i Tıbbiye (Medical School) and
Mekteb-i Harbiye (War College) were stated clearly in their phases of
establishment, the same was not done for the Darulfünun (E. İhsanoğlu,
1992: 335-395).

The Darulfünun was established and functioned in the particular
Ottoman style of that period. Following three failures, a fourth attempt
took place in 1900 during the reign of Sultan Abdülhamid II when the
necessary infrastructure and conditions for establishing the Darulfünun
were realized. Thus, the "Darulfünun-ı Şahane" was established as the
foundation of the present University of İstanbul.

This paper is an evaluation of the results of my previous two studies;
it also aims to clarify the development of the Darulfünun as an institution and as a concept as well as the reasons for the failure of the attempts
for its establishment.

Development of the Darulfünun as an institution: the first attempt

A temporary council for education, called *Meclis-i Muvakkat*, was set
up within the *Meclis-i Vâlâ* (High Council) in 1845 during Sultan
Abdülmecid's reign. It consisted of 7 persons and undertook the first
efforts towards the establishment of a "Darulfünun" in the Ottoman State.

The *Meclis-i Muvakkat,* which was assembled with the aim of reforming the educational system in the Ottoman State, submitted an official
report to the *Meclis-i Vâlâ* in 1846. For the first time in Ottoman literature, this report included the establishment of an educational institution
called the "Darulfünun". This term did not lend itself to a clear understanding as to what kind of an educational institution would be established and for what purpose. But, it appears that the first objective was
to train *münevver bendegân* (well-informed staff) so that the affairs of
the State would run more efficiently. The report also required that
the Darulfünun be immediately established at the Sublime Porte or in
another appropriate place.

This report, prepared by the temporary council, led to a more serious
interest in educational reform in the Ottoman State, upon the encouragement of Sultan Abdülmecid. It was decided in 1846 that a permanent

council of education under the name of *Meclis-i Maarif-i Umumiye* (Council of Public Instruction) be established with the aim of conducting more efficient studies on this subject. This council was charged with the undertaking of numerous studies related to the Ottoman Educational System. Among its duties were such crucial functions as the establishment of the Darulfünun and the schools of secondary education where the students of the Darulfünun would be educated[1].

The planned locality allocated for the establishment of the Darulfünun included the old ammunition store in the vicinity of the St. Sophia Mosque and the building land which belonged to the Sultan's Palace. The construction of the building, designed by the Italian architect G. Fossati, who was in İstanbul, began during the same year. The building of the Darulfünun was conceived after the pattern of European universities on a vast scale with 3 storeys and 125 rooms. It was constructed in nearly twenty years.

An Encümen-i Daniş (Society of Knowledge) was established in 1851 in order to prepare the books to be used by those who attended the classes in the Darulfünun. The Encümen-i Daniş, which was instituted in a pompous way, however, was not able to fulfill the expected duty, i.e., the preparation of textbooks.

While the building of the Darulfünun was still in construction, education started in some rooms of the same edifice in 1863 through the initiative of Fuad Paşa, the grand vizier of the time. The courses were taught in the manner of public lectures without any curriculum, mainly for the purpose of educating the public. Since, as stated above, there was no regulation of the Darulfünun nor how many years the period of instruction was planned to include. In practice, Ottoman scholars, some of whom were educated in modern Ottoman institutions of higher education and others in Europe, gave lectures on physics, chemistry and natural sciences as well as history and geography between 1863-1865 (E. İhsanoğlu, 1990: pp. 706-709). Moreover, a library and a laboratory were established in the Darulfünun for the benefit of those who attended the classes. Probably, after some time the lectures continued as regular courses. The public as well as the ministers and high officials attended the classes with great interest.

The courses in the Darulfünun, which regularly continued for more than two years until March 1865, were interrupted when its building was

---

[1] Başbakanlık Osmanlı Arşivi, İrade Dahiliye, nᵒ 6634.

constructed and allocated to the Ministry of Finance. Edhem Paşa, the Minister of Education of the time suggested that a new and smaller edifice be built for the Darulfünun. In order not to interrupt the education, Nuri Paşa's mansion near Çemberlitaş was rented temporarily and the courses began on 19 April 1865 in this building. They were again interrupted when this building burned in the great fire of Hoca Paşa on 8 September 1865.

The second attempt: Darulfünun-i osmanî

Following the failure of the first attempt to establish the Darulfünun, Ottoman statesmen took more serious measures about Ottoman educational reform. "Maarif-i Umumiye Nizamnamesi" (Regulation of Public Instruction), based on the experiences which the French educational system went through after the French Revolution of 1789, was prepared during the period when Safvet Paşa, a well-known figure of the Tanzimat period was Minister of Education. This regulation was adapted to the Ottoman educational system in 1869 and remained effective until the Period of the Second Constitution. It marked the beginning of an important period for the history of education in the Ottoman State. 51 items of this regulation, which comprised a total of 198 items, were related to the Darulfünun-i Osmanî. They dealt with the administrative, financial, internal and legal rules pertaining to this institution. Moreover, this regulation is particularly important by virtue of being the first one on this subject (*Düstur*, 1289, pp. 198-204).

At first, the Darulfünun was planned to have five departments like European universities. But, since the departments of medicine and theology were already present in the medical schools and medreses, only three departments were established in the Darulfünun. These were Law, Literature and Philosophy; Mathematics and Natural Sciences. The regulation stated that a library, museum and laboratory would be established and each department would have separate curricula. A brief examination of some important articles of this regulation related to the Darulfünun shows that it had some shortcomings. First of all, it did not specify where the graduates would be employed and what kind of professions they would pursue. The financial sources of this institution depended merely on subsidy, students' fees and donations. The courses would be open to the public and given in Turkish. At the end of the third year, the students would take a general final examination. Those who were successful would prepare a *müsvedde* (draft) which may be called a graduation thesis. In order to become a *müderris* (professor), the students had to study one more year, submit an *ilmî müsvedde* (thesis) and defend it before a jury.

It was decided that instruction in the Darulfünun-i Osmanî would start in 1869 in its second building which was completed at the same time as the Regulation of Public Instruction was made effective. In October 1869, 450 students were admitted from nearly a thousand candidates, mostly medrese students, by passing an examination. The instruction in the Darulfünun began, one year later than the planned time, on 20 February 1870 with a ceremony which the Grand Vizier Âlî Paşa, Minister of Education Safvet Paşa and Tahsin Efendi, the director of the Darulfünun attended. At first, it was planned that the Darulfünun would have the departments of Philosophy and Literature, Law, Natural Sciences and Mathematics. Since neither teachers nor books were available, however, these departments were united in a single curriculum and all the students attended the same classes. These courses were mainly, french, law, literature, economics, geography, history, philosophy, mathematics, physics, astronomy and chemistry.

Since the beginning of the second academic year coincided with Ramadan (October 1870), a series of public lectures were organized during this month. Cemaleddin Afganî, who was in İstanbul at that time and who had delivered a speech in the opening ceremony of the Darulfünun, gave a lecture on arts where he mentioned the prophethood as an art in his discussion of arts and their branches. This caused violent objections of the şeyhülislam's office. After this event, Afganî was exiled from İstanbul and rumours spread that the Darulfünun would be closed.

Instruction in the Darulfünun-i Osmanî continued between 1870-1872 without any interruption. It is not clear, however, whether there were any graduates of the Darulfünun or not and exactly how its activities ended. Some of the reasons for its failure were the following: the lack of a sufficient number of students educated in secondary schools and the deficiency of the educational staff. The Ottoman State tried to solve this problem by hiring native Muslim and non-Muslim teachers from various other educational institutions. It was not envisaged to invite a foreign teacher from Europe as the head of the Darulfünun-i Osmanî as it was done during Mahmud II's reign at the establishment of the Medical School. Similarly, although laboratory materials were brought from Europe to conduct physics experiments in the Darulfünun, no expert was invited in order to use them. Some other reasons for the failure of this institution were the high fees, the fact that all classes were open to the public, the regulation was not applied properly and no clear objective had been set.

The Darulfünun-i Sultanî and its departments

Following this second attempt, which was a failure, the high officials of the Tanzimat period did not give up their efforts in this direction since they believed in the need to establish a western type university. Therefore, Safvet Paşa, the Minister of Education of the time, charged Sava Paşa, director of the Mekteb-i Sultani (Lycée Impérial Ottoman de Galata-Séraï) in 1874 with the task of establishing a new Darulfünun in the building of the Mekteb-i Sultanî provided that it will not be a financial burden on the treasury. Based on previous experience, this time the dignitaries of the Tanzimat period attempted to establish the Darulfünun on the foundations of the Mekteb-i Sultani which was founded in 1868. Thus, they aimed to implant a new sprout of higher education into the body of this institution of secondary education.

Instruction in the Darulfünun-i Sultanî started in the 1874-75 academic year. It was planned to include three schools, i.e., the schools of law, literature and civil engineering. These three were called "Mekâtib-i Aliyye" (Schools of Higher Education). In the first year, the schools of Law and Civil Engineering became operational. The name of the School of Civil Engineering was changed at the end of the year, as the School of *Turuk-u Maabir* (Ecole des Ponts et Chaussées).

The two schools were opened more prudently in a less pompous way as compared to the previous inaugurations and the public was not informed about them until 1876. There were very few news about these schools in the newspapers and official documents of that period. At the same time, this institution was not openly named as the Darulfünun. Sava Paşa, the director of the Darulfünun-ı Şahane pointed out that the reason for this shyness was to take the necessary measures against any possible reaction to the image of the Darulfünun and that he was asked, in particular, to act prudently on this matter. Considering the fact that the schools were put into good order and the students had passed to their third year in 1876, there was no objection to informing the people about the inauguration of the Darulfünun. The "Regulation of the School of Law" was published the same year and it was stated that this regulation would be applied to all three schools with small modifications (*Düstur*, 1923: pp. 439-443).

According to its regulation, the Darulfünun-i Sultanî admitted the graduates of the Mekteb-i Sultanî and other high schools as well as the students with an equivalent level of knowledge. The period of instruction was four years. Both law students and those in the School of *Turuk-u Maabir* would graduate with the "grade de docteur" if they defended a thesis successfully. They could receive the "grade de licencié" by passing an easier examination than that of doctors and without submitting a thesis. Law graduates with the "grade de docteur" would be assigned posts in the Ministry of Justice, while those with the "grade de licencié" could work as lawyers. Graduates of the School of *Turuk-u Maabir* with the "grade de docteur" would be employed in the Ministry of Public Works and those with the "grade de licencié" could work as conductors or in similar jobs. The graduates of the School of Literature would teach in state schools or private schools.

Law School

The Law School in the Darulfünun-i Sultanî was established in order to educate well-informed staff who would hear cases and preserve the supremacy of law. The instruction in this school began during the 1874-75 academic year and the courses on Western as well as Islamic law were taught at the same time. Among them, there were some particularly important ones such as fıkh (Muslim canonical jurisprudence), civil code, Roman law and commercial law. 21 students entered the final examinations at the end of the first year. By the end of the third year, the total number of students rose to 61; thus, nearly 20 students were admitted to the Law School every year.

The building of the Darulfünun-i Sultanî was in the vicinity of Gülhane and the courses were given here. In 1876, the matter of moving the Darulfünun to its old building in Galatasaray came up. Students of the Law School, who were civil servants at the same time, objected to this change under the pretext that going and returning would be difficult if this school were moved to Beyoğlu. Some statesmen also supported this view and suggested that the Law School be separated from the Darulfünun and established somewhere near the Sublime Porte. Since Sava Paşa also objected saying that the separation of the Law School would be inconvenient, this school was moved to Beyoğlu together with the Darulfünun. Ali Suavi, who was appointed director of the Mekteb-i Sultani and the Darulfünun in 1877, observed that there were many educational and administrative irregularities in these schools. He informed the Ministry of Education and even the Sultan about this situation. The outbreak of the Ottoman-Russian War (the war of 1293/1876-77), the cautious attitude of Sultan Abdulhamid towards the minorities and the above report of Ali Suavi must have been some of the reasons for the interruption of the courses at the Darulfünun. On 16 October 1878, students were admitted to this school again and instruction began. But, two years later another Law School, attached to the Ministry of Justice, was opened in İstanbul, which indicates the preference to transfer the function of the education of law students from the Darulfünun to that Ministry.

The first students graduated from the Darulfünun-i Sultanî during the 1879-80 academic year. The number of graduates from the Law School was seven during the first year and six in the second year. They were assigned posts in the Ministry of Justice or in the courts of justice.

Due to the lack of information about the Law School of the Darulfünun-i Sultanî and other schools after 1881, presumably the activities of this institution ceased after this date.

School of Turuk-u Maabir

Turuk-u Maabir ("Ecole des Ponts et Chaussées") was the second school established within the Darulfünun-i Sultanî. At first, this school was opened under the name of Mülkiye Mühendis Mektebi (School of Civil Engineering) at the same time as the Law School in 1874. At the end of the first year, however, its name was changed into the School of *Turuk-u Maabir*. Students also studied mathematics and natural sciences in this school which aimed, rather than limiting the instruction to science only, to educate engineers who would render service in the construction of public works and the field of communications and transport in the vast imperial lands of the Ottoman State. The graduates of this school would be employed in the Ministry of Public Works.

Instruction in the School of *Turuk-u Maabir* started at the same time as the Law School during the 1874-75 academic year. The number of students who attended the former school and took the final examination at the end of the year was 26. This number rose to 43 at the end of the third year. During the 1877-78 academic year, instruction in the School of *Turuk-u Maabir* ceased, as was the case in the Law School, and the courses began one year later. The first students graduated from both schools in 1880. Although the exact number of the graduates in the first year is not clear, two engineers from this school were employed in the Ministry of Public Works. There were no graduates from the School of *Turuk-u Maabir* in 1881. Since there is no information about the activities of this school after 1881, courses must have been interrupted as in the Law School. Besides, civil engineering courses were taught separately in the Mülkiye Mühendis Mektebi which was opened by the state and was attached to the Ministry of Public Works in İstanbul in 1880.

School of Literature

There is rather limited information about the School of Literature. Contrary to the previous two schools, it was called "faculté" instead of "école" in the literature in French. The School of Literature, established with the aim of training teachers, depended on the same regulation as the Law School and the School of *Turuk-u Maabir*. However, it is not clear what kind of an education and instruction was in practice in the School of Literature. Courses in this school were obligatory for the first-year and second-year students in the Law School and the School of *Turuk-u Maabir*. Thus, the School of Literature was considered as a school which was different from the above-mentioned two.

*

*   *

Instruction of law and engineering continued separately from the Darulfünun in the Law School and Mülkiye Mühendis Mektebi, respectively, after the activities of the Darulfünun-i Sultanî ended. The success of instruction in these schools was due to the organic ties which they established with the ministries concerned.

The higher institutes of Ottoman education, which comprised a few departments and were established within a complex, did not achieve the desired success, while they approached the objective when established separately. Besides other reasons, one may state that the attempts to establish the Darulfünun during the 55 years between 1845-1900 were unsuccessful because these attempts of establishing institutions of higher education were not conceived in harmony with Ottoman social and state structure.

Development of the concept of Darulfün

This part of the paper examines how the idea of the Darulfünun developed. First of all, it will discuss where the need to establish a Darulfünun

VII

836

arose from, how the concept matured as a model, the administrative and financial changes it underwent and how the level of scholarship developed in this institution until the beginning of the twentieth century.

As stated in the first part of this paper, the term "Dârulfünun" was first mentioned in Ottoman literature at the *Meclis-i Muvakkat* (Temporary Council for Education) which convened in 1845. Such a term did not exist in Ottoman educational life until that day and its origin is not known. Most probably, however, it was put forward for the first time by the members of the *Meclis-i Muvakkat*. There is no source indicating that towards mid-nineteenth century, Western universities were taken as a model in establishing an institution for teaching modern sciences of Western origin. However, Engineer Ferik Emin Mehmed Paşa[2], who studied in France for long years and was a member of the *Meclis-i Muvakkat* and later of the *Daimî Maarif Meclisi* must have been influential in this development.

Schools of higher education such as the medical school, schoool of engineering and war college gave special training in these fields for military purposes. The graduates of these institutions were directly employed in the army, while the medreses mostly met the religious and legal needs of society. The Darulfünun must have arisen from the need to establish a civilian school of higher education besides the medrese. The official authorities deemed it necessary to establish such a school mainly for educating the people. Sultan Abdülmecid himself suggested that the success of the Ottoman reforms depended on the education of people. Thus, the state was compelled to reform education by improving the present schools and opening new ones.

Within the framework of this reform, an institution of higher education by the name of "Darulfünun", which did not exist until that day, was almost invented in a short period. This institution differed in structure from the classical Ottoman system and Western schools of higher education. The fact that Emin Mehmed Paşa, member of the *Meclis-i Muvakkat* was educated in Europe is the only plausible connection of the Darulfünun with the West.

[2] Ferik Emin Mehmed Paşa (d. 1851) Engineer. He was trained in the Tophane and was sent to Paris in 1833, where he studied language for two years. He became adjutant major upon his return to İstanbul. He was sent to Paris again in order to complete his education and superintend the Ottoman students in Paris where he stayed until 1840 (A. Şişman, 1983: p. 8).

In 1863, instruction in the Darulfünun started with public lectures organized in a few rooms of the big building, still under construction, allocated to this institution. For three years, these public lectures were given on natural sciences such as physics and chemistry as well as social sciences such as history and geography for the purpose of enlightening the people with some basic information. When the Darulfünun was established, some decisions appeared about this institution in the reports of the council. These are generally not sufficient for understanding all aspects of this institution. Therefore, it is not clear what the statesmen meant by the Darulfünun and the characteristics of the institution which they aimed to establish. It was stated, however, that the main objective of this institution was to train "well-informed servants", i.e. knowledgeable civil servants.

Attention was paid to realize the scholarly and administrative autonomy of the Darulfünun during the first attempt to establish this institution in 1869. By the official memorandum of the Minister of Education and an imperial rescript, it was decreed that this institution would be administered by a *nazır* (similar to the post of a president of a university). The regulation of the Darulfünun specified a period of instruction of three years and research programs which included the preparation of graduation and professorship theses. It also contained a curriculum which included courses on modern Western science as well as classical Ottoman culture with the aim of reconciling Islam and the West. The activities of the Darulfünun continued, until 1872, but its regulation was not applied thoroughly. A new attempt to establish the Darulfünun appeared in 1874.

During the first attempt which took place in 1845, a big building was planned for this institution after the model of Western universities, besides a library and a laboratory. It was not clear, however, what the administrators meant by the Darulfünun and what kind of an institution they planned to establish. In the second attempt, the first Ottoman Regulation of Public Instruction was made effective in 1869 and the French model was chosen for the Darulfünun which ended the discussions on this matter. Due to financial constraints, however, this regulation was not applied completely. Therefore, this attempt, too, failed in a few years.

The Darulfünun-ı Sultanî was conceived within the framework of a statute which was different from the one planned in the first two attempts. It was established in the building of Lycée Impérial Ottoman de Galata-Séraï in 1874, based on the infrastructure of an institution of secondary education. Darulfünun-i Sultanî which had three higher institutes emerged mainly as an institution of higher education for training the graduates of secondary schools. Although it was not specified in which fields the graduates of the Darulfünun-ı Osmanî would be employed, this was clearly stated in the case of the Darulfünun-ı Sultanî. Thus, the two previous Darulfünuns did not aim to give special training for professionals who would then be employed in particular fields. The objective of the third attempt was to realize specialization in particular fields according to the needs of the state at that time.

The financial state of the Darulfünun

In order to understand the legal statute of the Darulfünun and the reasons for its failure, one must study its financial aspects and sources of income.

During the first attempt, it was not possible to reserve a source for the Darulfünun from the present budget since it was a newly formed institution. Facing the necessity for creating new sources of income for new expenditures, Ottoman administrators turned to temporary measures and tried to use the small amount of money reserved from the sources for implementing educational reform. By a new practice, the use of bank notes in dealings and business became compulsory. The state tried to obtain the necessary means for the construction of the building of the Darulfünun by the revenue which accrued from issuing of bank notes. Among the other financial sources allocated to the Darulfünun were the new income from declaring a tender for the construction of the İzmir wharf for a higher amount and the revenue from the surplus customs tariffs as agreed in the trade agreement between Russia and the Ottoman State. When the Darulfünun was opened in 1863, the teachers who gave courses were not paid extra salaries. Therefore, the accumulated sum of money was spent only for the construction of the Darulfünun building.

During the second attempt, the financial sources of Darulfünun were determined and the necessary income was supplied through a *sandık* (cash-chest) where money would be collected. The main sources of income of the Darulfünun were, first of all, students' fees, followed by others such as donations by the state, the support by the waqfs and various grants. The fact that the Darulfünun had a financial regulation may be seen as a positive development. But, in a vicious circle, the

number of students, hence the revenues of the Darulfünun declined when
the students' fees were kept high in order to increase income. The fact
that the courses were open to the public and the majority of attendants
followed them without paying any fees was another negative influence.

The Darulfünun was regarded within the statute of an official state
institution because its building was founded by the state and the salaries
of the staff were augmented most of whom were already teaching in
other schools. Although there was no budget allocation for this institution,
it was dependent on the state. Its failure was entirely due to financial
constraints, lack of teachers and books, the small number of well-educated students and the level of scholarship which was rather weak.

The establishment of the Darulfünun-i Sultanî was planned to take
place in 1874 without causing any financial burden on the state. In the
beginning, it had sufficient means under the patronage of Lycée Impérial Ottoman de Galata-Séraï (Mekteb-i Soultanî). In the following years,
however, the Darulfünun-i Sultanî was faced with financial difficulties
mainly due to the fact that the number of students who received scholarships gradually increased vis à vis the students who paid tuition in the
Mekteb-i Sultanî which largely depended on the income resulting from
the fees of room and board. Consequently, the Darulfünun-i Sultanî had
to depend on state support and the patronage of the Sultan.

Ali Suavi was the director of the Mekteb-i Sultanî during the year
1877. His report about the students who were allocated scholarships
clarifies the situation. He submitted a petition to Sultan Abdülhamid II
on 25 August 1877 where he explained that 539 students studied in this
school, including 162 Muslims and 377 non-Muslims. Only 26 students
paid full tuition (25-30 liras), however, and more than half of them were
Muslims. He also presented in his petition that if the number of students
who benefitted from scholarships was reduced, the income of the school
would increase and a sizeable annual allowance reserved for this school
would be abolished gradually. The Darulfünun-i Sultanî was taken under
the patronage of the state after 1878. Its activities ended in 1881 mainly
because of political reasons and the disputes among statesmen rather
than financial difficulties (I. Doğan, 1991: 329-341).

The development of the Darulfünun, the problem of language, translation activities, lack of students and the level of instruction are subjects
which may be examined separately in the field of history of education
and science. During the first attempt, the question of the language of
instruction in the Darulfünun was not stated as a separate item of the

agenda. Besides, the writing and translation of textbooks for the Darul-fünun was specified as the main duty of the Encümen-i Daniş in which the memberships were distributed to statesmen of high-rank as honorary grades.

Turkish was accepted as the language of instruction during the second attempt. Therefore, the problem of language would be solved merely through translation of the textbooks of the courses taught by teachers who did not know Turkish. A translation society was set up within the Darulfünun aiming at the preparation of textbooks, but, as observed in the case of the Encümen-i Daniş, it did not produce any works.

The problem of the language of instruction first came to light in the Darulfünun-i Sultanî. Due to the "lack of books in Turkish and teachers who would give the courses in Turkish, until the necessary conditions were fulfilled for giving the courses in this language", the discussion as to whether they would be given in Turkish or French came to an end with the decision that some courses would be held in Turkish and others in French. Thus, a translation committee was set up which would undertake the translation of four books in total, two for the Law School and two for the School of *Turuk-u Maabir*.

The Darulfünun must be evaluated within the framework of the Ottoman reception of modern science and education as a whole. In the pre and post-Tanzimat periods, the Ottomans had the chance of direct contacts with Western science. But, Ottoman educational and scientific institutions were not completely successful in introducing the new knowledge based on observation and research and could not establish an indigenous tradition of science. Evidently, the Darulfünun failed to fulfill its expected function of contributing to the formation of new Ottoman scientific understanding based on European scientific thinking.

There were three attempts to establish the Darulfünun during the fifty-five years from the first attempt until the beginning of the twentieth century. The instruction was actually carried out only for thirteen years and the courses were either too advanced for the attendants to understand or too elementary. Although essentially the curricula were well prepared for an institution of higher education, some great disorders occurred in practice. As stated above, the first attempt took place without any program. During the second attempt, though the regulation specified the number of courses in all departments of the Darulfünun as thirty-seven, only ten courses were taught in practice. This number was slightly higher for the Darulfünun-i Sultanî, but only some parts of its program were applied.

Some books were purchased from Europe for the library of the Darulfünun. But, scientific periodicals which contained the results of new research were omitted from the list which shows that the scientific thinking in the Darulfünun was not research-oriented. In regard to the Darulfünun-i Sultanî, its main purpose was to train a staff of professionals.

A petition submitted by the Grand Vizier Said Paşa during the reign of Sultan Abdülhamid II expressed the idea of establishing a Darulfünun on the same level as Western universities. In his petition dated 14 February 1895, Said Paşa stated that the Ottoman institutions of higher education, particularly the Darulfünuns were professional schools, but did not fulfill their real function. He also emphasized the need to establish a Darulfünun with five departments, which would operate as American and European universities, in order to train scholars (Said Paşa, 1328: pp. 572-580).

On the basis of the above experiences, the first modern Ottoman university with five departments was established under the name of Darulfünun-i Şahane on 31 August 1900, the twenty-fifth anniversary of Abdülhamid ll's accession to the throne. It was the foundation of the present Turkish universities. Thus, there were two examples, i.e. the Darulfünun-i Sultanî and the Darulfünun-i Şahane, established in the beginning of Sultan Abdülhamid ll's reign and on the twenty-fifth anniversary of the sultan's accession to the throne, respectively, which marked a failure and a success in the same period. Apparently, the Ottoman thinking as well as experience of education and science had developed to the point of maintaining such an institution.

The Darulfünun is a notable example of the Ottoman reception of Western science and educational system. The Ottomans established this institution in order to meet the needs of the day without completely developing and understanding the idea of the university. Hence, it developed and prospered to a certain level through trial and error method.

BIBLIOGRAPHY

DOĞAN (İ.), 1991, *Tanzimatın İki Ucu, Münif Paşa ve Ali Suavi*, İstanbul, 429 p.
Düstur, 1. tertib, vol. 2, Matbaa-yı Amire, 1289.
Düstur, 1. tertib. vol. 3, Matbaa-yı Amire, 1293.
İHSANOĞLU (E.), 1990, "Dârulfünûn Tarihçesine Giriş: İlk İki Teşebbüs", *T.T.K., Belleten*, n° 210, 699-739.

—, 1992, "Tanzimat Öncesi ve Tanzimat Dönemi Osmanlı Bilim ve Eğitim Anlayışı, in H.D. Yıldız ed., 150. *Yılında Tanzimat*, Türk Tarih Kurumu, Ankara, 335-395.

—, 1993, "Dârülfünun", *Türkiye Diyanet Vakfı İslâm Ansiklopedisi*, vol. 8, Türkiye Diyanet Vakfı, İstanbul, 521-525.

—, 1993, "Dârülfünûn Tarihçesine Giriş (II) Üçüncü Teşebbüs: Dârulfünûn-ı Sultani", T.T.K., *Belleten*, n° 218, 201-239.

—, 1994, "Darülfünun", *Dünden Bugüne İstanbul Ansiklopedisi*, vol. 2, Kültür Bakanlığı, Tarih Vakfı, İstanbul, 559-562.

—, 1994, "Dârulfünûn": Mefhum ve Müessese Olarak Sultan II. Abdülhamid Dönemine Kadar Gelişmesi", *Sultan II. Abdülhamid ve Devri Semineri* (27-29 Mayis 1992). Ayrı Basım (Offprint), Edebiyat Fakültesi Basımevi, İstanbul, 173-190.

SAİD PAŞA, 1328/1912, *Said Paşa'nın Hatıratı*, Sabah Matbaası, Dersaadet (İstanbul) 2 vol.

ŞİŞMAN (A.), 1983, *Tanzimat Döneminde Fransa'ya Gönderilen Osmanlı Öğrencileri*, unpublished Ph. D. diss., İstanbul, 168 p.

THE GROUP OF SCHOLARS KNOWN AS MEMBERS OF BEŞİKTAŞ CEMİYET-İ İLMİYYESİ (BEŞİKTAŞ LEARNED SOCIETY)

This paper deal with some of the cultural and educational activities in the Ottoman State during the period covering the last years of Sultan Selim III and the reign of Sultan Mahmud II.

Although there are a number of studies and researches on the rapid dissemination of the Western culture after the Tanzimat period, the culture, education and science in the life of the Ottomans prior to the Tanzimat period need serious investigation.

In the period under investigation, in addition to the old type of institutions of education, science, and culture, establishments which followed the Western education systems were founded

The teachers who were appointed to these establishments were medrese graduates but some of them advanced their education and learned European languages. They consciously played active roles in transferring the Western sciences to Turkey. For this reason, we can consider this period as an epoch of transition and transformation.

It was in this period also that the intelligentsia were getting to know Western sciences and culture through various channels, and the State tried to adopt and apply what these new movements contributed to different aspects of life.

In the classic Ottoman education system, the most important role was played by *medreses* and the palace school, *Enderun Mektebi*. However, education in Ottoman society was not confined to these formal institutions. Educational and cultural activities were also carried out in the mosques, *tekkes*, libraries and in the mansions of the Vezirs and the rich[1]. These activities took place within this tradition and completed the formal education. At the beginning of the 19th century, scholars taught their

---

[1] Osman Nuri Ergin, *Türkiye Eğitim Tarihi*, Vol. II, İstanbul 1977 p. 376.

subjects of speciality to those students who applied of their own accord, without expecting any benefits in return. An example of this form of education is Kethüdazade Mehmet Arif Efendi, a well-known man of learning who lived during the Nizam-ı Cedit, the pre-Tanzimat and the Tanzimat periods. He is one of the students who received and later gave this type of lessons.

Kethüdazade preferred to teach in his own mansion instead of assuming an official post in the formal schools, and succeeded in forming students who later occupied important offices[2]. Besides following this tradition which was directed to Muslim students, and reflected his unique personality, wordly outlook and tolerance, he also taught Islam to students of other nationalities and religions[3]. The following example is worthy of on attention because it illustrates this point.

Emin Efendi, one of Kethüdazade's students, recalls their first day in class. While attending the lessons given by a scholar in Beşiktaş, Emin Efendi and a friend heard about Kethüdazade's courses and went to his mansion for lessons.

> One morning we woke up early, found out his address and eventually went to his mansion in Uzuncaova, Beşiktaş. We knocked at the door gently, for a scholar's door should not be knocked hard; he heard the knock looked through the corner window and asked what we wanted. When we asked for the teacher, he asked why we wanted him. Upon learning our wish to take lessons from him, he said, "Well, wait, I'll come". He himself came down to open the outside door for there was no rope to pull the latch. Then the students who entered the house went upstairs and sat in a room. Emin Efendi describes the room and the way the students sat: We entered the room, the teacher was seated on a cushion covered with *moscov* material and we sat on the floor opposite him. There were some books piled up on one side of his seat". When asked what he wanted to study, Emin Efendi

---

[2] *Menakıb-ı Kethüdazade Mehmet Arif Efendi*, compiled by Emin Efendi, 2nd ed. Istanbul 1305, p. 218. The first edition of *Menakıb-ı Kethüdazade* 158 pages and was compiled by Emin Efendi, one of Kethüdazade Mehmet Arif Efendi's student, as and was printed in 1294/1877. The 2nd enlarged edition was printed in 1305/1887-88 in 345 pages.

[3] *Menakıb*, 2nd ed. p. 282.

answered "pend". "He asked if we had books with us. We said 'yes' and took them out of our breast pockets and the course started[4].

As described above, because there were no public schools and high schools in those days, students used to go to the mansions of the scholars to learn the subjects not studied in the mosques and medreses[5], The educational activities practiced in the mosques and mansions evidently played an important role.

Within this traditional education, there were two different institutions, the mosques and the mansions. In the mosques, linguistics and theological sciences were taught to supplement the curricula of the medreses. These courses were usually concentrated in and around the Fatih quarter of Istanbul[6]. However, in the mansions the following subjects known as *Ulum-ı Cüziyye* were taught; mathematics, algebra, geometry, astronomy, philosophy, literature and history. Although we do not have sufficient information, we have reason to believe that these latter activities took place in the mansions of the scholars who settled in the Beşiktaş region[7].

The main objective of the students who took these courses instead of the official education, was to be graduate from the medrese earlier, to improve their education in subjects like mathematics and philosophy, to gain more knowledge and culture and thus to secure better positions in life[8].

After touching briefly on the tradition of mansions that constituted an important part of the education system at the beginning of the 19th century, we now turn to the main topic of this paper namely, the study of an intellectual circle consisting of a group of open-minded and far-sighted eminent scholars in Beşiktaş.

Since the time of Kanunî Sultan Süleyman, Beşiktaş was in great demand as a residential area, and some of the important statesmen and scholars had settled there[9]. In the notes of a traveller, written towards the end of the 18th century, it is stated that particulary on the seaside in

[4] *Menakıb,* 2nd ed. p. 190.
[5] *Menakıb,* 2nd ed. p. 218.
[6] Ahmed Cevdet Paşa, *Tezakir,* prep. by Cavid Baysun, part 40, Ankara 1967, pp. 9-11.
[7] *Menakıb,* 2nd ed. p. 218.
[8] *Tezakir,* Part 40, pp. 9-11.
[9] M.Tayyib Gökbilgin, "Boğaziçi", *İslam Ansiklopedisi,* Vol.II, Istanbul 1970, p. 676.

Beşiktaş, there were residences (yalı) mainly belonging to scholars who occupied the posts of *müftü* and *kazasker*[10]. There were many non-Muslim subjects living in this area and the scholars who lived there became aware of the concepts that were developed in the West[11], most probably through cultural exchange,

During the reign of Sultan Mahmud II, four eminent scholars came together and formed an intellectual group which we shall mention hereafter as the "Beşiktaş Group". Because these scholars had the opportunity to become acquainted with both classic Islamic culture and the new Western culture, this group played an important role in this period.

There are three main sources of information about this group. The first is *Menakıb-ı Kethüdazade* (Istanbul, 1294-1305/1877-1887) compiled by Emin Efendi[12]. The second is *Tarih-i Cevdet*[13] and the third is *Tarih-i Lütfi*[14]. In addition to these, a document was recently found explaining how the group Melekpaşazade Abdülkadir, as well as one of the members of the Beşiktaş group, were regarded by the State in 1826[15].

In the *Menakıb*, a student of Kethüdazade's called this group of ulema the Ortaköy Mezhebi, The Ortaköy Sect hinting that those who had seaside mansions in Ortaköy, such as Ferruh İsmail Efendi, Melekpaşazade Kazasker Kadri Bey, Kethüdazade and the others who were their constant visitors had leanings towards worldly enjoyments, for these mansions of Ferruh İsmail Efendi and Melekpaşazade Kadri Bey were the meeting place of the men of understanding and a centre for men of eloquence[16].

However, Ahmed Cevdet Paşa calls this group formed in Beşiktaş by Melekpaşazade Ferruh Efendi and Şanizade Ataullah Efendi, the Cemiyet-i İlmiye 'Learned Society'. The members of this society undertook to

[10] "Beşiktaş", *Istanbul Ansiklopedisi*, Vol.V, p. 2564.
[11] *Menakıb*, 2nd ed. p. 130.
[12] *Menakıb*, 2nd. ed. p. 224.
[13] Ahmet Cevdet Paşa's *Tarih-i Cevdet* is in 12 volumes. Thi first volume of the first edition was printed in 1271/1856, and the last volume in 1301/1883-84.
[14] The first volume of Ahmet Lütfi Efendi's work *Tarih-i Lütfi* was printed in 1290/1873.
[15] By Şeyhülislam Kadızade Mehmet Tahir Efendi's initialy dated 4 *Zilhicce* 1241/10 July 1826 is draft decree is dated *Evail-i Zilhicce* 1241/7-16 July 1826: Başbakanlık Osmanlı Arşivi, Cevdet Adliye, No.2002, 2002/1.
[16] *Menakıb*, 2nd ed. p. 224

teach courses to enthusiastic students. Kethüdazade also joined this group and taught philosophy and literature on certain days of the week [17].

Ahmet Lütfi Efendi, the members who owned seaside mansions in Ortaköy and their togetherness "in fact deserve to be called *Cemiyet-i İlmiye* (Learned Society)" [18].

The historians of the next two generations referring to Cevdet Paşa as a source, regarded this group as a Society of scholars. This viewpoint which is shared by present day historians, appears as occasionally fabrication. We come across some comments which present it as an academy of scholars that never existed in Turkey. These historians consider the activities of this group as the first stage in the formation of formal institutions such as the *Encümen-i Daniş* and *Cemiyet-i İlmiye-i Osmaniye* that were established much later, thus creating an institution which never existed in the cultural life of Istanbul in that period.

The most reasonable explanation given so far is that the activities of the group were that of a "salon" [19].

In fact, the activities of this group were not completely different from the education given in mansions during the same period. Nevertheless, their collaboration made the group appear as a kind of corporate body that had not existed in this field until then.

In order to have some information about the joint venture, it will be useful to touch upon the biographies of each member.

Ismail Ferruh Efendi originally came from Crimea and was appointed ambassador to London in 1797 with the honorary rank of Suvari Mukabelecisi. He returned to Istanbul in 1800. Later he became Director of Financial Administration of the third division of the Ottoman Empire and devoted his time to scholarly activities [20]. Ferruh Efendi made the first printed Turkish translation of the Qur'an entitled *Tefsir-i Mevakıb* [21]. He died in 1840 and was buried in Ortaköy. Ferruh Efendi had a seaside

---

[17] *Tarih-i Cevdet*, Vol. XII, p. 212

[18] Ahmet Lütfi Efendi, *Tarih-i Lütfi*, Vol.I, Dersaadet, 1290, pp. 168-69

[19] Şerif Mardin, *The Genesis of Young Ottoman Political Thought*, Princeton 1962, p. 231.

[20] Ercümend Kuran, *Avrupa'da Osmanlı İkamet Elçiliklerinin Kuruluşu ve İlk Elçilerin Siyasî Faaliyetleri 1793 - 1831*, Ankara 1968, p. 41

[21] *Sicill-i Osmanî*. Vol. IV, p. 14; *Menakıb*, 2nd ed., p. 38, 144.; E.Kuran, op. cit., p. 39 and p. 41; Carter V. Findley, *Bureaucratic Reform in the Ottoman Empire (The Sublime Porte 1789-1922)*. Princeton 1980, pp. 130, 132.

mansion in Ortaköy but the date of purchase is not known. The four available registry books of Bostancıbaşı do not contain a record of this mansion belween 1802 and 1815.

Melekpaşa 3 ade Abdülkadir Bey was the son of Melek Mehmed Paşa. They owned a seaside mansion in Ortaköy in 1795. He served as a *müderris, kadı* and *kazasker* and then became a member of the Supreme Court of Justice. According to the registry books of Bostancıbaşı of 1815, he had two seaside mansions in Ortaköy. One of these mansions was quite close to that of Şanizade Ataullah Efendi. Melekpaşazade died on 27 October 1846[22].

Şanizade Ataullah Efendi was born in Istanbul and lived in Ortaköy. He died in exile in Tire, 1826[23]. He was a great man of science, and in a way he was an encyclopaedist of his generation. He served as an annalist and also wrote a number of books on medicine and mathematics[24].

Kethüdazade Mehmet Arif Efendi was one of the most renowned scholars of his time. Far-sighted, and well-educated, he lived in Uzuncaova in Beşiktaş and died in 1847[25].

From this brief information it is clear that the group first started to meet in 1815 and continued doing so until 1826 when the members were sent into exile after having been accused of being Bektaşi. However, in a letter to his son, written apparently between 1835 and 1837, Kethüdazade says that they were gathering and having conversations in Ferruh Efendi's mansion[26].

[22] Mehmed Süreyya, *Sicill-i Osmanî*, Vol.III, p.350, Vol.IV, p. 510

[23] İbnül Emin Mahmut Kemâl İnal, *Son asır Türk şairleri*, Vol. I., İstanbul 1930, p. 111; Feridun Nafız Uzluk, "Şanizade Mehmed Ataullah Efendi", *14 Mart 1951 Tıb Bayramı dolayısıyla Monografi*, pp. 3, 4; Şanizade Mehmed Ataullah, *Tarih-i Şanizade*, Vol. IV., İstanbul (1292), p. 34; F. Babinger, *Osmanlı Tarih Yazarları ve Eserleri*, Trans. Coşkun Üçok, Ankara 1982, p. 376.

[24] Kethüdazade resided at his mansion in Uzuncaova for a long time. Upon the query of Sultan Mahmud II, "Shall I renovate his house or shall I buy him another one", he replied that he was not overfond of his present house and would prefer a new one. Thus the mansion near the Arab Quay was presented to him. *Menakıb*, 2nd ed. p. 6. When Kethüdazade married, his bride came to his father's house, *Menakıb*, 2nd ed. p. 86.

[25] İn the letter that he sent to his son in Baghdad Kethüdazade says, *Nadire-i devran İsmail Ferruh Efendi hazretlerinin Sahilhane-i saadetlerinde idare-i giruse muhabbet ve güft ve şunud ve lezzet yafte-i tabakçe-i serair ve şühud iken Bağdat Heştabad'ta oğlum Beyefendiye terkim-i rakime-i muhabbet edeceğim dedim. Buyurdular ki bende dahi Yaver Beğ ve Derviş Paşazade Beğe selâm yaz.* Kethüdazade Mehmet Arif Efendi, *Eşar-ı Kethüdazade*, compiled by Ragıb Hulusi, Dersaadet 1271/1855, p. 3.

[26] Bernard Lewis, "The Impact of the French Revolution on Turkey", *Journal of World History*, Vol. 1/2, 1953, p. 113.

Looking at the way the members of this group were brought up, the type of education they received and the scientific studies they conducted, it is obvious that they shared some common characteristics. It should be kept in mind that these four people were in touch with Europeans and in their form influenced those who came into contact with them. They were well-versed in Islamic culture, at the same time receiving education in philosophy and mathematics, they claimed to be neglected in the *medreses* of the time. Moreover, they were open to the culture and novelties of the West. These personalities were able to combine, as much as possible, Islam and the new sciences and culture which they acquired. It is certain that the newly established educational institutions and schools of engineering somehow had an effect in the formation of this group. It may also be noted that they were, to a certain extent, influenced by the non-muslim subjects of the Ottoman State who were educated in Europe, particularly in Padua, Italy[27].

Kethüdazade's own opinions on the matter provide us with new ideas about the effect of the West on this group. To put it briefly, the behaviour and the ideas of Kethüdazade demonstrate how Islam and Western sciences could be reconciled.

Clarification of this point will be useful. Kethüdazade was open-minded towards liberal ideas, modernization and progress, to the point of making friends with people from every nationality and religion[28]. He even went to church[29] and took part in balls organized by the English, in his robe, and *kavuk* (turban). He knew a lot about life in Europe. For example, he pointed out that the dog carts which the military school cadets wanted to use did not resemble the originals because the dogs used in Europe were specially bred and trained for this purpose. This idea was dropped after his intervention[30].

While appreciating the Westerners for some of their practices, Kethüdazade made also comparisons and suggestions about the limit to the relationship with the West and about what should be observed in these relations[31]. While he disapproved imitating European styles of dress,

[27] *Menakıb,* 2nd ed. pp. 41, 148.
[28] *Menakıb,* 2nd ed. p. 148.
[29] He writes that Europeans were eating ice cream, *Menakıb,* 2nd ed. pp. 119, 185.
[30] *Menakıb,* 2nd ed. p. 296.
[31] *Menakıb,* 2nd ed. p. 130.

furniture and building, he was in favour of adopting arms manufacturing industry and other useful techniques and sciences[32]. About teaching foreign languages to students, he was of the opinion that they should first be taught their religious beliefs and cultural values, because otherwise, under the influence of the language they were learning, they might adopt another identity. Differing from his contemporaries and the statesmen of later generations and men of thought, Kethüdazade put forward his ideas about Western sciences and technology. His idea was that in addition to teaching these sciences and practising them, new discoveries should be aimed at, which he called "experiment", and that just as in Europe, the State must support people who conducted these studies[33]. Kethüdazade approved and even supported the new system established by Sultan Mahmud II[34].

The Beşiktaş group, though rudimental, appears to have been an organization as is clear by the following:

1. Many scholars who came together and gave lessons were generally farsighted intellectuals, familiar with natural and mathematical sciences and acquainted with Western ideas.

2. The members gathered in certain places and gave lessons free of charge; their conversations on science, literature and politics were free.

3. The regular expenses of the group were met by the members in a friendly manner, and each "partaker"[35] contributed even when he was away from Istanbul[36].

4. Most important of all, in 1826, when the Janissary forces were being disbanded and the Bektaşi were sent into exile, all the members of the group were accused of being Bektaşi and were also exiled[37].

---

[32] *Ibid.*

[33] Eşar-ı Kethüdazade, p. 3.

[34] Ebüzziya Tevfik Bey states that *"Arifane lafzının galat olarak kullanıldığı, aslının hârifane-arkadaşça-olması gerekir"*. *Mecmua-ı Ebüzziya*, Vol.IV, part 39, 1 Safer 1302, p. 1238.

[35] *Tarih-i Cevdet*, Vol. XII, p. 213.

[36] *Tarih-i Cevdet*, Vol. XI, p. 212.

[37] Arif Efendi (Damadzade) was Kazasker Murat Efendi's son and Şeyhülislam Mekkizade's son-in-law. After he became a *müderris*, he received honorary degrees in Yenişehir in 1214/1799, in Bursa in 1223/1808, received the rank of Molla of Mekka in 1231/1816 and an honorary degree in İstanbul in 1236/1821. Lastly in 1241, when he received the honorary degree of Anadolu, he was accused of being a Bektaşi and was sent to exile in Güzelhisar, *S.O.*, Vol. III, p. 271, 350.

From Cevdet Paşa's work and other sources as well, it is clear that these people in fact were not Bektaşi but drew attention and were accused of being Bektaşi[38] because of their liberal ideas and the attitude of envious people. It may be said that these men of learning were unjustly treated by Sultan Mahmud II who was influenced by some bigoted scholars benefitting from the turbulence and disorder in Istanbul. The situation which led the Sultan to take this decision was prompted by the disbandment of the Janissary forces whose members were exiled. It is also possible that the State took such an attitude towards the members of this group because they were suspected of being dissidents[39].

In conclusion, it may be said that during the period under examination, apart from the *medrese* and the (Palace School) *Enderun Mektebi*, teaching and other cultural activities conducted in some mansions became quite influential. The intellectual group founded in Beşiktaş by four men of learning was quite effective in this period, possibly because they had the chance to be familiar with both Islamic and the new Western culture. However, it should be kept in mind that this period is difficult to understand because this group wa known as the Learned Society *Cemiyet-i İlmiye* and there were some unfounded ideas about them.

This group of scholars in Beşiktaş believed that it were spread possible to reconcile Islam and the West in non-religious matters, and that, as to the religious matters, an understanding based on tolerance could be established; fields of science, knowledge and technology, which made life easier for human beings, might be accepted without any conflict.

One of the reforms introduced by Mahmud II was based on his preference to establish new Western-style educational institutions, instead of reforming the system that had been practised within the Ottoman tradition until then. Thus, two different systems appeared which were based on two opposite philosophies and world views. At first, these systems were in accordance with one another, but later on they became contradictory.

The concept presented by scholars of the Beşiktaş Learned Society in the pre-Tanzimat period, trie to reconcile Islamic-based Ottoman cul-

---

[38] *Tarih-i Cevdet*, Vol. XII, pp. 212, 213; *Menakıb*, 2nd ed., pp. 34, 5; F.N. Uzluk, *op. cit.*, p. 3, Mehmed Cevad, *Maarif-i Umûmiye Nezareti Tarihçe-i Teşkilat ve Icraatı*, Vol. I, İstanbul 1338, p. 69.
[39] *Menakıb*, 2nd ed. pp. 143-144; Ş. Mardin, *op. cit.*, p. 230.

ture and the Western civilisation/science in the belief that modern science did not contradict Islam, it should be treated as a new ideological movement, entirely separated from the two contradicting ideological movements, one of which accepted the West exactly as it was while the other rejected it completely. One of the first representatives of this movement was Ismail Hakkı of Erzurum (1730/1780).

If the reconciling philosophies of these scholars who were accused by their opponents of "not paying attention to their religious duties, but mixing and having familiar intercourse with atheists and heretics and thus having inclinations towards this kind of vicious sects" had received support in that period, perhaps another type of intellectual, more influential and more powerful in the society, would have come into existence. Instead, in 1848, one or two generations later, a generation arose which rejected the fundamental elements of the culture to which they belonged and which became alienated from their environment. Their example is the student from the School of Medicine who said, "Eh! Monsieur, ce n'est pas au Galata Serai qu'il faut venir chercher la religion" (Oh! Sir, you do not look for religion in Galata Saray)[40].

---

[40] Charles MacFarlane, *Turkey and its Destiny*, London 1850, p. 268.

GENESIS OF LEARNED SOCIETIES AND PROFESSIONAL ASSOCIATIONS IN OTTOMAN TURKEY

Dedicated to the esteemed memory of Aydın Sayılı,
pioneer of research on the history of
scientific institutions in Islam

The Ottomans founded their civilization on the basis of the Islamic cultural heritage and within the Islamic scientific tradition, and brought this culture and science to a climax in the 16th century. In the meantime, despite their feeling of superiority over Europe, they followed the technical developments and geographical discoveries which occurred in the West, with a selective attitude. However, their feeling of absolute superiority and autarky, which prevented them from realizing the importance of the newly developing intellectual and scientific trends of the Renaissance and the Scientific Revolution, did not continue in the seventeenth century.[1] The first indications of their changing attitude can be traced in the works of the famous Turkish bibliographer and scholar Kâtip Çelebi (1609-1657) who gave information on European scientific and artistic institutions in his work titled *Levamiü n-nur* translated from Latin. In the same work, he also described the high-level cultural milieu that had developed in Europe. In addition to this information which he derived from Mercator, Kâtip Çelebi gave the first indications that the Ottomans were backward in sciences and that Europeans had achieved great progress[2].

---

[1] E. İhsanoğlu, "Ottomans and European science", *Science and Empires*, ed. P. Petitjean et al., Dordrecht: Kluwer Academic Publishers, 1992, pp.37-48; "Introduction of Western science to the Ottoman world: A case study of modern astronomy (1660-1860)", *Transfer of Modern Science and Technology to the Muslim World*, ed. E. İhsanoğlu, IRCICA, İstanbul 1992, pp. 67-120; "Some remarks on Ottoman science and its relation with European science up to the end of 18th century", *Journal of the Japan-Netherlands Institute*, Vol. 3, Tokyo 1991, pp. 45-73.

[2] O. Şaik Gökyay, *Katip Çelebi'den seçmeler*, Istanbul 1968, p. 180.

162

The present article reviews the history of establishment of learned societies and professional associations in Ottoman Turkey.[3] By taking into account the contacts and interactions which took place between Ottomans and Europeans, it examines the processes whereby modern science and scholarly activities were institutionalized in these two distinct worlds, which had different cultural, social, economic, and administrative structures, and which were ruled by different legal systems, one based on Islamic law and the other, on Roman law. While it deals with the Ottoman learned societies and professional associations to determine their impact on Ottoman scientific and cultural life, their features as pioneers, and their successes and failures, the study also takes into consideration the increasingly complex structure of modern societies and the dependence of these institutions upon specialized knowledge in various fields of inquiry. The sequence of events under review illustrates the birth of institutions in response to new needs, and the dependence of their development on the conditions connected with their establishment as well as on the existence of an appropriate milieu to encourage their continuation. The distinction between learned societies and professional associations is made in order to differentiate the societies which were formed by people from all circles and all occupations for the purpose of encouraging science and culture, from those that people of one profession formed in order to encourage professional solidarity or scientific studies in that particular field. Manufacturers' and merchants' guilds established for the purpose of meeting common interests, as well as political societies and parties, are beyond the scope of this study. However, as an exception, the paper does cover the *Cemiyet-i Tedrisiye-i İslamiye,* later known under the name of *Daruşşafaka* society, which continues to function until today, and whose statute did not correspond exactly to any one of the categories mentioned previously. On the other hand, the learned societies and professional associations are taken up in chronological order according to their dates of

---

[3] A general picture of the movement to institutionalize science in the Ottoman Empire was drawn at the symposium on "Ottoman Scientific and Professional Associations" joinly organized by the Faculty of Letters of Istanbul University and the Research Centre for Islamic History, Art and Culture (IRCICA), in Istanbul, 1987. The papers of this symposium were published the same year, under the title *Osmanlı ilmî ve meslekî cemiyetleri*, I. Millî Türk Bilim Tarihi Sempozyumu Bildirileri, 3-5 Nisan 1987 (papers presented to the First National Symposium on the History of Sciences in Turkey, 3-5 April 1987), edited by Ekmeleddin İhsanoğlu (Faculty of Letters Publishing House, Istanbul 1987, 264 p.), hereafter referred to as *OIMC*. Review articles of this book were written by Carl Max Kortepeter in 1991 (Karl Max Kortepeter, *The Ottoman Turks: Nomad Kingdom to World Empire*, The ISIS Press, Istanbul 1991, review article with an outline, p. 291), and Carter Vaughn Findley in 1992 (*International Journal of Middle East Studies*, Vol. 24, Feb. 1992, pp. 140-141).

establishment, instead of being reviewed separately. This allows us to describe the environment which surrounded these institutions, as well as their mutual influences in different periods of history.

SOME LEGAL AND CULTURAL ASPECTS OF INSTITUTIONALISATION

Corporate bodies such as waqfs, guilds, mutual assistance funds, and religious orders, which brought together persons or properties, did exist in practice in the Muslim world. But there were no bodies uniting real persons around specific objectives, in a kind of legal organization that we would call "society". Although Islamic law had no special provision on juridical personality, it did not prevent the establishment of corporate bodies such as "associations" or "societies" as we would call them today. There was nothing contrary to the establishment of such bodies in the laws issued by Ottoman Sultans either. In 19th-century Ottoman Turkey, such juridical persons as those formed by people having common objectives were called "*cemiyet*".[4] However, legal arrangements providing for the establishment of "societies" with juridical personality in the real sense of the term were activated within the framework of the movement to establish "societies" which arose practically under western influence and developed with the modernization of Ottoman society. It seems that at least for the time being, it is not possible to explain fully the legal foundations of the movement to establish learned societies and professional associations which appeared in the second half of 19th century. However, it can be stated with certainty that the attempt followed the European example.

It can be observed that in both Europe and the Ottoman world, movements to institutionalize science and learning started in the aftermath of some scientific and cultural reforms. Following the academies of the Renaissance which had humanistic and literary purposes, institutions conducting scholarly and scientific activities with socially recognized roles started to appear in France and England in the 1660's. Hence, the *Académie Française,* which specialized in linguistics, and the *Académie des Sciences de Paris,* were established in 1635 and 1666 respectively, both on the initiative of the

---

4 In addition to being used in its literal sense to designate any association of people, this word was also used with a negative connotation in the literature of Ottoman classical history, labelling the trouble-making associations formed by rebellious subjects as "*fesat cemiyetleri*". Another argument about associations is that one may question whether the state would allow the establishment of bodies other than itself organized in a similar way, "with a head, a secretary, an accountant, administrators and staff"; Hüseyin Hatemi, *Medeni hukuk tüzel kişileri*, İstanbul 1979, pp. 165-184.

state, while *The Royal Society of London* was established in 1660 by individuals. In Ottoman cultural life, the first attempts to institutionalize learning
and science in a way different from the traditional Islamic institutions took
place on the initiative of the state, in both the classical period and the period
of modernization.

The academies established in Europe were first mentioned in Ottoman
literature as early as the first half of the 17th century. Kâtip Çelebi, in his
work titled *Keşfü z-zünun,* which is one of the best known works of Islamic
bibliography, refers for the first time in Ottoman literature, and perhaps in
Islamic literature in general, to the Renaissance academies, expressing his
own critical views. While he enumerates the books on philosophy under the
title *"Ilm al-hikmah",* he draws attention to the dearth of books on "natural,
spiritual and mathematical philosophy" in Islam and to their abundance in
Greek and Latin literature. He also points out that works written in these
fields by Islamic scholars were not numerous compared to those written by
scholars from "European academies". Kâtip Çelebi calls the members of the
academies *"Ehl-i Aqademya"* and describes the "Academy" as "a place
similar to the Medrese in our countries, where those concerned with sciences
assemble"[5]. However, he does not notice that these academies were established in addition to the European universities which were the counterparts
of Ottoman *"medreses",* in order to meet the needs that universities could
not fulfill and to pursue disciplines which were not welcomed in scholastic
academic circles. It is much later that similar needs started to be felt in the
Ottoman society.

THE UNIQUE ATTEMPT IN THE CLASSICAL PERIOD

The Ottomans' first systematic attempt to assemble scholars officially was
made for the purpose of carrying out the translation activity which started
under the orders of Grand Vizier Nevşehirli Damat İbrahim Paşa in 1720,
during the reign of Sultan Ahmet III, a period known in Ottoman history as
the "Tulip Age". This translation activity was undertaken by a group of
scholars, poets, writers, high-ranking civil servants, and Sufi sheikhs, which
can be described as an *ad hoc* group, whose number sometimes reached
thirty, and whose remunerations came from the state treasury[6]. Although this
group of scholars did not have any juridical status, their activities were successful, especially when taking into account the circumstances of the time,

5 Katip Çelebi, *Keşfü z-zünun,* Vol. I, Maarif Matb., n.p. 1941, p. 684.
6 Mehmet İpşirli, "Lâle devrinde teşkil edilen tercüme heyetine dair bazı gözlemler", *OIMC,*
   pp. 33-42.

and also the fact that they translated several works of history and philosophy into Turkish as well as some books of Aristotle from Greek into Arabic. These translations contributed to the revival of classical learning and cultural activities during the Tulip Age. This group of scholars represented the first attempt of the state to set up an Ottoman intellectual institution without an organic structure.

According to our source of information, the idea to establish this *ad hoc* group was formulated by Nevşehirli Damat İbrahim Paşa, who was motivated by his own interest in history, as well as his intention to submit translations of history books to Sultan Ahmet III. The same reference also suggests that there was a resemblance between the work produced by this group and the *Rasail* written by the *Ikhwan as-safa,* a group of scholars of the 10th century, because each work was translated by a different person, as in the case of the compilers of the *Rasail.*[7] Still, it appears from the work of this group that there was a conscientious effort to revive Ottoman culture by emulating an earlier example deriving from the same Islamic background. In the early years, each member of the group was assigned to translate a whole book, but starting with 1725, different chapters of the same book were distributed among the members. The group was composed of *kadıs* and professors awaiting appointment to their next post, dervishes, high-ranking officials, members of religious orders, and "learned people and scholars who worked, and were known, in the fields of history and literature". The sources do not explain the factors which had prompted the Grand Vizier to establish this group[8].

EARLY INSPIRATIONS FROM EUROPE

In the 18th century, Ottoman ambassadors sent to European capitals such as Vienna, Paris, and St. Petersburg made visits to some academies and institutions of learning. For example, Yirmisekiz Mehmet Çelebi, who headed an Embassy to France, returned to Istanbul in 1721 after having seen some new French institutions dealing with science, the botanic garden, the anatomy museum, the Observatory of Paris, and met some scholars and scientists. Ambassador Şehdi Osman Efendi visited the Russian Academy of Sciences

---

7 The source of information is *Çelebizade Asım Tarihi,* Istanbul 1282, pp. 358-361. The fact that Çelebizade Asım does not mention the *Bayt al-Hikmah* and *Dar al-Hikmah* institutions of classical Islamic history, where groups of scholars met and worked together, indicates that these institutions, revealed to us by recent studies were not known well enough to be taken as examples at that time. On the history of these institutions: George Makdisi, *The Rise of Colleges: Institutions of Learning in Islam and the West,* Edinburgh University Press, Edinburgh 1981.

8 İ. H. Uzunçarşılı, *Osmanlı tarihi,* Vol. IV, Part 1, 2nd ed., Ankara 1978, pp. 153-155.

in St. Petersburg, the Natural History Museum (which he called the *"acaib-hane"* — house of wonders), the library, and the printing house. In his travel report, he did not mention that the Academy of Sciences was a research-oriented institution. Having examined sixteen official travel reports (*"sefa-retnames"*) written by ambassadors, we found that these travels abroad helped to acquaint the Ottoman State and society with some new practices and techniques, such as the printing technique which sought to spread culture to a larger readership. However, they were not influential in leading to the establishment of organized scholarly activities or institutions of learning and research similar to the European academies and learned societies[9].

TRANSFORMATION OF THE OTTOMAN EDUCATIONAL TRADITION

During the classical Ottoman period, the medrese was the main educational institution in both religious and non-religious sciences (such as natural and mathematical sciences), where intellectuals were trained. In addition to the medrese, there was the *Enderun* School in the Ottoman Palace, which provided training to the elite ruling class. Physicians and astronomers graduated from the medrese. Moreover, a separate medrese of medicine called *Darüttıb* had been established within the Süleymaniye complex for the purpose of training physicians. Therefore, like other graduates of the medrese, physicians and astronomers too were considered members of the *"ilmiye"* (learned) class. Other professional groups were trained traditionally within the master-apprentice system in order to meet the needs of state and society. Architects and engineers were the best known examples of professionals trained in this system. In Ottoman society, the official education given in the medrese and the Enderun was supplemented by some educational and cultural activities carried out in mosques, dervish lodges, libraries, and residences of some statesmen and scholars. This practice had become a tradition which was in a way parallel and complementary to public education, and whereby well-off, authoritative ulema gave lessons to students who applied for it[10].

In the early period of modernization, new types of Ottoman intellectuals were educated in institutions such as the *Mühendishane* (Engineering

9 E. İhsanoğlu, "Tanzimat öncesi ve Tanzimat dönemi Osmanlı bilim ve eğitim anlayışı", *150. yılında Tanzimat*, ed. Hakkı Dursun Yıldız, TTK, Ankara 1992, p. 347.

10 For a brief account of scientific activities in the Ottoman period, *see* E. İhsanoğlu, "Ottoman science in the classical period and early contacts with European science and technology", *Transfer of Modern Science and ...*, pp. 1-48 and "Tanzimat öncesi ve Tanzimat dönemi ....", pp. 335-395; for non-official education, *see* "19. asrın başlarında — Tanzimat öncesi — kültür ve eğitim hayatı ve Beşiktaş Cemiyet-i İlmiyesi olarak bilinen ulema grubunun buradaki yeri", *OIMC*, pp. 44-45.

School) and the *Mekteb-i Tıbbiye* (Medical School) which were established under Western influence starting from the last decades of the 18th century. These new institutions were structurally different from the medreses, and modern sciences had great weight in their curricula. Their graduates were to make key contributions to the developments in Ottoman cultural and scientific life in the 19th century.

At the beginning of the 19th century, the scholars in some quarters of Istanbul started to teach various sciences in their residences, introducing another new practice into this atmosphere of novelties. The case of the ulema group which assembled in the Beşiktaş district is a vivid illustration of such activities. Learned men who were renowned particularly in mathematics, astronomy, and science had settled in this quarter where a group of high ranking ulema had formed an intellectual circle of free and progressive thinkers. Although this circle did not have the status of a society in the Western sense, it may still be considered the first attempt of individuals to establish societies in Ottoman Turkey.

The members of this group, Kethüdazade Mehmet Arif Efendi, İsmail Ferruh Efendi, Şanizade Ataullah Efendi and Melekpaşazade Abdülkadir Efendi, who were the four leading scholars of the period, had joined in a scholarly and intellectual togetherness. They assembled in their seaside residences between the Beşiktaş and Ortaköy districts of Istanbul, where they gave free lessons and held literary and political conversations. The customary expenses of the group were met according to each member's ability to pay; when a member was sent to the provinces on duty, he forwarded his share of payment regularly. This "solidarity" and "cooperative behaviour" of the members of the group indicates that they had the spirit and the drive of an organization or an institution, which also shows that theirs can be considered the first such practice in Ottoman society[11].

Besides being knowledgeable in Islamic culture, this intellectual group was also acquainted with Western science and culture which were newly introduced into Ottoman society in the beginning of the 19th century. They were open to Western novelties and sought to achieve a synthesis between Islamic and Western cultures. Therefore, they played an important role in this period. The group believed that it was possible to reach a harmony between Islam and the West on non-religious matters, to establish a mutual understanding based on tolerance regarding religious matters, and to acknowledge the findings of science and the usefulness of some Western techniques which facilitated daily life. According to our findings, the attempt

---

11 E. İhsanoğlu, loc.cit, *OIMC*, pp. 43-48.

168

made by this group to establish a society in the beginning of the 19th century remained the only endeavour of its kind[12].

MODERNIZATION AND THE FORMATION OF LEARNED SOCIETIES AND PROFESSIONAL ASSOCIATIONS

As in Europe, movements to establish societies in Ottoman Turkey appeared after the implementation of some scientific and cultural reforms. The cultural transformation of the Ottoman society in the first half of the 19th century and especially the *Tanzimat* reforms of 1839, well-known in Ottoman history, brought about several notable developments towards the modernization of Ottoman society. The first attempts to set up societies were made in this period in order to respond to arising needs. In 1851, a council named *Encümen-i Daniş* was established by the state. Its members were high ranking statesmen. The Encümen-i Daniş was the first example of its kind in Islamic history. In addition to some formal and ceremonial similarities, this council resembled the Académie Française, set up in 1635, as regards aims and objectives. It constituted the nucleus of the learned societies and professional associations of Ottoman Turkey.

The idea to establish the Encümen-i Daniş was formulated by the *Meclis-i Muvakkat* (a temporary council set up by the state and entrusted with the task of planning the educational reform) in the context of its decision to set up a *Darulfünun* (an institution of higher learning separate from the medrese). The Encümen-i Daniş, which was established soon after this decision, is unprecedented in Islamic law as well as in Sultanic laws. An examination of its regulations (*"Nizâmname"*) reveals that it enjoyed the status of a public institution having its own legal framework.

Though it has not been possible to determine the original motives for establishing the Encümen-i Daniş, there is much probability that Meclis-i Muvakkat members, as well as Engineer and Division General Emin Mehmet Paşa, head of the *Daimî Maarif Meclisi* (Permanent Council for Education established in 1846) who had been educated in France for years, were influential. Calling this council Encümen-i Daniş, a compound of two Persian words which literally means "Assembly of Knowledge" implies the intention of avoiding other suitable terms which would have the same meaning, such as *"cemiyet-i ulum"*, *"heyet-i ulum"*, or *"cemiyet-i ilmiye"* but which include the Arabic word *"ulum"* whose meaning encompasses both religious and non-religious sciences. This reflects, in our opinion, the wish to set up an institution different from the traditional type, which is also true for the

---

[12] *Ibid.*, pp. 73-74.

Darulfünun which was set up later and whose founders used the word *"fünun"* (sciences) instead of the usual term *"ulum"*, to emphasize that it was different from the "medrese" and that it focused on modern Western sciences.[13]

A report prepared by the *Meclis-i Maarif-i Umumiye* (Permanent Council for general education) following the decision of the Meclis-i Muvakkat was submitted to the Supreme Council *(Meclis-i Vala)*, which decided to establish both the Darulfünun and the Encümen-i Daniş. The reports concerning the establishment of these institutions were sent to the *Bab-ı Meşihat* (Office of the Sheikhulislam) for approval. In his answer, Şeyhülislam Arif Hikmet expressed the opinion that it would be appropriate to set up the Encümen as soon as possible.[14]

The inaugural ceremony of the Encümen-i Daniş took place on 18 July 1851 in the presence of Sultan Abdülmecit, and was addressed by Grand Vizier Mustafa Reşit Paşa. Though it was officially announced that the objective was to "set up a society to bring together people possessing the necessary knowledge and qualifications to produce, as soon as possible, the books which are needed by the public", Encümen member Ahmet Cevdet Paşa emphasized, in his inaugural declaration, the objectives of preserving the Ottoman Turkish language and setting rules of grammar. This suggests that the functions of the Académie Française were taken into consideration. In fact, Sir James Redhouse, who was a member of the Encümen-i Daniş, referred to it twice in the dictionary he was preparing at that time, as "The Academy of Science of Constantinople" in one place and as "The Literary and Scientific Academy of Constantinople" in another.[15]

There are points of similarity between the regulations of the Encümen-i Daniş and the initial regulations of the Académie Française. The latter was set up as a group of scholars and literary men which was recognized and regulated by the State; its membership was limited to 40 and its objectives included the preservation of the language and observance of grammar rules. However, unlike the Académie Française which continued to function from 1635 until our time, the activities of the Encümen-i Daniş ended in 1861. Though it was expected to perform great tasks, it was not able to realize its objectives, except the publication of a few works by Ahmet Cevdet Paşa[16].

---

13 E. İhsanoğlu, "Tanzimat döneminde İstanbul'da Darulfünun kurma teşebbüsleri", *150. yılında....*, p. 397.

14 Kenan Akyüz, *Encümen-i Daniş*, Ankara 1975, p. 36.

15 Sir James W. Redhouse, *A Turkish and English Lexicon*, Constantinople 1890, p. 213, p. 885.

16 Akyüz, *op. cit*, p. 63.

170

The regulations of the Encümen can be examined under three aspects. The first one is its organizational structure. There were two kinds of membership to the Encümen: internal membership, and external membership. The number of internal members was limited to 40, while external membership was left unlimited. It was directed by two presidents who were elected from among the members and ranked as first and second head. Membership was conferred as an honour and members were not remunerated. The second aspect is the conditions of membership. Internal members were nominated by the Supreme Council, i.e. by the state, at the establishment of the Encümen and just for once; subsequent members were to be elected by the internal members, by secret vote. Qualifications required for membership included the following: being well-versed in learning and science, having a perfect knowledge of one of the European languages, and being knowledgeable enough to write a book in one discipline. As to the third aspect, it is related to the field of activities of the Encümen, which was the preparation of books useful for the public by means of editing and translation. In addition, it would encourage the use of a plain Turkish language, give awards to those who wrote books, and its members would collect information about the places they visited on mission, among others.[17]

Although the Encümen-i Daniş was affiliated to the Council of Education, that is to the state, its regulations provided an example to learned societies and professional associations which were set up later. New intellectual trends and new types of occupations emerged in the Ottoman Empire in the 19th century, and new circles of people were formed around them. This established a precedent for new forms of institutions and organizations differing from the conventional type. The model of the Encümen provided an answer.

SUCCESSIVE DEVELOPMENTS

Regular attempts to establish learned societies and professional associations in Ottoman Turkey started after the establishment of the Encümen-i Daniş. Activities of the first learned societies and professional associations which were set up by resident foreigners in the country were later continued by graduates of the modern Ottoman educational establishment. In the period following the Encümen's establishment up to the dissolution of the Ottoman Empire in the first quarter of the 20th century, numerous societies were formed in various fields of learning and in different professions.

---

[17] *Ibid.*, pp. 51-57.

The *Société Orientale de Constantinople* was the first learned society established by resident foreigners who lived in Istanbul. The aim of this society, founded by European orientalists in 1852, was to collect and disseminate information on eastern countries and especially the Ottoman Empire, examine the nature, geography, history, language, literature, ancient works, science, and art of these countries. The society followed the models of the Paris Oriental Society and the German Oriental Society, and according to our knowledge, it is the first learned society established as a result of individual efforts. Its regulations were signed by twenty-three foreigners. There is not enough information on the activities carried out by this society which had forty members.[18]

The second society established by foreigners in Istanbul was the *Société Médicale de Constantinople.* It was founded on 15 February 1856, after a preparation which lasted six months, by P. Pincoffs, an English physician working in the allied forces which were stationed in Istanbul during the Crimean War and a group of friends. Its main purpose was to assist the physicians who were treating the soldiers wounded in the war. They constituted themselves a circle for discussing medical subjects. The society had forty members, all of whom were foreigners. According to our knowledge, no obstacle was faced in its establishment. With the assistance of Grand Vizier Fuat Paşa, the Sultan conferred the title *"Şahane"* (Imperial) upon the society on 22 May 1856, and also allocated a monthly sum of fifty gold coins to meet its expenses. The name of the society was changed to *Société Impériale de Médecine de Constantinople,* known as *Cemiyet-i Tıbbiye-i Şahane* in Turkish literature. The second honorary member of the society, after P. Pincoffs, was Fuat Paşa[19]. The establishment of this society created a new case in legal procedures, bringing the requirement that societies founded by individuals be officially recognized. Hence, it set an example for later societies, which obtained an official authorization after their establishment. This practice was adopted as a legal rule if not as an official procedure, and there is no evidence of any objection to it by the official authorities.

The society had honorary, internal, and corresponding members. The condition that only physicians, surgeons, and graduates of the pharmacy and veterinary schools were accepted as members gave it the appearance of a professional association. The society's regulations were made up of 21

---

18 H. L. Fleischer, "Die Morgenländische Gesellschaft in Constantinopel", *ZDMG*, Vol. VII, 1853, pp. 273-278, in *OIMC*, p. 7.
19 Hüsrev Hatemi, Aykut Kazancıgil, "Türk Tıp Cemiyeti (Derneği), Cemiyet-i Tıbbiye-i Şahane ve tıbbın gelişmesine katkıları", *OIMC*, pp. 110-120.

clauses and had points of similarity with those of the Encümen-i Daniş. In addition, there were provisions for the payment of membership fees. It was stipulated that the society would publish a journal titled *Gazette Médicale d'Orient* and that its working language would be French. This society carried out its functions until our time with some changes. It is known today as the *"Türk Tıp Derneği"*. Its activities served to introduce new concepts and approaches to Ottoman society. The exclusion of Turks from this society, with the exception of a few high ranking officials who were offered honorary membership, led Turkish intellectuals to set up similar societies in subsequent years.[20]

The first society founded by Turkish intellectuals for the purpose of spreading science and culture was the *Cemiyet-i İlmiye-i Osmaniye*. It was established in 1861 under the leadership of Münif Efendi (later known as Münif Paşa), and headed by Halil Bey, the Turkish Ambassador in St. Petersburg. As did the Cemiyet-i Tıbbiye-i Şahane, the society applied for official authorization as soon as it was opened and its members nominated. Official permission was given to its activities, but unlike the Cemiyet-i Tıbbiye-i Şahane, this society was not granted its request for a monthly allocation. Its membership structure included internal members, honorary members and other similar features, but no distinction was made on the basis of religion, language, or ethnic origin. People of all national origins were ad-mitted to membership.[21]

The Cemiyet-i İlmiye-i Osmaniye was functional for six years, until 1867. It has been suggested that the Cemiyet-i İlmiye-i Osmaniye had points of similarity with the Royal Society of London, on the grounds that both were founded by individuals and not by the state. However, there is no re-semblance between these two societies in either the circumstances under which they were set up or the objectives and activities they were expected to fulfill. The Cemiyet-i İlmiye-i Osmaniye, whose aim was, briefly, to ac-quaint intellectual and educated circles with European science, was rather similar to the clubs that Italian humanists formed in the Renaissance period. During the 14th, 15th, and 16th centuries, Italian humanists had found that their new literary and critical interests were not welcomed by the universities or by scholastics and sought places outside of schools, in the chanceries of city states, at courts, and in the entourages of despot and merchant princes such as the Medici. Throughout the 1400s, such men organized clubs, most of which were unspecialized in nature and few of which paid any attention to

---

20 Ekrem Kadri Unat, "Osmanlı Devletinde tıp cemiyetleri", *OIMC*, pp. 86-87.
21 E. İhsanoğlu, "Cemiyet-i İlmiye-i Osmaniye'nin kuruluş ve faaliyetleri", *OIMC*, pp. 203-204.

science or even mathematics. But they did help to popularize the idea that learning could and ought to be revived, and aroused the interest of some of the social elite in this goal, which was generally associated with the belief that the material, social, and political conditions of life could be changed for the better.[22] Similarly, the interest that the Cemiyet-i İlmiye-i Osmaniye took in science was not directed to the practice, application, or development of science. What the society did was limited to understanding and explaining science as a cultural phenomenon. Therefore, it did not make any effort towards transferring science or promoting scientific activities. Probably, it was not aware of such requirements.

The main activity of the Cemiyet-i İlmiye-i Osmaniye was to introduce the new class of intellectuals, and graduates of Western-type educational institutions, to modern science. The society tried to realize this aim by publishing a journal titled *Mecmua-i Fünun*. Altogether forty-seven issues appeared, containing articles randomly chosen from European popular periodicals and books. The society carried out its activities under the administration of its founders, i.e. bureaucrats trained within the process of modernization which had led to the society's establishment. It excluded the new modern scientists educated in that period, such as chemists, physicists, astronomers, and mathematicians. Members of this society were not very eager to reconcile modern science with Islam as it was the case of the group of ulemas in Beşiktaş mentioned above. Unfortunately, instead of contributing to the establishment of a new scientific tradition and organization in Ottoman Turkey, the society became the first representative of "scientism".[23]

During the first decade of the move to establish institutions in the Ottoman Empire, societies having scholarly, professional, and cultural purposes enjoyed freedom and were also recognized by the state. Although their rules of operation, nature, and characteristics were different, their regulations were similar to those of their predecessors, with additional provisions to meet new requirements. Taking into account this similarity between the regulations of different societies, it would be more appropriate to examine the learned societies established later with a focus on their objectives rather than their statutory features. All the more so because the learned societies established after those mentioned above also started to set some economic and political goals for themselves.

22 Roger L. Emerson, "The organisation of science and its pursuit in early modern Europe", *Companion to the History of Modern Science*, Ed. R. C. Olby, G. W. Cantor, Routledge, London, New York, 1990, p. 969.
23 E. İhsanoğlu, "Cemiyet-i İlmiye-i Osmaniye'nin...", pp. 218-220.

174

The first society set up by Turkish physicians was the *Cemiyet-i Tıbbiye-i Osmaniye,* established secretly in 1862. It was founded by graduates of the *Mekteb-i Tıbbiye-i Şahane.* Its purpose was limited to the translation of works on medicine and dentistry into Turkish and the publication of a monthly journal. The society applied for official authorization in 1865. Its establishment was approved and it was given a monthly allowance of 1000 piastres. The society made great efforts to implement medical education in the Turkish language, thus involving itself in an on-going dispute with the Cemiyet-i Tıbbiye-i Şahane which argued that medical education should be in French. The Cemiyet-i Tıbbiye-i Osmaniye won the case and achieved its aim in 1870. Its second successful step was the publication of the first modern Turkish medical dictionary titled *Lugat-ı Tıbbiye.* The first edition of this dictionary appeared in 1873 and the second one, in 1901. Discussions which took place between the members of this society and those of the Société Impériale de Médecine, who defended the premise that medical education should continue in French, are noteworthy from the viewpoint of the history of medicine and Ottoman culture.[24]

The first pharmaceutical society was established in Istanbul in 1863 by non-Muslims under the name of *Société de Pharmacie de Constantinople.* It was closed shortly after its foundation, but reestablished in 1879 under the same name; it remained active for twenty years. Its publication was the *Journal de la Société de Pharmacie de Constantinople.* The society was active in the following areas: enacting a law to regulate the work of the pharmacists and pharmacies, putting into effect a price list for medicines, limiting the number of pharmacies, forbidding the sale of drugs outside of pharmacies, and publishing a codex. As to Turkish pharmacists, they united in the *Osmanlı Eczacı İttihad Cemiyeti* (Union Pharmaceutique Ottomane) in 1908. The objective of uniting the eight pharmaceutical societies which were established in the Ottoman period in the framework of a particular programme was realized only after the proclamation of the Republic, with the foundation of the *Türkiye Eczacıları Cemiyeti* (Turkish Society of Pharmacists) in Istanbul, in 1924.[25]

In 1864, Ottoman intellectuals established a new type of society, the *Cemiyet-i Tedrisiye-i İslamiye* (Islamic Society for Education). This society differed from those established previously by the State or by individuals,

---

24 Nil Sarı, "Cemiyet-i Tıbbiye-i Osmaniye ve tıp dilinin türkçeleşmesi akımı", *OIMC,* pp. 121-142; E. İhsanoğlu, Feza Günergun, "Tıp eğitiminin türkçeleşmesi meselesinde bazı tesbitler", First International Congress on the History of Medicine and Deontology, Istanbul, 1993.
25 Turhan Baytop, "Osmanlı imparatorluk döneminde eczacılık cemiyetleri", *OIMC,* pp. 143-154.

with respect to its statute as well as its activities, because it did not work like a Western-type academy or a club for the elite, but it was mindful of the general public, and functioned in accordance with Islamic traditions. This contributed to its duration up to the present times. It is known today as the Daruşşafaka. It provided the example of an institution which preserved its vitality because it was in conformity with the norms of its social environment.

The Cemiyet-i Tedrisiye-i İslamiye was founded my middle-class civil servants and bureaucrats with the purpose of educating Muslim children. Its activities did not cover non-Muslim children. A few years after its establishment, the society set up a school named Daruşşafaka-i İslamiye as a free boarding school for Muslim students, as a counterpart of the Galatasaray secondary school where most of the students were Christians and Jews. The fate of this school illustrates the dispute between conservative intellectuals and those who were more inclined to adopt the European culture. Following his appointment as Minister of Education in 1867, Münif Paşa, who was one of the founders of the Cemiyet-i İlmiye-i Osmaniye, closed this school out of 'jealousy and rivalry'[26].

Another society established by a group of conservative Ottoman intellectuals was the *Cemiyet-i İlmiye* (Learned Society). According to available information, this society was functional for eleven months only, between the years 1879-80. Its purpose was to spread science and technology in the Ottoman Empire and to provide assistance to some schools. Among its members, there were those who belonged to the ulema class and were receptive to modern sciences. Moreover, considering the cultural climate of the time, the society took a conciliatory attitude between the two extremes of total acceptance or rejection of Western culture. The society functioned for a very short period, about which the only available source is its journal *Mecmua-i Ulum,* seven issues of which were published.[27]

PROCLAMATION OF THE FIRST CONSTITUTION
AND THE LEGAL STATUS OF SOCIETIES

Although the *Mecelle* (Islamic Civil Code), which was drafted by the *Mecelle Cemiyeti* (Mecelle council) set up in 1868 under the supervision of Ahmet Cevdet Paşa, referred to "a multitude of persons unified in the status of one" (article 423), the *Kanun-ı Esasi* (first Ottoman Constitution) proclaimed in 1876 did not have a special provision for societies. At that time, the concept of legal personality had not developed yet in the framework of

---

26 *Daruşşafaka, Türkiye'de ilk halk mektebi,* İstanbul 1927, pp. 3-5.
27 E. İhsanoğlu, "Cemiyet-i İlmiye ve Mecmua-ı Ulum", *OIMC*, pp. 221-245.

Ottoman legislation. However, it is observed that during the reign of Sultan Abdülhamit II, and until the proclamation of the Second Constitution in 1908 and the enactment of the law of societies in 1909, the establishment of learned societies and professional associations slowed down, with a few exceptions. The reason was that some societies which acted like political parties within and outside the country were regarded with suspicion.[28].

ESTABLISHMENT OF SOCIETIES AFTER THE SECOND CONSTITUTION

The establishment of societies in the Ottoman Empire accelerated especially after the proclamation of the Second Constitution in 1908. One reason for the remarkable increase of societies after this date was the freedom that the Constitution had brought to the country. Yet the primary reason was the enactment of the law of societies, in 1909. With this law, Ottoman societies gained a legal basis. Instead of requesting authorization, they were required to notify the authorities of their establishment.[29] Another characteristic of the increased activity to establish societies after the 1908 Constitution was the variety of professional associations founded in this period. Until 1908, professional associations had been established only in the fields of medicine and pharmacy. After this date, there appeared societies in the fields of agriculture, veterinary medicine, dentistry, engineering, and architecture as well.

The first society of dentistry was established after the School of Dentistry started to function in 1909 as part of the Faculty of Medicine. This society was founded by the students of this school in 1914, under the name of *Dişhekimleri Mezunin ve Talebe Cemiyeti* (Society of Graduates and Students of the School of Dentistry). There were four societies in this field between 1914-23, but the *Türk Diş Tabipleri Cemiyeti* (Society of Turkish Dentists) founded in 1922, is the first society that lasted long.[30]

Among the societies established after the Second Constitution, there were two founded independently by the graduates of the schools of agriculture and forestry, in addition to a third society founded under the name of *Osmanlı Ziraat Cemiyeti* (Ottoman Agricultural Society).[31]

Education in veterinary medicine first began in the Military Veterinary School which was founded in 1842. But the first veterinary society called

---

28 E. İhsanoğlu, "Modernleşme süreci içinde Osmanlı Devletinde ilmi ve mesleki cemiyetleşme hareketlerine genel bir bakış", *OIMC*, pp. 11-12.

29 Hüseyin Hatemi, "Bilim derneklerinin hukuki çerçevesi (Dernek tüzelkişiliği)", *OIMC*, pp. 83-84.

30 E.İhsanoğlu, "Modernleşme süreci...", p. 12.

31 *Ibid.*, p. 11.

Osmanlı Cemiyet-i İlmiye-i Baytariyesi (Ottoman Learned Veterinary Society) was established in 1908; at that time, there were five other veterinary societies in different parts of the country. Among them, the *Taşra Baytari Cemiyeti* (Provincial Veterinary School) was founded in Adana, but dissolved before it became active.[32]

Attempts to establish societies in the field of applied sciences such as engineering and architecture began after 1908. Although engineering was the oldest applied science taught according to modern methods of instruction, engineers did not become organised until a very late date. The main reason for this delay was that the graduates of military and civil institutions, such as the *Mühendishane-i Berr-i Hümayun* (Imperial School of Engineering), the *Hendese-i Mülkiye* (Civil School of Engineering) were charged with duties in the army or state service as soon as they completed their education. Moreover, the small number of engineers who worked on their own in those years may have been a factor which delayed the establishment of societies in this field. The regular instruction of civil architecture began only in 1882 upon the establishment of the *Sanayi-i Nefise Mekteb-i Alisi* (Higher School of Fine Arts). This was also a reason for the delay in the formation of a body of Turkish architects and the establishment of a society in this field.

The first society active in the field of engineering and architecture was established in 1908 under the name of *Osmanlı Mühendis ve Mimar Cemiyeti* (Society of Ottoman Engineers and Architects). Its founders were engineer bureaucrats who graduated from the School of Civil Engineering and architects most of whom worked on their own. The most remarkable activities of the society focused on the solution of problems related to the instruction and application of engineering in the Ottoman Empire. Its publication titled *Osmanlı Mühendis ve Mimar Cemiyeti Mecmuası* (Journal of the Society of Ottoman Engineers and Architects, published in 1909-1910) created the necessary milieu to fulfill this purpose. Following this first attempt, non-Muslim Ottoman engineers and architects united with some members of foreign nationality in a society which they called *Association des Architectes et Ingénieurs en Turquie* (Association of Engineers and Architects in Turkey). Engineers and architects founded independent societies in the early period of the Turkish Republic.[33]

The *Osmanlı Coğrafya Cemiyeti* (Ottoman Geographic Society) established in 1908 by a few Ottoman men of letters and culture was the first society active in this field. In addition, an examination of its objectives

---

32 *Ibid.*, p. 17.
33 Feza Günergun, "Osmanlı mühendis ve mimarları arasında ilk cemiyetleşme teşebbüsleri", *OIMC*, pp. 154-196.

shows that it may be regarded as a scholarly society rather than a professional association. The purpose of the society was to undertake research in the field of geography, collect and collate statistics, follow the developments regarding geography, and establish contacts with counterparts in other countries. Another society which was active in the field of geography starting from 1913 was the Istanbul branch of the *Société de Géographie de Paris* (Geographical Society of Paris)[34].

It appears that no societies were established in fields of basic sciences such as mathematics, natural, and physical sciences in the Ottoman period. Attempts to establish a society of chemistry began in 1919-20, but the first society in this field was set up in the period of the Republic, in 1924, under the name of *Türk Kimyagerler Cemiyeti* (Society of Turkish Chemists). The fact that chemistry is partly an applied science may be the reason why the first society in the field of basic sciences was a society of chemistry. This society was established at a relatively late date, only after the formation of an independent body of chemists in the first decades of the 20th century. This had become possible after the establishment of an independent department of chemistry in the Faculty of Science of the Darulfünun (Istanbul University), whereas chemistry education was previously provided as an ancillary to the curricula of medicine and other scientific disciplines. Physicists too endeavoured to establish a society in 1930, but plans to set up societies of physics, astronomy, and mathematics did not materialize until the 1950's.[35]

The *Türk Bilgi Derneği* (Turkish Society of Knowledge) established in 1914, after the Second Constitution, aimed to conduct research and accumulate information in the following areas, which were beyond popular knowledge: Turkology, Islamic studies, biology, philosophy and sociology, mathematics, materialism, and Turkism. It was established for the purpose of forming the nucleus of a future academy. The publication of the Society was titled *Bilgi Mecmuası* (Journal of Knowledge) and contained articles on a wide range of subjects, from history, language, literature, philosophy, political science, sociology, economics, education, pedagogy, and medicine to mathematics.[36] Like the previously mentioned *Cemiyet-i İlmiye*, activities of this institution were short-lived and only seven issues of its periodical could be published.

[34] E. İhsanoğlu, "Modernleşme süreci...", p. 14.
[35] *Ibid*, p. 15.
[36] Zafer Toprak, "Türk Bilgi Derneği (1914) ve Bilgi Mecmuası", *OIMC*, pp. 247-254.

INSTITUTIONALISATION IN COMPARISON WITH THE EUROPEAN EXPERIENCES

It can be noted that an analytical examination of the formation process of learned societies and professional associations reviewed in this paper enables us to acquire a perspective within the context of the history of science.

As is known, scientific societies in Europe were established from the 17th century onwards by amateur scientists who were interested in the "new science" that was born as a result of the Scientific Revolution. These societies were prompted by a reaction to the medieval universities' curricula in scholastic science. In addition to these societies, academies were the main institutions where modern science was studied until it became part of the teaching program of universities in the 19th century.[37]

The establishment of the Royal Society of London and the Académie des Sciences in Paris aroused the "learned society movement" which gained pace in the 18th century. In consequence, from the first years of that century, state academies were established in Berlin, St. Petersburg, Uppsala, Bologna, and Prag, in addition to the numerous learned societies which were founded by individuals in various European cities. The Ottomans were hardly influenced by this movement which occurred in Europe in the 18th century. The establishment of Ottoman learned societies started only around mid-19th century. It is only after this date that the scholars who were trained in various branches of science reached a sufficient number to organize a society in their field of specialisation.

There was a parallel trend in the Ottoman world: while the old (Islamic) science was taught traditionally in the medreses, which were institutions of classical learning, modern science was taught in Western-type educational institutions. As in Europe, the founders of Ottoman learned societies were individuals who had the chance to get acquainted with the "new science", or in other terms, modern sciences. Members of the ulema class, who actively took part in the early phases of the move to establish societies, were later replaced by the new intellectuals as a result of the modernization process.

Despite some formal similarities between them, the most important difference between societies established in Ottoman Turkey and Europe lie in their programmes and activities. Societies founded in Europe aimed to develop science through research, and to encourage inventions. The program of the Royal Society, which is a leading example, emphasized experiments and practical applications. The Society's work was directed to improving the "useful arts", manufacture, machines, and inventions. Moreover, it was open

---

37 Stephen F. Mason, *A History of the Sciences*, Collier Books, New York 1962, pp. 256-266.

not only to men of science, but also to merchants, navigators, and specialists of mechanical arts.

Although the programs of some Ottoman societies included such activities as "scientific research" and "encouraging certain crafts and professions, disseminating the sciences, technique, agriculture and industry", they were not able to fulfill these aims. The majority of members of these learned societies were graduates of the new Ottoman institutions where modern sciences were taught; other intellectual groups, professional or business circles did not join them. This may be due to the heterogeneous composition of the Ottoman society.

The development patterns of the societies described above show that an awareness of the need for innovations was not a sufficient condition for the realization of the desired objectives. The existence of an appropriate environment to accommodate these innovations was an equally important necessity. The complex scientific milieu of Europe, which was a result of the scientific and technological progress encouraged by economic and social developments, did not exist in the Ottoman world. Therefore, scientific productivity was only a secondary objective for Ottoman learned societies. There were particular political, social, and economic conditions behind the objectives of scientific activity in Europe. In Ottoman Turkey, the circumstances were completely different and even opposite to those in Europe. This fact was reflected in the objectives of the Ottoman societies. The Ottomans aimed to regain their military and political power as soon as possible by transferring new technology and acquainting the public with it. Thus, their learned societies had rather general objectives, aiming basically to introduce the public to modern science in a simple way. An examination of their publications clearly indicates this purpose. These short-lived publications did not contain original articles; in general, they gave information or printed popular articles which could appeal to the general reader in Europe. Meanwhile, new professionals such as physicians, pharmacists, and engineers who were trained in the newly established educational institutions attempted to establish professional associations which would enable them to reinforce their professional solidarity and strengthen their position in society.

Another obstacle to the development of learned societies and professional associations may have been the absence of a relationship based on mutual interests between the state and these societies. The state did not see itself deriving substantial benefits from the experiences of societies in the technical field. Hence, it did not oblige them to fulfill certain functions. The income of these societies came from membership fees. They did not have sponsors like those who provided material support to European societies.

Nevertheless, these societies, and especially the professional associations looked to the state for financial support. In a few instances, the state provided financial assistance to their activities. The most important example is the support given by the state to the *Cemiyet-i Tıbbiye-i Mülkiye* (Society of Civilian Physicians) founded by Turkish physicians, for its endeavours to develop a Turkish medical terminology.

The Ottoman professional associations were more numerous than those qualified as learned societies, and all of them dealt with applied sciences. This situation may be explained by the following factors: the instruction of applied sciences started much earlier than basic sciences, because of the objective to attain the technological level of the West as soon as possible and to train the manpower required for the army, the new state institutions, and society in general. Consequently, there was a group educated in this field from an early date. Societies started to be established relatively early in professions of civilian nature, in other words, in professions which were practised outside official and military services. Similarly, societies appeared earlier in professions having a wider social scope. Thus, the earliest societies were those working in the fields of medicine and pharmacy because the professionals in these fields were numerous, had the opportunity to work on their own, and were able to establish wider social relationships.

The learned societies and professional associations of the Ottoman period were established mainly in Istanbul, the capital city of the Empire, with a few exceptions. Although these societies admitted members from outside of Istanbul as well, their activities were limited to the capital.

An examination of the body of founders and members shows that the first attempts to establish professional associations in the fields of medicine and pharmacy were undertaken by non-Muslims. Very few of their members were Turkish. In later periods, as the number of Turkish professionals increased, they united among themselves and tried to organize professional associations. The case of engineers and architects was just the opposite. Five years after the establishment of societies by Turkish architects and engineers, non-Muslim architects organized their own society. Meanwhile, it was mostly Turkish professionals who organized societies in the fields of agriculture and veterinary medicine.

As it is known, the Ottoman state established contacts with the West at an early date in many fields, and endeavoured to transfer European knowledge and techniques. The Ottomans brought technicians from Europe and sent students to European institutions for education. They founded institutions similar to their counterparts in the West, where technical instruction was given. The Ottomans even abandoned the system of measures and weights that had been used in Ottoman society for centuries, and adopted the metric system.[38] All of these examples indicate that if they were willing to do so, the Ottomans would have been able to transfer scientific developments from the West to Ottoman lands, to learn and apply Western science and technology, as in the example of Peter the Great in Russia. It appears, however, that the Ottomans were too hasty in bridging the gap with the West. For this reason, they were content with short-term remedies to their problems. And this being the objective, they gave priority to establishing institutions for technical education which would meet the needs of state and society instead of academies and scholarly institutions.

In conclusion, we might state that contrary to the precedents in the West, Ottoman societies that were established under the name of "learned society" did not seek to investigate and examine nature and man. They generally started and ended as very short-lived cultural movements with limited activities. The founders of these societies did not endeavour to guarantee their permanence by endowing them with a system such as that of the "waqf" institutions previously founded in the Ottoman-Islamic tradition. These societies were established at a time when neither the socio-economic conditions of the newly developing scientific community, nor the accumulation of indigenous scientific knowledge and experience, had reached a potential necessary for sustaining their activities. Although the expectations of state and society were similar in this respect, and despite the fact that the necessary legal framework was formed with time and through a development process, it is evident that a social and cultural environment to accommodate these bodies and encourage their continuation had not yet reached the necessary level of maturity. Only in the case of professional associations, which had attained such maturation, do we witness continuity until the present.

---

[38] Feza Günergun, "Introduction of the metric system to the Ottoman State", *Transfer of Modern Science ...*, pp. 297-316.

LEARNED SOCIETIES AND PROFESSIONAL ASSOCIATIONS ESTABLISHED IN OTTOMAN TURKEY IN THE 18th-19th CENTURIES

This list includes, in addition to the learned societies and professional associations, some cultural associations which were important for their activities and were interested in the subject of science, as well as official councils which dealt with educational and cultural subjects. The dates on the left of their name are establishment dates. Dates which are not certain are indicated with the sign [ ]. The names of the societies are followed by the title of their publication if any, and their ordering numbers in two cumulative catalogues. The first is the catalogue of the periodicals in Arabic script printed 1828-1928, briefly IRCICA-AHSYK* and the second, the catalogue of the periodicals published in French in Turkey from 1795 to our time, briefly ISIS-PFDT**. References which give information on the societies are also cited.

DATE OF ESTABLISHMENT, NAME OR DESIGNATION, REFERENCES

1720-30   TRANSLATION COUNCIL
           M. İpşirli: "Lâle devrinde teşkil edilen tercüme heyetine dair bazı gözlemler", *OIMC*, p. 33

CA. 1800   THE GROUP OF SCHOLARS IN BEŞİKTAŞ, known as the "BEŞİKTAŞ CEMİYET-İ İLMIYESİ"
           E. İhsanoğlu: "19. asrın başlarında — Tanzimat öncesi — kültür ve eğitim hayatı ve Beşiktaş Cemiyet-i İlmiyesi olarak bilinen ulemâ grubunun buradaki yeri", *OIMC*, p. 43

1845   MECLİS-İ MUVAKKAT (TEMPORARY COUNCIL)
           E. İhsanoğlu: "Cemiyet-i İlmiye-i Osmaniye'nin kuruluş ve faaliyetleri", *OIMC*, p. 197

1846   "DAİMÎ MAARİF MECLİSİ" (PERMANENT COUNCIL FOR EDUCATION)
           E. İhsanoğlu: loc.cit., p.198

1851   ENCÜMEN-İ DANİŞ
           Kenan Akyüz: *Encümen-i Daniş*, Ankara 1975; Cahit Bilim: "İlk Türk bilim akademisi: Encümen-i Daniş", *Hacettepe Üniversitesi Edebiyat Fakültesi Dergisi*, Vol. 3, no. 2, Dec. 1985

1853   SOCIÉTÉ ORIENTALE DE CONSTANTINOPLE
           H. L. Fleischer: "Die Morgenländische Gesellschaft in Constantinople", *Zeitschrift der Deutschen Morgenländischen Gesellschaft* (ZDMG), Vol. VII, 1853, p. 273-78

---

*   *İstanbul kütüphaneleri Arap harfli süreli yayınlar toplu katalogu 1828-1928,* prepared by Hasan Duman, published by the Research Centre for Islamic History, Art and Culture (IRCICA), Istanbul, 1986.
**   G. Groc ve İ. Çağlar, *La presse française de Turquie de 1795 à nos jours, Histoire et catalogue,* ISIS, İstanbul 1985.

184

1856	CEMİYET-İ TIBBİYE-İ ŞAHANE (Société Impériale de Médecine de Constantinople) (Publication: *Gazette Médicale d'Orient* (Şark Tabib Gazetesi), 1857-1925, ISIS-PFDT, 203) Hüsrev Hatemi and Aykut Kazancıgil: "Türk Tıp Cemiyeti (Derneği) Cemiyet-i Tıbbiye-i Şahane ve tıbbın gelişmesine katkıları," *OIMC*, p. 111 E.K. Unat: "Osmanlı Devletinde tıp cemiyetleri", *OIMC*, p. 86 *Cinquantenaire de la Société Impériale de Médecine de Constantinople*, Constantinople (1907-1908)
1861	CEMİYET-İ TIBBİYE Başbakanlık Osmanlı Devlet Arşivi (Prime Ministry's Ottoman Archives), İrade Dahiliye, no. 31742 CEMİYET-İ İLMİYE-İ OSMANİYE (Publication: *Mecmua-i Fünun*, 1862-1867, *İstanbul kütüphaneleri Arap harfli süreli yayınlar toplu katalogu, 1828-1928*, préparé par Hasan Duman, éd. E. İhsanoğlu, IRCICA, 1986-IRCICA-AHSYK) E. İhsanoğlu: "Cemiyet-i İlmiye-i Osmaniye'nin kuruluş ve faaliyetleri", *OIMC*, p. 200
1862	CEMİYET-İ TIBBİYE-İ OSMANİYE — Date of establishment mentioned in the regulations: 1866 (1283) E.K. Unat: loc.cit., *OIMC*, p. 88; N. Sarı: "Cemiyet-i Tıbbiye-i Osmaniye ve Türk dilini türkçeleştirme akımı", *OIMC*, p. 121 *İkdam*, 22 Zilhicce 1328 (24 Kanun-i evvel 1910), no. 280, p. 5, line 5; 24 Zilhicce 1328 (26 Kanun-i evvel 1910), no. 282, p. 3, line 6
1864	CEMİYET-İ TEDRİSİYE-İ İSLAMİYE (Publication: *Mebahis-i İlmiye*, IRCICA AHSYK, 0900) Hüseyin Hatemi: "Bilim derneklerinin hukuki çerçevesi (Dernek tüzel kişiliği)", *OIMC*, p.82; *Medeni hukuk tüzelkişileri*, Vol. I, Istanbul 1979, p. 192-193; *Darüşşafaka. Türkiye'de ilk halk cemiyeti*, İstanbul 1927 *Takvim-i Vekayi*, 16 Eylül 1284, no. 893; 8 Safer 1285, no. 971; 7 Şevval 1285, no. 1042 *İkdam*, 12 Muharrem 1327 (4 Feb. 1909), no. 5278 *Yeni İkdam*, 7 Rebiül-evvel 1328 (19 March 1910), no. 66; 6

Rebiül-ahir 1328 (16 April 1910), no. 34; 19 Rebiül ahir 1329 (19 April 1911), no. 398
Tanin, 9 Şaban 1237 (25 August 1909), no. 352

1866   TRANSLATION COUNCIL
Mahmud Cevad: *Maarif-i Umumiye Nezareti tarihçe-i teşkilât ve icraatı,* Vol. I, Matbaa-ı Amire
1338, p. 90-94

1869   ASSOCIATION DES MÉDECINS DE SCUTARI
E.K. Unat: loc.cit., *OIMC*, p. 96

1879   CEMİYET-İ İLMİYE
(Publication: *Mecmua-i Ulûm*, 1879-80, IRCICA- AHSYK, 0931)
E. İhsanoğlu: "Cemiyet-i İlmiye ve Mecmua-i Ulûm", *OIMC*, p. 221

SOCIÉTÉ DE PHARMACIE DE CONSTANTINOPLE (Cemiyet-i Eczaciyan der Asitane-i Aliyye)
(Publication: *Journal de la Societe de Pharmacie de Constantinople*, 1879-1880)
T. Baytop: "Osmanlı İmparatorluğu döneminde eczacılık cemiyetleri", *OIMC*, p. 144

(1885)   ZIRAAT VE SANAT TERCÜME-İ FÜNUN ODASI
(Publication: *Ziraat ve Sanat Tercüme-i Fünûn Odası*, 1885, IRCICA-AHSYK, 1792)

1887   SOCIÉTÉ DE MÉDECINE DE SALONIQUE
E.K. Unat: loc.cit., *OIMC*, p. 97

1903   CLUB MÉDICAL DE CONSTANTINOPLE
(Publication: *Comptes-rendus du Club Médical de Constantinople*, 1903-1907)
E.K. Unat: loc.cit., *OIMC*, p. 98

1907   ASSOCIATION DES PHARMACIENS ETRANGERS
Turhan Baytop: loc.cit., *OIMC*, p. 147

1908   OSMANLI MÜHENDİS VE MİMAR CEMİYETİ
(Publication: *Osmanlı Mühendis ve Mimar Cemiyeti Mecmuası,* 1909-1910, IRCICA-AHSYK, 1205)
F. Günergun: "Osmanlı mühendis ve mimarları arasında ilk cemiyetleşme teşebbüsleri", *OIMC*, p. 156

ETIBBA-I MÜLKIYE-I OSMANIYE CEMIYET-I İTTIHADIYESI (Association des Médecins Civils Ottomans)
E.K. Unat: loc.cit., *OIMC*, p. 98

OSMANLI COĞRAFYA CEMIYETİ
İkdam, 20 Şevval 1326/14 Teşrin-II 1908, no. 5198, p. 4., line 4; 9 Zilkade 1326/3 Kanun-I 1908, no.5217, p. 3, line 2

OSMANLI ZİRAAT CEMİYETİ
(Publication: *Resimli Çiftçi*, 1909, IRCICA-AHSYK, 1276)
İkdam, 20 Şevval 1326/14 Teşrin-II 1908, no. 5198, p. 2, line 5; 15 Muharrem 1327/7 Feb. 1909, no. 5281, p. 3, line 5; 28 Muharrem 1327/21 Feb. 1909, no. 5294, p. 3, line 6

BAHRİYE TERAKKİ CEMİYETİ (Zabıtân-ı Bahrî Terakki Cemiyeti)
*İkdam,*20 Ramazan 1326/16 Teşrin-I 1908, no. 5171, p. 4, line 6; 24 Şevval 1326/19 Teşrin-II 1908, no. 5203, p. 4, line 3

OSMANLI CEMİYET-İ İLMİYE-I BAYTARİYESİ
(Publication: *Mecmua-i Fünûn-ı Baytariye*, 1908-10, IRCICA-AHSYK, 0913)
Ferruh Dinçer: "Türkiye'de kurulan veteriner dernekler ile bugüne kadar olan gelişmeleri", *Türk Veteriner Hekimleri Derneği Dergisi*. no. 11-12, 1964, p. 487-502

OSMANLI ECZACI İTTİHAD CEMİYETİ (Union Pharmaceutique Ottomane)
T. Baytop: loc.cit., *OIMC*, p. 148

BİLUMUM ECZACI MÜSTAHDİMİNİ İLE ECZACI MEKTEBİ MÜDAVİMİ CEMİYETİ
T. Baytop: loc.cit., *OIMC*, p. 152

ETIBBA-İ ASKERİYE KULÜBÜ
İkdam, 24 Şevval 1326/19 Teşrin-II 1908, no. 5203, p. 4, line 3; 19 Zilkade 1326/13 Kanun-I 1908, no. 5227, p. 4, line 4

(1909) OSMANLI ZİRAAT MEKTEB-İ ALİSİ MEZUNİN CEMİYETİ
(Publication: *Osmanlı Ziraat ve Ticaret Gazetesi*, 1909-14, IRCICA-AHSYK, 1211
Yeni İkdam, 5 Safer 1330 (25 Kanun-II 1912), no. 675, p. 4, line 6

1909 CEMİYET-İ UMUR-I TIBBBİYE
İkdam, 24 Safer 1327 (17 March 1909), no. 5319, p. 3, line 2

(1909)   BERRÎ-BAHRÎ ETÎBBA-İ ASKERİYE CEMİYETİ
            İkdam, 22 Muharrem 1327 (14 Feb. 1909), no. 5288, p. 4, line 5

            DEVLET-İ OSMANİYE ECZACILARI CEMİYETİ (Société des Pharma-
            ciens de l'Empire Ottoman)
            T. Baytop: loc.cit., *OIMC,* p. 149

            ASSOCIATION DES PHARMACIENS DE SMYRNE (Farmakeftiki Enosis
            Smirneon)
            T. Baytop: loc.cit., *OIMC,* p. 151

            MÜLKİYE BAYTARLARİ İTTİHAD VE TEAVÜN CEMİYETİ (Baytar İtti-
            had ve Teavün Cemiyeti)
            F. Dinçer: loc.cit.

1910   BAYTAR MEKTEB-İ ALİSİ MEZUNİN CEMİYETİ
            (Publication: *Risale-i Fenn-i Baytarî, "Recueil Vétérinaire",*
            1328-29 (1931-32) IRCICA-AHYSK, 1301)
            F. Dinçer: loc.cit.

            OSMANLI ZİRAAT CEMİYETİ
            Yeni İkdam, 7 Rebiül-I 1328 (19 March 1910), no. 6, p. 4, line 3;
            26 Rebi'ül-I 1328 (6 Mai 1910), no. 54, p. 5, line 6

            OSMANLI TIP FAKÜLTESİ ŞEFLER CEMİYETİ
            E.K. Unat: loc.cit., *OIMC,* p. 100

1910-11   YENİ MUHİT ÜL-MAARİF CEMİYETİ
            (Publication: *Yeni Muhit ül-Maarif Gazetesi,* 1911, IRCICA-
            AHSYK, 740)
            Osmanlı Mühendis ve Mimar Cemiyeti Mecmuası, no. 12 (Eylül
            1326/1910), p. 337-39
            Yeni İkdam, 7 Muharrem 1329/8 Kanun-II 1911, no. 295, p. 2,
            line 3; 4 Safer 1329/3 Feb. 1911, no. 321, p. 5, line 6; 23 Safer
            1329/22 Feb. 1911, no. 342, p.5, line 3; 23 Receb 1329/20 July
            1911, no.491, p. 5, line 1

1911   TAŞRA ETİBBA-I MÜLKİYE CEMİYETİ
            E.K. Unat: loc.cit., *OIMC,* p. 99

            ETİBBA-İ MÜLKİYE CEMİYETİ
            E.K. Unat: loc.cit., *OIMC,* p. 100
            Yeni İkdam, 23 Muharrem 1330/13 Kanun-II 1912, no. 663, p. 5,
            line 1; 25 Muharrem 1330/15 Kanun-II 1912, no 665, p.4, line 6;
            21 Safer 1330/10 Feb. 1912, no. 691, p. 4, line 6

188

TAŞRA BAYTARİ CEMİYETİ (Adana)
F. Dinçer: loc.cit.

(1913) ORMAN MEKTEB-İ ALİSİ MEZUNİN CEMİYETİ
(Publication: *Toprak*, 1913-26, IRCICA-AHSYK, 1590)

1913 ASSOCIATION DES ARCHITECTES ET INGÉNIEURS DE TURQUIE
F. Günergun: loc.cit., *OIMC*, p.179

1914 TÜRK BİLGİ DERNEĞİ
(Publication: *Bilgi Mecmuası*, 1913-14, IRCICA- AHSYK, 0169)
Z. Toprak: "Türk Bilgi Derneği (1914) ve Bilgi Mecmuası", *OIMC*, p. 247

OSMANLI TABABET-İ AKLİYE VE ASABİYE CEMİYETİ
E.K. Unat: loc.cit., *OIMC*, p. 102

1918 TABABET-İ RUHİYE CEMİYETİ
E.K. Unat: loc.cit., *OIMC*, p. 102

VEREMLE MÜCADELE OSMANLI CEMİYETİ
E.K. Unat: loc.cit., *OIMC*, p. 102

1919 İSTANBUL EMRAZI CİLDİYE VE EFRENCİYE CEMİYETİ
(Publication: *İstanbul Emraz-ı Cildiye ve Efrenciye Cemiyeti Mecmuası / Bulletin de la Société de Dermatologie et de Syphilographie de Constantinople*, 1921, IRCICA-AHSYK, 703)
E.K. Unat: loc.cit., *OIMC*, p. 103

ETİBBA MUHADENET CEMİYETİ
E.K. Unat: loc.cit., *OIMC*, p. 104

ETİBBA TEAVÜN CEMİYETİ
E.K. Unat: loc.cit., *OIMC*, p.105

(1919) İLİM VE UMRAN-I SINAİ CEMİYETİ
İlim ve Umran-ı Sınaî Cemiyeti Nizamnamesi, İstanbul, Tıbbiye-i Şâhâne Matbaası, 1335

1920 TÜRK BAYTARLAR BİRLİĞİ (Ankara)
F. Dinçer: loc.cit.

(1920) ZİRAAT CEMİYETİ
(Publication: *Yeni Çiftlik*, 1920-26, IRCICA-AHSKYK, 1713)

(1920)   ZİRAAT CEMİYETİ
              (Publication: *Yeni Ziraat Gazetesi,*1920-24, IRCICA-AHSYK,
              1759)

1921   DERSAADET ECZA TÜCCARANI CEMİYETİ (Association des Droguis-
              tes de Constantinople)
              T. Baytop: loc.cit., *OIMC,* p. 152

(1921)   SOCIÉTÉ DES SCIENCES MÉDICALES
              (Publication: *Pages Médicales de Constantinople* (1921), ISIS-
              PFDT, p. 367)

1922   TÜRK DİŞ TABİPLERİ CEMİYETİ
              "Türk Diş Tabipleri Cemiyeti ve Teavün Sandığı nizamnamesi",
              Hüsnü Tabiat Mecmuası, Istanbul, 1938

X

*Modernization Efforts in Science, Technology and Industry
in the Ottoman Empire (18-19th centuries)* *

The Ottoman Empire, which was established as a small principality at the turn of the 14th century, gradually expanded into the lands of the Byzantine Empire, both in Anatolia and the Balkans. After 1517, when its sovereignty reached Arab lands, and the empire became the most powerful state in the Islamic world and one of the greatest states in Europe, covering a vast area extending from Central Europe to the Indian Ocean. For centuries the Ottoman Empire maintained a delicate equilibrium in their struggles with Europe. When the balance of power turned definitely against them, the Ottomans started to lose their lands. However, it is only following their defeat in the First World War that the empire disintegrated and, after 1923, disappeared from the world scene.

The empire was once very wealthy. Situated on the crossroads of the great trade routes between Asia and Europe, rich in mineral resources, it was influential in world trade and regulated the economic balance between the East and the West. In addition to the material superiority acquired after the conquest of Egypt in 1517, the Ottoman sultans also became the spiritual leaders of the Islamic world by assuming the caliphate, that is, the highest authority in Islam.

By the reign of Süleyman I (1520-1566) continuous military success in an area stretching across three continents had made the Ottoman Empire a world power. Yet, in the 17th century, fortune turned in favor of Europe. To illustrate the image the Ottoman sultans had of themselves and their empire during this period, let us quote the following lines by Sultan Süleyman the Magnificent:

«I am God's slave and sultan of this world. By the grace of God I am the head of Mohammad's community. God's might and Mohammad's miracles are my

* This paper is the expanded version of the presentations that were submitted at the Congress on «Scientific and Philosophical Thought in the Greek Intellectual Area» organized by the Institute for Neohellenic Research, National Hellenic Research Foundation in Athens on 19-21 June and the international symposium on «Introduction of Modern Science and Technology to Turkey and Japan» organized by the Research Centre for Islamic History, Art and Culture (IRCICA, Istanbul) in cooperation with the International Research Center for Japanese Studies (IRCJS, Kyoto) and the Turkish Society for History of Science (TBTK, Istanbul), held on 7-11 October 1996, Istanbul.

companions. I am Süleyman, in whose name hutbe is read in Mecca and Medina. In Baghdad I am the shah, in Byzantine realms the Caesar, and in Egypt the sultan; who sends his fleets to the seas of Europe, the Maghrib and India. I am the Sultan who took the crown and throne of Hungary and granted them to a humble slave. The voivoda Petru raised his head in revolt, but my horse's hoofs ground him into the dust, and I conquered the land of Moldavia»[1].

The administrative and bureaucratic practices of the Ottoman Empire were a progressive continuation of the ancient traditions of the pre-Ottoman Islamic states. The sharing out of functions within the administration was in accordance with these traditions. Islamic political theory recognized the «Men of the Pen», *Kalemiye*, besides the «Men of the Sword», *Seyfiye*, and the «Learned Men of Religion», İlmiye, as three pillars of the administration. Above all these three groups were the sultan and the household of the palace who served him. Thus, the state was built upon these four pillars.

Among the auxiliary institutions of the central government, a number of civil commissionerships such as those for the mint, customs and cereals, and a number of military organizations, such as the Janissary corps, gun-foundry and arsenal, had their own offices. In the provinces, governors, *kadıs'* courts and numerous important *vakıfs* had their own secretarial staff. A commissioner, with a secretary to assist him, was appointed for all state undertakings, whether construction, mining, manufacture or agriculture. Thus, the total number of secretaries was far greater than the limited body who worked in the offices of the central government.

The Men of Religion were educated in the medreses, the Men of the Pen were trained in the scribal offices, and the Men of the Sword in the *ocaks* (Corps) and the *enderun* (the palace school); these three state pillars all received professional training together with a general education. The aim of this type of education based on Islamic tradition was to give the best possible training to the administrators and experts who governed state affairs. Several successful statesmen, lawyers, physicians and astronomers were educated in these institutions.

These men were cultured, knowledgeable, and well-versed in Arabic and Persian as well as Turkish. They also had a thorough knowledge of classical Islamic literature, and contributed to the literary activities of their time. In addition to formal education, nearly all Ottomans, including the sultan, were affiliated to Sufi orders. The knowledge and experience provided by Sufism enabled them to become open-minded, self-controlled, and to conduct themselves with religious sincerity and tolerance. All these elements of cultural education and spiritual discipline were integral to the *adab-ı Osmanî* (The Ottoman way of life).»[2]

Charitable institutions in Islamic society, were established as *vakıfs*. This ensured the continuation of public services, since a *vakıf* devoted in perpetuity the profits from a particular source to a charitable purpose, without touching the capital. In the *vakfiye* –the deed of endowment, recorded in the *kadi's* register – the founder of a *vakıf* defined the endowment, its purpose, conditions and forms of management, and appointed the *mütevelli*, the endowment's chief trustee.

In general, the *vakıf* was a complex that included the following buildings grouped around a mosque: a medrese, an imaret for cooking and distributing food, a hospice (*tabhane*), a hospital, (asylum and *daruşşifa*), a bathhouse (*hamam*), caravanserai, and shops. This complex had the nature of a religious, cultural, and social institution. These vakıfs played a significant role in the construction of new cities, and reestablishment of the *mahalles* (quarters) in the old cities. They helped build houses for the staff of the vakıfs, and funded municipal services such as water and sewerage systems. Each shop was allocated to masters of a particular trade and craft, such as bakery, mills, candle manufacturing, dyeing and butchery. So was the marketplace, which provided perpetual secure income for the vakıf[3].

Ottoman Social and Economic Background

The Ottoman state had a powerful central authority and an effective bureaucracy. The state saw to it that the people lived in an order manner by enacting örfî (customary) laws, by enforcing the principles of the *shari'a* (Islamic law) and by applying the töres, the social rules and customs. In other words, the Sultan's central power organized economic, social and religious life. The people worked in trades, industry and agriculture in accordance with the needs and policies of the state.

During the Ottoman classical period, people of different ethnic origins and belief systems coexisted peacefully. The sultan held absolute authority over the state and, at the same time, he assumed the function of the caliph, i.e. the religious leader of the Muslims who constituted a majority of the society. As the absolute ruler of the empire, the sultan insured that the subjects of various religions and ethnic groups would live together in peace under the social order known as the *Nizâm-ı Alem* (Pax Ottomana).

Legal and juridical problems in society were solved with the Islamic and the customary laws. The members of the learned-religious institution, who represented one of the four pillars of the state, ran the judicial system. Compared with the other three classes, the *ilmiye* received a longer education, and they organized Ottoman religious and educational life. The *kadı*, a member of the *ilmiye* class, supervised legal, municipal, and civil matters in both the cities, and the administrative and juridical districts of the provinces. Another important member of the *ilmiye* class was the *müderris*, who influenced educational and

X

scholarly life. The *şeyhülislam*, who headed the hierarchy of the *ilmiye* class, and the *müftis*, his representatives in the provinces, issued *fetvas* regarding religious and administrative matters, and also on subjects related to the daily life of the people.

Briefly, the *ilmiye* class played a key role in regulating law, administration and education. All their members were educated in the medreses, the main educational institution.

As the Ottoman state expanded and spread across three continents, its financial institutions gradually developed to meet the empire's growing needs. Though the essential characteristics of the financial system were inherited from the previous Turkish-Islamic States, the Ottomans developed them further[4]. One of the key principles of the classical Ottoman system was *provisionism*. Its purpose was to ensure that all goods and services would be cheap, plentiful, and of good quality. With respect to foreign trade, provisionism sought to maximize the supply of goods and services to the internal market. Exports were discouraged by prohibitions, quotas, and taxes. Imports, by contrast, were encouraged and facilitated.

A second important principle of Ottoman economic policy was *traditionalism*, namely the impulse to preserve existing conditions and look to the past for models, rather than search to accommodate changing circumstances. Traditionalism found its expression in the time-honored motto that one should not work against what comes from the olden times (*Kadimden olagelene aykırı iş yapılmaması*). It remained unchanged as a vital component of Ottoman economic thought until the end of the 18th century.

Fiscalism was the third principle, and its goal was to maximize treasury income, and to prevent it from falling below already-attained levels. Influenced by the production capacity of the Ottoman economy and the degree of monetization, the growth in treasury income was achieved only slowly and with difficulty. Ottoman fiscalism thus developed mainly in the direction of preventing drops in income and reducing expenses. A rigid fisco-centrism evolved which viewed all economic activity only in terms of the tax income they would yield[5].

Towards the end of the 17th century, facing European military and economic superiority, the Ottomans came to recognize the inadequacy of their organization and the need to reform classical institutions. As Professor İnalcik rightly observes, the Ottomans were unable to adapt themselves to the changed conditions. They failed to understand modern economic problems, and remained bound by the traditional formulae of the near-eastern state. Against the mercantilist economics of contemporary European powers, Ottoman statesmen adhered to the policy of free markets, their main concern being to provide the home market with an abundance of necessary commodities. Unable to formulate a comprehensive economic policy for the empire, they saw no danger in extending the capitulations, so that from the second half of the 16th century Europeans even began to control the transport between the empire's Mediterranean ports. The Ottoman government, bound by traditional concepts, encouraged the import of goods into the empire, but discouraged exports. They taxed imports and exports at the same rate, and prohibited the export of certain goods which could cause shortages in the home market. By keeping restrictions on corporations they hindered development in some branches of industry and exports[6].

The weakness of the Ottoman army became apparent when it was defeated at Vienna in July 1683. Following this fiasco, the Ottomans began grudgingly to accept the superiority of the West in some fields. From then on, they entered a period during which they looked more attentively to the West for solutions to some of their problems[7].

The Ottoman scientific tradition and the impact of the West during the 18th and 19th centuries

Significant changes occurred in Ottoman scientific and educational life within a broad span of time. Therefore, we cannot label the radical changes in Ottoman history as the immediate consequences resulting from specific needs or initiated at particular fixed dates but rather an outcome of a long process of transformation. Generally, «the old and the new» were practiced concurrently.

Ottoman science emerged and developed on the basis of the scientific legacy and institutions of the Seljukid Turks. It greatly benefited from the activities of scholars who came from Egypt, Syria, Iran, and Turkistan, which were homelands of some of the most important scientific and cultural centers of the time. The Ottomans preserved and enriched the cultural and scientific heritage of the Islamic world, giving it new dynamism and vigor. Thus, the Islamic scientific tradition reached its climax in the 16th century. Besides the old centers of the Islamic civilization, new cultural and scholarly ones flourished in such places as Bursa, Edirne, Istanbul, Skopje, and Sarajevo. The Ottoman cultural and scientific heritage which developed in this period constitutes the cultural identity and scientific legacy of present-day Turkey as well as of several Middle Eastern, North African, and Balkan countries.

It was in their own scientific traditions that the Ottomans found the solutions to the intellectual issues they faced. The medreses which were the principal institutions of education and learning in the Ottoman Empire were organized in a manner to meet the needs of the state and society. Indeed, the Ottoman educational system was self-sufficient in every respect. Therefore, the Ottomans did not consider the transfer of science from contemporary European states necessary during their heyday[8].

X

Because of their feeling of moral and cultural superiority as well as their economic self-sufficiency, the Ottomans did not feel the need to follow up the scientific and intellectual developments of the Renaissance and the Scientific Revolution. These did not seem essential to them. It would be an anachronistic evaluation to say –as some modern historians claim– that they did not realize that these developments would challenge them in the future. But this did not mean that they were not aware of the technical developments and geographical discoveries of the 15th and 16th centuries. The developments that Europe witnessed were followed by the Ottomans in different ways and means, selectively in the fields of warfare, mining technology, geography and medicine.

Geography and close historical ties made the Ottoman world the first nonwestern environment to which Western science spread. Proximity allowed the Ottomans to learn early on of European innovations and discoveries. The Ottomans began, already in 15th century, to transfer Western technology (especially firearms, cartography, and mining), as they also had some access to Renaissance astronomy and medicine through emigrant Jewish scholars. The interests of the Ottomans remained selective, however, because of their feelings of moral and cultural superiority and the self-sufficiency of their economic and educational system[9].

As a result of our studies, it appears that from the 17th century to the beginning of the 19th century, Ottoman knowledge of the West came through three main channels: 1. translations made from European languages; 2. personal observations of Ottoman ambassadors who paid official visits to Europe; 3. modern educational institutions.

Translations and adaptations from European languages

From the 17th century onwards, we find many translations of Western scientific writings. Here, we shall attempt to follow the introduction of modern scientific concepts to the Ottoman scientific milieu. As far as we can establish, the first work of astronomy translated from European languages was the Ephemerides Celestium Richelianae ex *Lansbergii Tabulis* (Paris, 1641), astronomical tables by the French astronomer Noel Duret (d. ca. 1650). The translation was made by the Ottoman astronomer Tezkereci Köse İbrahim Efendi (Zigetvarlı) in 1660 under the title, *Secencel el-eflak fi gayet el- idrak*. This translation was also the first book in Ottoman literature to mention Copernicus and his heliocentric system. The initial reaction of the chief astronomer of the Sultan was to declare the book a «European vanity», but after learning of its use and checking it against Uluğ Bey's *Zic* (astronomical tables), he realized its value and rewarded the translator. His first reaction, however, typifies Ottoman reluctance to accept Western scientific superiority.

Heliocentricity, which stirred such controversy in Europe, was viewed by Ottoman astronomers just as an alternative technical detail, and it didn't become a subject for polemics. One probable reason is that the heliocentric theory of the universe didn't conflict with any religious dogmas. Astronomy books subsequently translated from European languages also dealt mostly with astronomical tables.

Among the translations completed between the 17th and 18th centuries was a major book on modern geography by Abu Bakr b. Bahram el-Dimashki, which was based on Janszoon Blaeu's Latin work, *the Atlas major* (1685). In the 18th century, Müteferrika translated from Latin Andreas Cellarius' *Atlas coelestis* (1708). He completed the translation in 1733, and called it *Mecmuatü'l- hey'eti'l- kadime ve'l- cedide*, meaning the collection of old and new astronomy. In 1751, Osman b. Abdulmennan translated Bernhard Varenius' work from Latin, and called it *Tercüme-i kitab-ı coğrafya* (Translation of geographia generalis). In addition to these translations, classical Ottoman astronomy and geography and related scientific activities continued within their traditional framework.

A survey of scientific literature in Turkish thus shows that after overcoming their feelings of superiority, Ottoman scholars readily accepted new concepts, information and techniques. The administration had a positive outlook, and the religious scholars (*ulema*) were not particularly hostile – as their response to the heliocentric system shows. There was no obvious conflict between religion and Western science at this stage[10].

From the 16th century onwards, the arrival of physicians and diseases from the West introduced new medical ideas and methods of prophylaxis and treatment. From the 17th century onwards, the medical doctrines of Paracelsus and his followers began to appear in Ottoman medical literature under the names of *tıbb-ı cedid* (new medicine) and *tıbb-ı kimyaî* (chemical medicine)[11].

One of the most prominent followers of this trend was Salih b. Nasrullah (d. 1669). In his work entitled *Nuzhatu'l-abdan*, he quoted from various European representatives of the new medicine, and gave the compositions of remedies. Al- ézniki (18th century), likewise cited Arab, Persian, Greek and European physicians together in his *Kitab-ı kunuz-i hayat al-insan kavanin-i etibba-yı feylesofan*, and presented new medicines alongside the old. Ömer Şifai (d. 1742), too, in his work *al-Cevher al-ferid* states that the remedies he gives were taken from the books of Latin doctors, and that he translated them from European languages into Turkish. In this way, recent medicine of European origin was practiced side-by-side with traditional medicine until the beginning of the 19th century[12].

A characteristic of the 19th century scientific translations or compilations –some clear examples of which can be seen at the turn of century– is the coexistence of modern and Turkish-Islamic traditions. Examples can be found of works where the geocentric and heliocentric systems of the universe were introduced together; similar cases can be traced in medical writing. 18th century Ottoman medical works transferred practical medical knowledge from Europe, but old concepts (such as the concept of humors) still prevailed in physiology and anatomy. Modern physiology was introduced to the Ottoman world through the work translated from Italian by Hekimbaşı Mustafa Behçet Efendi (1774-1834) under the title *Terceme-i fizyolocia* (Translation of physiology), ca. 1803. The Ottomans became acquainted with modern anatomy through *Hamse-i Şanizade* (1820-1826), a five-volume work, by the Ottoman physician Şanizade Ataullah (1771-1826)[13]. During the second half of the 19th century, particularly after 1870, when it was decided to conduct medical education in the Imperial School of Medical Sciences in Turkish, several books on medicine were translated into Turkish and published, starting with *Lugat-ı tıbbiye* (Medical dictionary), 1873[14].

Towards the end of the 18th century, teachers at the *Mühendishane-i Berri-i Hümayun* (Imperial School of Engineering), which was established to teach modern sciences to the army officers, started to translate and compile books from European scientific literature. In general, the instructors relied on the textbooks used in European military technical schools. Among the first scientific books, which were published at the turn of the 19th century, were about ten books on mathematics and engineering compiled and translated by Hüseyin Rıfkı Tamani (d. 1817), the first chief instructor of the *Mühendishane*. ishak Efendi (d. 1836), another chief instructor of the Mühendishane, published thirteen volumes based on Western sources, especially French. Among these, the four-volume *Mecmua-i ulum-ı riyaziye* (Compendium of mathematical sciences; Istanbul, 1831-1834) is particularly important, since it was the first large-scale attempt to present in one of the languages of a Muslim nation, a comprehensive textbook of the various sciences, such as mathematics, physics, chemistry, astronomy, biology, botany and mineralogy. İshak Efendi's efforts to find Turkish equivalents for new scientific terms and to transfer modern science had an influence that extended well beyond the borders of Ottoman Turkey to other Islamic countries[15].

Instructors at the *Mühendishane*, as well as the graduates and instructors at the *Mekteb-i Tıbbıye-i Şahane* (Imperial School of Medical Sciences; reformed and opened in 1838), continued to translate European scientific books. After the proclamation of the Tanzimat (1839), modern education became widespread, civilian education was reorganized, and new scientific and technical books were printed. The mid-19th century thus witnessed an increase in both the number of printed books on modern science and techniques, and in the variety of subjects introduced. Earlier, between the establishment of the first Turkish printing press in 1727 and the proclamation of the Tanzimat in 1839, 28 books of science were printed; during the Tanzimat period (1840-1876) the figure jumped to 242.

There was a numeral increase in the books printed on mathematics and medicine, but a decrease in the number of works published on geography, military sciences, engineering, astronomy and navigation[16]. On the other hand, after the Tanzimat, some different subjects were tackled. For example, Derviş Paşa published the first chemistry book in Turkish entitled *Usul-i kimya* (Elements of chemistry, Istanbul 1848) and Hekimbaşı Salih Efendi printed *İlm-i hayvanat ve nebatat* (Zoology and botany, Istanbul 1865), the first book on these subjects in Turkish. Moreover, during the first three decades of the Tanzimat period (1839-1869) four books were printed every year, but during the last seven years (1870-1876) this number rose to eighteen annualy. These figures reflect the growing Ottoman interest in Western science.

The drop in publications on sciences such as geography, military sciences, and engineering indicates a shift of interest from the military to the civil realm. A parallel result can also be inferred from the prefaces of works compiled on the same subjects during the 19th century (before and after the Tanzimat). Ishak Efendi, in his *Mecmua-i ulum-i riyaziye*, mentions the importance of chemistry for the war industry, and Kırımlı Aziz Bey in his *Kimya-yı tıbbî* (Medical chemistry, Istanbul 1868-1871) pointed out that chemistry, in addition to its usefulness for medicine, was the basis of several industries and techniques mainly of non-military character.

During the 18th and 19th centuries, the languages of Ottoman scientific literature were Turkish and Arabic. Although there were also Persian works, these were rather rare, less than 1%. However, the ratio of these languages differed in manuscripts and printed works. While nearly all of the books about modern science and technology, printed during these two centuries (with a few exceptions) were in Turkish, most of the manuscripts were produced in Arabic. In the 18th century, for example, 72.1% of the manuscripts were written in Arabic and 24.7% were written in Turkish. In the 19th century, the great majority of astronomical works were in manuscript form beside the printed works. Of these, 53.8% were in Arabic and 41.4% in Turkish. Thus, Turkish was increasingly used for writing on astronomy, but Arabic manuscripts remained more numerous, as usual. Although we do not have statistical figures for other branches of science, we may presume a pattern parallel to that found in astronomy. In sum, the scientific manuscripts of the 18th and 19th centuries were in Arabic and Turkish, while most of the printed literature was in Turkish[17].

Broadly speaking, then, our survey of the scientific literature of this period may be summed up thus: with some exceptions, theory, experimentation and research -the major constituents of Western science-were largely neglected by Ottoman statesmen and scholars. Most of the modern scientific literature of this time was limited to textbooks for high schools and higher educational institutions.

New educational institutions and the westernization of military technical training

The second main channel through which the Ottomans became acquainted with Western science and transferred it into their world were the institutions of military technical training established from the first half of the 18th century. During the last decade of the 17th century, the rapid development of European military techniques led to a series of Ottoman defeats. Until then, the Ottomans had managed to manufacture cannons, guns, and ships based on techniques imported from the West. It was not difficult early on for the Ottomans to keep up with European technology, for it changed relatively slowly. Large state enterprises such as the *Tersane* (Maritime Arsenal), *Tophane* (Arsenal of Ordnance and Artillery), *Baruthane* (Powder mill) and Darphane (Mint), established by the state, functioned fairly successfully to meet the needs of the military, for the technological gap between the Ottomans and Europe was not yet great.

Forced into constant retreat in Central Europe, the Ottomans gave up their policy of conquest and began to follow European developments closely, turning their attention to the cultural and technical sources of European superiority. Thus commenced a period of affluence, called the Tulip Age (1718-1730); under Western influence, new developments emerged, not only in the technical fields, but also in art and architecture. However, disagreements arose between the followers of a peace policy and their opponents, and erupted in a reaction against Nevşehirli Damat İbrahim Pata, the Grand Vizier. Extravagance and luxury incited this reaction which led to the rebellion of Patrona Halil (1730) ending the Tulip Age.

The Tulip Age saw a change in Ottoman-European relations, and innovations such as the fire pump and the printing press were established in the Ottoman capital. Ottoman administrators who learnt about European daily life via European ambassadors also developed a great interest in nonmilitary European inventions. New ways of thinking and understanding emerged.

During the 18th century, innovation in European war technology began to accelerate, and, compared with the earlier periods, it became harder for the Ottomans to keep pace. The Ottomans compared their armies with the military powers of Europe and they decided to transfer European technology in order to redress the imbalance in power. Naturally, they had to consult European specialists to apply this technology. Until the 1770s, the Ottomans tried to meet this need by employing European specialists who had converted to Islam.

The Ottomans sought gradually to import Western military science and to modernize their army. A significant change occurred in Ottoman military education, with the establishment of new *ocaks* (corps) within the military organization and new drills were put into practice. A first attempt was the creation in 1735 of the *Ulufeli Humbaracı Ocağı* (Corps of Bombardiers), under the supervision of Comte de Bonneval of French origin. He took refuge with the Ottoman state in 1729, took the name Ahmed after having embraced Islam and became known as Humbaracı Ahmed Pasha.

The corps of bombardiers, with a new arrangement securing a more stable salary payment, was established around the beginning of 1735, and was regulated differently from the other corps of Ottoman military organization both in terms of military and administrative aspects. Besides undergoing drills, the bombardiers in this corps received theoretical training in geometry, trigonometry, ballistics and technical drawing[18].

As a result of the changing European trade policy and the close contacts between the Ottomans and Europe in the second half of the 18th century, relations were set up with France in order to transfer military technologies through the import of European experts. Thus, a group of French experts came to Istanbul within the framework of military aid agreements. Since most of them were invited by the Ottoman State, they were not expected to change their religion. An increasing number of experts and officers flooded to Istanbul as the Europeans working in the higher positions of the Ottoman State were not asked to convert to Islam. This situation can be considered as a type of brain drain from Europe to Ottoman Empire.

Among the French experts who came to Istanbul was Baron de Tott who was of Hungarian origin. He was employed in building fortifications and in teaching new European military techniques. Firstly, in the autumn of 1770, he built new fortifications and artillery batteries on the Dardanelles. Thus, he proved himself to the Ottoman statesmen who were still hesitant about the recruitment of Christians in the army. Baron de Tott stayed in Istanbul until 1776. During his stay, he supervised the establishment of a new foundry where French workers cast new types of cannon balls. He also installed a European device called St. Maritz, which was used for boring holes and polishing cannons, a system which was used in Europe at that time. Another of his contributions was the founding of the *Sür'at Topçuları Ocağı* (known in French as the Corps de Diligents), where artillerymen were trained in the European manner. De Tott further played a role in establishing a school (Ecole de Mathématiques or Ecole de Théorie, 29 April 1775), where courses were given for the first time on theoretical mathematics and military techniques. In this period, the Ottomans began to import guns and gunpowder as well as warships from Europe, and they gradually became more dependent on European technology.

Baron de Tott introduced European techniques to the Imperial Maritime Arsenal as well. Thus, due to his efforts, new techniques, which were not used in Ottoman Imperial Maritime Arsenals and Imperial Foundries beforehand, were applied. Despite Baron de Tott's personal ambitions, the policies of the French government and the objectives and needs of the Ottoman state were very different, perhaps even opposed. Baron's presence in the Ottoman lands and his cooperation with the Ottomans lasted six years, and he returned to France when local, French, and personal interests ceased to overlap.

France and England stood on opposite sides in the American War of Independence (1775-1783). The czarina Katerina II took advantage of their conflict vis à vis their European policies to promote Russian expansion in Eastern Europe and the Ottoman lands. When the war in America ended, a new policy emerged in Western Europe to support the Ottoman state through military supplies, technical aid and assistance for the modernization of its army. This policy reflected the European perception that the Ottoman state could halt Russian expansion in the region. For their part, the Ottomans were greatly alarmed when they lost Crimea, and started preparing for a war against Russia to regain this territory.

The Ottoman Grand Vizier Halil Hamid Paşa (1782-1785) requested technical assistance and a fleet from Europe and also enacted laws for the training and discipline of the army. Between 1783-1788, numerous French military experts and officers came to Istanbul within the framework of military aid and cooperation to work on various technical projects and the fortification of the Ottoman borders. The first task was to fortify the Bosphorus against a possible Russian attack. Over fifty engineers, supervisors and workers, with experience in gunnery, bombarding and shipbuilding, came to Istanbul for this purpose. Among them, two engineers by the names of Lafitte-Clavé and Monnier have a special place.

Besides their basic work, Lafitte-Clavé and Monnier also taught fortification at the Imperial School of Engineering which was opened at the Maritime Arsenal. The gunner and military engineer Saint Rémy arrived during this period, and persuaded the Ottoman statesmen in 1785 to establish larger and more modern gun casting furnaces instead of the old ones. However, the new furnace, built at great expense, proved unprofitable, and Saint Rémy was sent back to France. Between 1785-1787 the French shipbuilding engineer Le Roy and his team of ten people were employed at the Maritime Arsenal and built more than a hundred warships of different sizes.

When the Ottomans went to war with Russia between 1787-88, an alliance was formed between Russia and France. All of the French experts and masters thus left Istanbul, and native Ottoman masters and workers were employed in their place. Likewise, the French teachers in military technical schools were replaced by Ottoman professors, the *müderris*, members of the *ulema* class, and technicians from other European countries such as Sweden. More French technicians came to Istanbul after the French Revolution, but they were sent back when Napoleon invaded Egypt[19].

X

The military engineering training which began in the 18th century underwent significant transformations during the 19th century. The second attempt towards a western type military education began with the training of physicians. All these new practices stemmed from the priority given to military needs. However, Western type elementary schools, high schools and civil higher educational institutions were established in the 19th-century Ottoman Empire. These new types of institutions replaced the old ones, particularly the *medreses* which were subsequently neglected as we shall elaborate in our following article.

In the 19th century, European technical knowledge continued to filter through previously established schools of engineering and also through the students sent abroad to study in various fields. The Imperial School of Medical Sciences, created in 1838, played a crucial role in the introduction of modern medicine into Ottoman Turkey.

Ottoman technological transfer and the establishment of modern industry during and after the industrial revolution in Europe

In Central Europe a period of mercantilism, the beginning and duration of which were different in various countries, preceded the Industrial Revolution which began at the end of the 18th century. Likewise, in the beginning of the 18th century, when the palace and the officials in İstanbul were more receptive to Western influences than before, there was a similar period of transition between traditional artisanal society and the foundation of factories in the 19th century[20]. During this «period of manufacture» the guilds mainly dealt with small-scale production[21].

The changes that took place in Ottoman industry in the 18th century were not comparable in any way to the explosive industrial growth then occurring in Western Europe. But neither was Ottoman industry totally stagnant, on the contrary, it displayed a complex pattern of development shaped by various regional influences and different industrial sectors.[22].

Around the beginning of the 18th century, the Ottoman state sought to meet the economic challenge of Europe and its own need for manufactured goods through industrial initiatives based on traditional technologies. As Mehmet Genç has noted, however, this situation changed after 1709:

«From 1709 onward, we come across two important changes. The first relates to the organization of the production: rather than being the exclusive prerogative of the state, management was entrusted to an entrepreneur with his own capital and liability along the lines of profit and loss. The second concerns the quality of woolen production: the decision was made to locate and import technological know-how from abroad because it was realized that such skills were not to be found within the empire»[23].

During the early decades of the 18th century, workshops for wool, silk and porcelain were established in Istanbul. These were: a workshop of wool-cloth equipped with machines and equipment, a workshop of silk cloth that was set up immediately afterwards and a workshop of porcelain which was founded in 1718. A *basmahane* (cotton printing manufacture) and a dye-house were established in the 1720s. A small paper workshop was founded in Yalova in 1744-1745. Although the wool and paper workshops were closed a few years later, the other enterprises continued to function for many years[24]. Genç observes:

«The inability to obtain technology and the increasingly abundant and cheap manufactured imports made it imperative that the state apply a strictly protectionist regime over the long term. On the one hand, the Ottoman Divan (imperial council) provided every sort of advantage possible in the area of interest-free credit, even long-term capital without repayment. It also helped in securing supplies of raw materials, locating and settling workers, and offering broad tax exemptions. On the other hand, the state's guiding principle of provisionism in no way permitted any form of mercantilistic protectionism, either by curtailing imports, or by imposing duties that would raise domestic prices»[25].

Thus, according to Genç, the industrialization attempt was a failure from the beginning because it didn't apply the principles of mercantilism.

Fostered by the reforms of Sultan Mustafa III (1757-1774) and Selim III (1789-1807), the workshops established in the 18th century were generally designed to serve the military. The attempts to establish large-scale mechanized factories stimulated the transfer of technology from the West.

Following his accession to the throne, Sultan Selim III initiated a reform movement called the *Nizam-ı Cedid* (New Order) which reorganised the military and implemented reforms in the army. During the 1790s, Sultan Selim III took a personal interest in improving the manufacture of weaponry. As early as 1793-1794, he introduced contemporary European methods and equipment for the production of cannons, rifles, mines, and gunpowder. As late as 1804, for example, he undertook the construction of elaborate buildings to house a woolen mill for uniforms and a paper factory near the Bosphorus, at Hünkar İskelesi. Following the overthrow of Selim III, in the first two decades of Sultan Mahmud II's reign, few, if any, industrial improvements seem to have been attempted, but a burst of activity followed this gap. A spinning mill was built near Eyüp in Istanbul in 1827, a leather tannery and boot works in Beykoz were improved early in the 1830s, a part of the paper factory located at Hünkar iskelesi was converted to a cloth manufacture in the same years[26]. After the abolition of the Janissary Corps (1826), the army adopted European-style equipment; this however, worked against domestic self-sufficiency. By 1841, the need for a massive industrial program became obvious.

X

A factory called Feshane was opened in 1832-1833 to manufacture the fez, the headgear for the soldiers of the *Asakir-i Mansure-i Muhammediye*, which replaced the Janissary Corps. The fez was also commonly worn by the general public. Originally imported from Tunisia, the fez began to be produced in Istanbul after 32 master craftsmen were brought from Tunisia. In 1839, when the building in Kadırga became too small, the *Feshane* was moved to its new building in the Defterdar quarter of Eyüp. In 1841, the Ottomans began to produce the fez with steam powered machines instead of the spinning wheel based on organic energy. This steam engine used was bought from Belgium upon the advice of Belgian experts administering the factory. A large part of the fezzes produced in the *Feshane* were bought by the state to meet the needs of soldiers and officials. The rest were sold to the people in the *Feshane* shops. Since they were of good quality, they were popular among the people. Yet from the beginning the factory faced stiff foreign competition and soon after its establishment, the French founded their own factory and started shipping fezzes to Istanbul and other Ottoman cities[27].

The competition faced by the *Feshane* contributed to its success, because the Ottomans felt compelled to keep up with new technology. Between 1847-48, they increased production with machines imported from England and Belgium. Profits were invested in the new machines. With the partial exception of the *Feshane*, early attempts to introduce European industrial methods focused exclusively on the manufacture of goods for governmental and military use.

Upon the proclamation of the Tanzimat, a series of laws and legislations were enacted to re-establish the political power and the new developmental policies were accepted. Thus, economic enterprises were established, most financed by the *Hazine-i Hassa* (the sultan's privy purse). The enterprises set up between 1847-48 can be called imperial factories. Among these, were the Zeytinburnu Iron Factory, Izmit Woolen Cloth Factory, Hereke Silk Cloth Factory, Veliefendi Printed Wool-Cloth Factory, Mihalic State Farms, the School of Iron Ore and Agriculture in Büyükada. There were also plans to open *talimhanes* (training courses) on «mines, geometry, chemistry and sheep-breeding»[28].

The Zeytinburnu Iron Factory was another large heavy industrial complex. To serve the army's needs, the Gun foundry and the Arsenal were installed with steam engines. The Arsenal at the Golden Horn (Haliç), provided with European equipment and personnel, typifies the attitude of the Ottoman state toward technology transfer. Like the Feshane, it employed numerous foreign (particularly English, French and American) workers and administrators. Until the end of the century, the Ottoman industries equipped with the Western technologies depended largely on foreign labor force[29].

The main purpose of founding and building imperial factories was to produce the necessary materials for the army, and to meet expenses with internal rather than external resources. Different kinds of factories multiplied in the imperial domains. It was essential that production increase, and since these factories produced goods, the state encouraged them by not buying similar goods from abroad in order to balance exports and imports. Above all, the state applied measures of frugality to reestablish this balance of trade. The bureaucrats of the Tanzimat were aware that it was as important to encourage exports as to limit imports[30]. Nevertheless, this objective went unrealized because as Donald Quataert states:

«Other government policies and attitudes clearly retarded the pace of technology transfer and industrial development. The Ottoman regime continued to impose and maintain tariff structures that were very unfavorable to industry, for example, retaining duties on the flow of goods within the empire until nearly the end of the century...»

European insistence on low import duties was also a critical factor[31].

As a Belgian worker in İzmit observed in 1848:

«It would be very odd if we could not turn out a piece of the finest cloth occasionally, seeing that we have the best machinery from England and France, that the finest wools are imported, via Trieste, from Saxony and the best wool countries, and that we Frenchmen and Belgians work it. You could not call it Turkish cloth; it is only cloth made in Turkey by European machinery, out of European material, and by good European hands»[32].

These words indicate the degree of Ottoman dependence on Europe. During the first half of the 19th century, modest steps towards industrialization were taken with the establishment of the above-mentioned factories in Istanbul and its surroundings. However, these attempts were interrupted due to the Crimean War (1854-1856) and the resulting financial crisis.

By building factories, the Ottomans aimed to stimulate the industrialization of the country. They thus attached high priority to spreading industrial education and they invited masters from Europe to teach modern techniques to domestic workers and masters. The state also sent students to Europe to bring back new industrial technology. It further tried to ensure that these factories ran profitably by regulating external and internal trade, changing the customs policy, and finally, applying subventions.

The industrial development program of the 1860s-1870s initiated by the official Industrial Reform Commission sought to mechanize several sectors such as tannery, shoe-making and silverwire manufacture. But the program hardly advanced beyond the planing stage. This Commission seems to be the last example of a concerted state effort towards factory formation in the Ottoman empire. Thereafter, the state shifted its emphasis from building factories to encouraging their creation by entrepreneurs, though we shall see that it continued to found some factories[33].

To reduce its economic dependence on Europe, the state reformed customs regulations in 1875, and enacted laws allowing the establishment of new factories. Machines and tools for factories using advanced technology were allowed to enter duty-free, and the yarn produced in these factories was declared exempt from all internal and export duties. Concessions were granted to individuals to found factories. The role of the entrepreneurs in industry gradually increased, and the number of factories grew significantly in the 1880s, when three-quarters of the Ottoman factories were established.

The first factory, the Paşabahçe Glass Factory, was founded in 1884, and has functioned successfully to the present day. It was established with modern equipment, and its workers were imported from Europe, particularly Bohemia. By contrast, the Hamidiye Paper Factory, founded in 1886, proved a failure despite its modern machinery. Profitable production was made impossible by intrigues and financial maneuvering and the need to import all raw materials. This is just one of many examples of insufficient planning in Istanbul which, combined with European competition, forced numerous factories run by entrepreneurs to shut down[34].

The civilian reforms undertaken by the state during the reigns of the sultans Mahmud II and Abdülmecid in the 19th century, aimed at the prosperity of the country and the well-being of the people. In addition to their efforts in the economic sphere, the Ottomans started to transfer and apply modern transportation and communication technology from the West. The goal here was both to strengthen state central authority and to answer the needs brought about by social and economic changes. In the cities, the old structures such as the khans and caravansaries were replaced by buildings such as railway stations, warehouses, quays, and hotels. Traditional customs gave way to new hobbies and forms of entertainment such as the theater[35].

During the Tanzimat period, when administrative and civilian reforms were planned, the Ottomans turned their attention to new objectives. In addition to these attempts towards industrialization, the Ottomans benefited to the utmost from the means provided by modern technology to implement the reforms in the field of public works which in turn greatly improved the people's life standards.

The attempts in the 19th century to industrialize, and to transfer Western technology didn't yield the expected results. This limited success may have resulted from mistakes in Ottoman policies and the pressure of foreign powers. To begin with, it was very hard for the Ottomans to find the necessary capital for industrialization; western capital investments, which entailed heavy conditions and difficulties, did not develop in the direction that they wished. Instead, western investments favored the interests of the non-Muslim subjects and ethnic groups who had cultural affinities with Europe. Moreover, the West quite naturally made its investment decisions with an eye, above all, toward profits.

X

62

Under these conditions, the Ottomans tried to cling to the suitable opportunities in world economy on the international platform. However, within the framework of the political and economic relations of the Ottomans with Europe in the 19th century their attempts to transfer modern technology and to found independent industrial enterprises were also hindered by deep-rooted European hostility.

Although the initiatives in heavy industry met with limited success, there was a rapid growth in low-level technology transfer. For example, the yarn and dye technologies were adopted quickly and quite extensively. These cheap and easy-to-use examples of contemporary technology supported lively production in small-scale workshops and homes. There was considerable innovation and adaptability here. As Quataert asserts, this low-level technology transfer suited the interests of the European economy[36].

From the establishment of the state at the turn of the 14th century until the end of the 17th century, The Ottomans were self-sufficient, and content with their cultural, scientific and technological background. During this time, the Ottomans selectively transferred elements of western science and technology, but, with their feeling of superiority, they didn't feel the need to accept western science as a whole. Nevertheless, as the military balance with Europe tilted against the Ottomans at the end of the 17th century, their attitude became more receptive. Let us note that, however, the purpose of the Ottoman transfer of technology was to use, rather than to produce. From the beginning of the 18th century onwards, the palace and its surroundings started to feel the need to learn from Europe. Innovations were introduced into military organization, and traditional and modern elements existed side-by-side in education, science and technology.

During the reign of Mahmud II, the hold of traditional factors grew weaker, while European influences gained strength. The Ottomans became more dependent on Europe in education, science, technology, and industry thus triggering a conflict between the old and the new in society, and fostering dual norms in cultural life. Among the factors that hindered Ottoman development were: the endeavours of the European countries to dominate Ottoman markets, flawed polices in Ottoman dealings with the West, and the support which ethnic and religious minorities received from outside the empire. But the most important obstacles were the absence of big capital accumulation, and Ottoman dependence on European sources.

The Ottomans' haste to bridge the gap with Europe and regain their old power led them to commit political errors. Moreover, it appears that following the Scientific and Industrial Revolutions, the Ottomans adopted modern science and technology mostly through «translations» and «purchase», and failed to produce science and develop a technology – failed, that is to establish an indigenous tradition in science and industry which would decrease their dependence on the West. I believe that this was the most crucial factor that made the Ottoman experience different from that of Russia and Japan.

Acknowledgements

I wish to thank Dr. Mehmet Genç for his valuable comments, constructive criticisms and suggestions. I am grateful to Dr. Mustafa Kaçar for his contribution to research, as well as to Dr. Semiramis Havuşoflu and Dr. Yumna Özer for translating the manuscript into English.

BIBLIOGRAPHY

Barkan, Ö. L., «İmaret sitelerinin kuruluş ve işleyiş tarzüna ait araştürmalar», i.Ü. İktisat Fakültesi Dergisi- *Revue de la Faculté des Sciences Economiques de l'Université d'Istanbul*, 23, 1-2 (Ekim 1962-Şubat 1963), 239-296.

Clark, E. C., «The Ottoman industrial revolution», *International journal of Middle East studies*, 5 (1974), 65-76.

Emecen, F., «Osmanlı siyasî tarihi: kuruluştan Küçük Kaynarca'ya» in E. İhsanoğlu, ed. *Osmanlı devleti ve medeniyeti tarihi* (History of the Ottoman state and civilisation),(Istanbul: Research Centre for Islamic History, Art and Culture/ IRCICA, 1994), pp. 5-63.

Genç, M., «Ottoman industry in the eighteenth century: general framework, characteristics, and main trends» in D. Quataert ed., *Manufacturing in the Ottoman empire and Turkey, 1500-1950*, (New York: State University of New York Press,1994), pp. 59-86.

Güran, T., «Tanzimat döneminde devlet fabrikalarü» in Hakkü Dursun Yüldüz ed., *150. yülünda Tanzimat*, (Ankara: Türk Tarih Kurumu, 1992), pp. 235-258.

İhsanoğlu, E., «Ottoman science in the classical period and early contacts with European science and technology» in E. İhsanoğlu, ed., *Transfer of modern science and technology to the Muslim world* (Istanbul : IRCICA, 1992), pp. 1-120.

—, *Büyük Cihad'dan Frenk fodulluğuna* (Istanbul: İletişim Yayüncülük, 1996).

—, «Ottomans and European Science» in P. Petitjean et al., eds., *Science and Empires*, (Dordrecht: Kluwer Academic Publishers, 1992), pp. 37-48.

—, «Başhoca İshak Efendi: pioneer of modern science in Turkey» in ⌐ ____ E. Farah, ed., *Decision making and change in the Ottoman empire* (Kirksville: The Thomas Jefferson University Press, 1993), pp. 157-168.

—, «Osmanlı döneminde astronomi literatürü tarihi» in Fatih'ten günümüze astronomi (Astronomy from the time of Sultan Mehmet the Conqueror up to the present day) (Istanbul: Istanbul Üniversitesi Fen Fakültesi,1994), pp. 17-25.

— (ed.), *Osmanlı devleti ve medeniyeti tarihi* (History of the Ottoman state and civilisation) (Istanbul: IRCICA, 1994).

—, «Osmanlı imparatorluğunda çağünü yakalama gayretleri» in E. İhsaoğlu and M. Kaçar, eds., *Hağünü yaklayan Osmanlı* (Ottoman's attempts to keep up with the age) (Istanbul: IRCICA, 1995), pp. v-xvü.

—, and Feza Günergun, «Tüp Eğitiminin Türkçeleşmesi Meselesinde Bazü Tespitler», *Türk Tüp Tarihi Yüllüğü=Acta Turcica Histoirae Medicinae, I. Tüp Tarihi ve Deontoloji Kongresine Sunulan Tüp Tarihi ile ilgili Tebliğler*, edited by Arslan Terzioğlu, Istanbul 1994, pp. 127-134.

X

inalcük H., *The Ottoman empire: the classical age 1300-1600*, translated by N. Itzkowitz and C. Imber (London: Weidenfeld and Nicolson, 1975).

Kaçar, M., «Osmanlü imparatorluğunda askeri sahada yenileşme döneminin başlangücü» in F. Günergun, ed., *Osmanlı bilimi araştürmalarü* (Studies on Ottoman Science) (Istanbul: Istanbul Üniversitesi Edebiyat Fakültesi, 1995), pp. 227-238.

—, *Osmanlı devletinde Mühendishane'lerin kuruluşu ve bilim ve eğitim anlayüşündaki değişmeler* (The foundation of «Mühendishane»s in the Ottoman state and the changes in the scientific and educational life), unpublished Ph.D. thesis, University of Istanbul, Departement of History Of Science (1996).

Kazancügil, A., B. Zülfikar, *XIX. yüzyülda Osmanlı imparatorluğunda anatomi* (Anatomy in the 19 th Century Ottoman empire) (Istanbul: Özel Yayünlar, 1991).

Kütükoğlu, M. S., «Osmanlü iktisadî yapüsü» in E. ihsanoğlu, ed., *Osmanlı devleti ve medeniyeti tarihi* (History of the Ottoman state and civilisation) (Istanbul: IRCICA, 1994), pp. 513-650.

Müller-Wiener, W., «15-19. yüzyüllarü arasünda istanbul'da imalathaneler ve fabrikalar» in E. ihsanoğlu,ed., *Osmanlülar ve batü teknolojisi, yeni araştürmalar ve yeni görüşler* (Ottomans and Western technology, new researches and opinions) (Istanbul: Istanbul Üniversitesi Edebiyat Fakültesi, 1992), p. 53-120.

Quataert, D., *Manufacturing and technology transfer in the Ottoman empire 1800-1914*, Etudes Turques USHS, Nr.2 (Istanbul-Strasbourg :The Isis Press, 1992).

Sarü, N., «The Paracelsian Influence on Ottoman Medicine in the Seventeenth and Eighteenth Centuries» in E. ihsanoğlu, ed., *Transfer of modern science and technology to the muslim world* (Istanbul: IRCICA, 1992), pp.157-179.

Yediyüldüz, B., «Osmanlü Toplumu» in E. İhsanoğlu, ed., *Osmanlı devleti ve medeniyeti tarihi* (History of the Ottoman state and civilisation) (Istanbul: IRCICA, 1994), pp. 441-510.

X

ENDNOTES

1. In order to comprehend the developments in the 18th and 19th centuries, which form the essential framework of the period under study, we present a summary of the historical background as well as the social and economic structure of the Ottoman Empire based on H. inalcük's work The *Ottoman Empire: the classical age 1300-1600*. In addition, we also refer to different parts of *Osmanlı devleti ve medeniyeti tarihi* (History of the Ottoman state and civilization), edited by Ekmeleddin ihsanoğlu. For the text quoted here see H. inalcük, p. 41.

2. Bahaeddin Yediyüldüz, «Osmanlı Toplumu,» pp. 464-465.

3. Ö. Lütfi Barkan, «İmaret sitelerinin kuruluş ve işleyiş tarzüna ait araştürmalar», p. 239.

4. Mübahat S. Kütükoğlu,»Osmanlı İktisadî Yapüsü», p. 513.

5. For a brief explanation of this concept, developed by Mehmet Genç, see M. Genç, «Ottoman industry in the eighteenth century,» pp. 59-86.

6. H. İnalcük, pp. 51-52.

7. Feridun Emecen, «Osmanlü siyasî tarihi: kululultan Küçük Kaynarca'ya», p.56.

8. E. ihsanoğlu, «Ottoman science in the classical period», pp. 23-24.

9. E. ihsanoğlu, *Büyük Cihad'dan Frenk fodulluğuna*, p. 36.

10. E. İhsanoğlu, «Ottomans and European science», pp. 41-42.

11. Nil Sarü, «The Paracelsusian influence on Ottoman medicine in the seventeenth and eighteenth centuries», pp. 167-169.

12. E.İhsanoğlu, «Ottoman science in the classical period, « p. 41.

13. Aykut Kazancügil, Bedizel Zülfikar, *XIX. yüzyülda Osmanlü imparatorluğunda anatomi.*

14. Due to the handicaps in conducting the medical education in French at the *Mekteb-i Tübbiye-i Tahane* (Imperial School of Medical Sciences), established in 1836, and the successful medical education at the *Mekteb-i Tübbiye-i Mülkiye* (Civil School of Medical Sciences), established in 1869, an attempt was made to conduct the medical education in Turkish at the Imperial School of Medicine as well. However, the non-Muslim instructors of this school opposed this intensely and claimed that it was impossible to carry out medical education in Turkish. Nevertheless, in 1870, the state decided to carry out instruction in Turkish also at the Imperial School of Medical Sciences. Most of the members of the *Cemiyet-i Tübbiye-i Osmaniye* supported Turkish instruction while members of the *Cemiyet-i Tübbiye-i Şahane*, most of whom were non-Muslim Ottoman and European physicians, had a negative attitude toward Turkish instruction. See E. ihsanoğlu & F. Günergun, «Tüp Eğitiminin Türkçeleşmesi Meselesinde Bazü Tespitler», *TürkTüp Tarihi Yüllüğü=Acta Turcica Historiae Medicinae*, edited by A. Terzioğlu, Istanbul 1994, pp. 127-134.

15. E. İhsanoğlu, «Başhoca İshak Efendi: pioneer of modern science in Turkey,» pp. 157-168.

16. Feza Günergun, «A general survey on Turkish books of science printed during the last two centuries of the Ottoman state», paper delivered at the «International Congress of History of Science» (18th: August 1-9, 1989, Hamburg-Munich).

17. E. İhsanoğlu, «Osmanlı döneminde astronomi literatürü tarihi,» p. 22.

18. M. Kaçar, «Osmanlı imparatorluğunda Askeri Sahada Yenileşme Döneminin Başlangücü», pp. 227-238.

19. M. Kaçar, *Osmanlı devletinde bilim ve eğitim anlayüşündaki değişmeler ve Mühendishane'lerin kuruluşu*, pp. 38-95.

20. Wolfgang Müller-Wiener, «15-19. yüzyüllarü arasünda İstanbul'da imalathaneler ve fabrikalar», pp. 65-73.

21. In the Ottoman Empire the guilds were based on a closed economic system that united their members together around such sound moral principles as the enjoyment of work, professional discipline, honesty and contentedness. Thus, they provided economic and social security for the society. The guilds assured that the professions would remain respectable and, at the same time,

X

maintained the standards and prevented unfair rivalry. In all these activities, the state did not intervene in the affairs of the guilds except for the quality, quantity and price of the goods produced. H. İnalcük, pp. 150-151, B. Yediyüldüz, p. 475-476.

22. M. Genç, p. 59.

23. Ibid.,, p. 71.

24. W. Müller-Wiener, pp. 66-67.

25. M. Genç, p. 73.

26. E. C. Clark, «The Ottoman industrial revolution», 65-66.

27. Tevfik Güran, «Tanzimat döneminde devlet fabrikalarü», pp. 235-237.

28. Ibid., p. 236.

29. Donald Quataert, *Manufacturing and technology transfer in the Ottoman empire 1800-1914*, pp. 29-30.

30. T. Güran, pp. 236-237.

31. D. Quataert, p.10.

32. E.C. Clark, p. 75.

33. D. Quataert, pp. 30-31.

34. W. Müller-Wiener, pp. 81-83.

35. E. İhsanoğlu, «Osmanlü imparatoluğunda çağünü yakalama gayretleri», pp. vüü-xvü.

36. D. Quataert, passim.

XI

Aviation: The Last Episode in the Ottoman Transfer of Western Technology

Aviation constitutes the last episode in the centuries long interest of the Ottomans in Western technology. An examination of this interest may shed light on some aspects of the relations between the Ottomans and the West as well as setting forth its regulating parameters and its paradigm in the last period.

Ottomans who closely followed the developments in Europe were eager to be acquainted with them, acquire what they deemed necessary and use them. The Ottomans devoted particular attention to the technical novelties in the military arts which they did not possess but considered beneficial such as fire-arms, the compass, shipbuilding, etc.; moreover, from the second half of the eighteenth century onwards, European educational and scientific institutions attracted the interest of the Ottomans.[1]

Until the nineteenth century the Ottomans transferred technology from Europe, particularly taking military needs into consideration. During the reign of Sultan Mahmud II (1808-1839), the state took the necessary steps for improving public works as well as the well-being of the civilian population. In this framework, from the beginning of the nineteenth century, the Ottomans started to transfer technology from the West in order to set up modern networks of communication and transportation in the Empire. The attempts to set up modern postal services in the 1830s was followed by the establishment of telegraph lines. The telegraph constituted the first technological transfer from the West in the field of communication. The telegraph network developed rapidly following the construction of the first telegraph lines in 1855, about ten years after they were built in Europe, and from the 1860s onward, the Empire established itself in the global communication network. Telephone lines were set up first in 1881 and became widespread after the proclamation of the Second Constitution (II. Meşrutiyet) in 1908.[2]

In the field of transportation, as the construction of modern highways began in 1830s, the transport by carriage developed greatly owing to the efforts of the large enterprises in the nineteenth century. Shortly after the first steamship was bought from England in 1827, the use of steam engines began in naval transportation with both military and civil purposes. Although the attempts to build a railroad network date to the beginning of the second half of the nineteenth century, the bad experiences resulting from the abuses by foreign contractors led the state to undertake its own investments in railroads. The Hamidiye Hijaz railroad, which was completed between 1900-1908, is the best example of the point which the Ottomans reached in the management of transfer of technology from the West. Despite political instability and economic difficulties, on the whole the Ottomans managed to spread the postal and telegraphic services, railroads, highways and maritime lines almost equally over all provinces of the Empire. However, it should be noted that this introduction of modern transportation and communication technology was a process of adoption rather than development of technology. A similar attitude can be seen in the beginning of the twentieth century in the transfer of aircraft technology.[3]

Throughout the nineteenth century and the first two decades of the twentieth century, Ottoman administrators exerted very intensive pressure in order to import Western technology and, in particular, war technology. However, this relationship between the Ottomans and the West did not develop as the Ottomans had hoped. Among the factors that limited Ottoman success was the fact that they adhered to their own understanding and policies, and that the economic and political interests of the European states conflicted with the Ottomans. In the transfer of aircraft technology, the Ottomans constantly depended on Europe mainly due to their impatient and hasty attitude, failing to follow the right procedures, taking wrong decisions, in addition to the efforts by the European powers to hinder the success of the Ottomans.

This study aims to examine, in terms of the relations between the Ottomans and Western technology, the history of Ottoman aviation which started with ballooning in the last quarter of the eighteenth century and ended with military aircraft in the first quarter of the twentieth century. Whatever may be the interpretation of the Ottomans' achievement in this field, it must be admitted that the Ottoman army, which started to gain victories by using the weapons of the Middle Ages, throughout history acquired the necessary knowledge to use new arms, ending with the airplanes invented in the West. The Ottomans who founded and expanded an empire on horseback for centuries, were finally able to defend it in the cockpit, entrusting their legacy to history and to their heir, the Turkish Republic.[4]

A general account of the Ottoman administrators' attitude towards modern technology in the nineteenth century shows that they managed to adopt several technological novelties such as steam engines, telegraphs and railways. Such modern technologies were widely applied in civilian and military fields without a large time gap between Ottomans and other European states. When we look at aviation we see that immediately after the first trials in Europe, balloon ascents were made by foreigners in the Ottoman capital. These ascents, mostly undertaken for entertainment only, aroused curiosity among the Ottomans, but since the potential of balloons for practical purposes seemed limited, ballooning did not impress the Ottoman administrators and the public much. One of the first examples for this, was the Egyptians' indifference to the trials undertaken by French military balloonists in Cairo during Napoleon's occupation of Egypt.[5] However, the Ottoman sultan Selim III took an interest in the success of French aviators in Europe and during his reign a military trial was made in Istanbul in 1801 with a balloon designed by İngiliz Selim Ağa who held the position of *halife* (assistant) in the Imperial School of Military Engineering, the Mühendishane,[6] but there is no evidence indicating that these trials were developed or that the Ottomans made further attempts in this regard.

Until 1908, the Ottoman statesmen and people considered the demonstrations made by foreign balloonists and particularly by the Italian Antonio Comaschi in Istanbul in 1844 and 1845 as good entertainment. Since balloons were not used against the Ottomans neither in the Balkans nor in Russia, they did not constitute a military threat to the Ottoman Empire. But in the period which followed the proclamation of the Second

Constitution, the balloon was used in warfare as a military facility together with aircraft in Europe; this led the Ottoman administrators to buy balloons. In a way, this situation indicates the attitude of the Ottomans towards the technological novelties in the West. What mattered to them was the practical applicability of an invention and this characteristic peculiar to the Ottomans was evident in several other technological transfers from the West. The disinterested attitude of the Ottomans on this subject stemmed from the above reasons, otherwise, we have not come across any historical facts which justify Oberling's views stating neither that 'Conservative elements in Ottoman society strenuously opposed the transfer of 'infidel' technology into the Empire, and their obtrusiveness made it difficult for individuals to do so on their own initiative'[7] nor that 'In the Ottoman Empire all innovation had long been regarded as taboo, especially if it originated in the heathen West.'[8] On the contrary, there are several historical examples on this subject which support quite the opposite of this view.

The Ottomans bought their first balloon in 1911, 126 years after the very first balloon ascents of 1785.[9] Ottoman attitude to aircraft technology differed completely from their standpoint on ballooning. The Ottomans bought and implemented the aircraft technology which originated in 1903 and was developed farther in 1906. This clearly indicates their close interest in this new discovery and manifests their pragmatic attitude.

On December 7, 1903, Orville and Wilbur Wright successfully undertook their first experimental airplane flight, thus arousing excitement and curiosity all over the world. Three years later, in 1906, experimental flights started in Europe, first in France, and subsequently in other countries. After a short while, air shows started in the Ottoman empire as well, but these were performed by foreign aviators. The first air show which took place in Istanbul in December 1909 was reminiscent of the aerobatics shows around the beginning of the nineteenth century. Contrary to the balloon, the airplane, which roused great interest among the people, did this time attract the attention of administrators.

The Ottoman press followed daily Louis Blériot's successful flight across the English Channel on July 25, 1909, stirring great excitement in the world as well as the air meet which was organized between August 23-29 in the French city of Rheims; thus, several aspects of the airplane were introduced to the Turkish public.[10] These articles aroused great interest in the public opinion. They provided information about this new invention and indicated that airplanes were more convenient and less expensive than balloons. What draws the readers' attention in these articles are the attempts of their writers to find a Turkish translation or equivalent for the French word *aéroplane*, So we find a series of attempts starting from using the Turkish word *uçurtma* for kites and the term *merakib-i havaiye* (air vehicles), compounded for the first time as a translation for the French word *aéroplane*. Just as the Europeans made up words for new inventions from Greek and Latin, the Ottoman Turks, too, referred to Arabic which they regarded as a classical language, thus naming this new invention *tayyare* after the verb 'to fly' in Arabic. Later on, this word became widespread among several muslim nations and has been in use until the present day.

The year 1909 was quite productive for Ottoman aviation. Following the balloon ascents in May, performed by the French balloonist Ernest Barbotte in Istanbul, for the first time an airplane flight was organized in the capital of the Ottoman Empire. Towards the end of November, Baron Pierre de Caters, who was the runner-up in the race held in Rheims around the beginning of August, came to Istanbul in his Voisin double-decker.

Baron de Caters was an amateur aviator and belonged to the Belgian aristocracy. He came to Istanbul immediately after the air meet held in Rheims. His purpose was to make several trials with his airplane hoping to break the European record with the aid of favorable weather conditions. While the Baron was trying to receive permission for his flight, he also met Mahmud Şevket Pasha, commander of Hareket Ordusu (Action Army). The Pasha decided to set up a commission of officers to follow and study the trials to be made, for the purpose of determining to what extent it would be possible to benefit from the airplanes in the military field.[11] Mahmud Şevket Pasha was a pioneer in introducing aviation to the Ottoman state. As will be seen below, his meeting with de Caters was influential in the development of Ottoman military aviation.

The Baron planned to make his first flight on November 30, 1909. He chose the Hürriyet-i Ebediye hill (Okmeydanı) for his takeoff. As the preparations were incomplete and the weather conditions were unfavorable, the flight was postponed. On the morning of December 1, 1909, as the weather was favorable he decided to make his first trial. The airplane was ready toward evening and although it was late for the flight, he moved his airplane twice in the twilight to try its engine. He moved a few hundred meters back and forth on the ground and then decided to postpone the trial and repeat it the next day at 2 p.m. in the same place and drove the aircraft into the shed.[12] On December 2, despite strong winds, he made his first flight toward evening in the presence of nearly two thousand spectators including the princes Ziyaeddin, Necmeddin, Ömer Hilmi and Selahaddin Effendis and princesses of the imperial family, the Italian ambassador and the high officials of the embassy.[13] After a few maneuvers, the airplane roughly landed in the vicinity of the Bulgarian Hospital. The airplane was slightly damaged and its wheels were bent a little. This first flight was partly successful; the foreign press in Istanbul criticized this famous pilot who had won several prizes in many competitions, but was somehow unsuccessful in completing this flight.

Baron de Caters also failed in the demonstration flights on December 3rd and 5th.[14] Since Ottoman public opinion did not appreciate him as much as he had hoped, he left Istanbul. These trials, although unsuccessful, gave the public the opportunity to have a closer look at the airplane for the first time and witness this greatest discovery of the century.[15] Till then, the Ottoman public had not seen an airplane, but knew about it through newspapers. However, the public in Istanbul at least had an idea about the airplane and observed how it flew. A negative effect of this incident was to show that in air transportation there was the risk of a crash.

One week after Baron de Caters left Istanbul, M. Louis Blériot arrived and his Blériot airplane was exhibited to the public in the Pera Palas Hotel at Beyoğlu.[16] Thousands of people came together on the morning of December 12th at Taksim Square where Blériot realized his flight. However, after ascending 40-50 meters and flying over the people, the airplane moved toward Beyoğlu and fell over a house near Pangaltı. M. Blériot, who was slightly injured, was sent to the hospital after preliminary medical treatment. The airplane was greatly damaged.[17]

Following these flights in 1909, the outlook and the activities in regard to aviation acquired a different direction in the Ottoman state. As aircraft assumed military functions, the subject immediately gained importance. The above-mentioned flights by the Caters and Blériot were not successful enough to satisfy the military authorities. Therefore, the Minister of War Mahmud Şevket Pasha sent two staff officers to Europe as military attachés with the duty of collecting information about modern armies. Enver Bey and Fethi Bey were sent abroad to examine the military organisation of the German and French armies respectively. Fethi Bey, the military attaché in France, recommended that two officers from the Turkish army be sent to France for flight training and that Blériot, R.E.P. and Depérdessin airplanes be bought.[18]

The year 1910 was one of great developments in aviation in Europe. To follow this progress more closely, the Ottoman administrators sent delegations to Europe and tried to find the way of benefitting from aviation as a third military force in addition to the land and naval forces.

On September 29, 1911, Italy declared war upon the Ottoman State and started to bomb Tripoli from the sea. Air raids started on October 22, 1911, and on November 1, 1911, Italian aircraft dropped bombs over Tripoli for the first time in the history of aviation.[19] This event caused great alarm among the Ottomans. Minister of War Mahmud Şevket Pasha attempted to buy airplanes and balloons from Europe, he also sent two Ottoman officers Captain Fesâ and Lieutenant Kenan Effendi to Paris for training. The Ottoman General Staff also took a serious interest in aviation and charged Lieutenant Colonel Süreyya Bey [İlmen] with this matter.

Süreyya Bey, who was in charge of the organisation of Ottoman air forces, founded the Tayyare Komisyonu (Commission for Aircraft). It consisted entirely of military officers and was the first Ottoman organization to deal with aviation. Mahmud Şevket Pasha presented a memorandum to the Meclis-i Vükela (the Cabinet) on the organization of an air fleet. This memorandum underlined for the first time the necessity to train officers in aviation and mechanics able to pilot aircraft. It was aimed at building a school and a center for aviation in Istanbul. At first, it was decided to establish the aviation school and the headquarters in the locality of Bulgurlu, but according to the investigations of a delegation headed by Süreyya Bey and two French experts, Ayastefanos (Yeşilköy), the location of the present airport of Istanbul, was more suitable to build an airport.[20] At this point, Ottoman aviation determined its own direction. This delegation dedicated its efforts to meet the needs of the army immediately and chose a short-term practical solution, thus influencing the destiny of Turkish aviation.

An aid campaign was started under the name of *iane-i milliye* (national assistance) to set up the Ottoman air force. An article written in a fervent style appeared in the newspaper *Sabah* encouraging the aid campaign, named after the three aircraft (*Vatan*, *Mesrutiyet* and *Ordu*) which the Ottomans wanted to purchase, following the practice in France. It arose great interest all over the Ottoman lands. First of all, the Minister of War Mahmud Şevket Pasha donated 25 gold liras and one fourth of his six month salary every month as a contribution to this campaign. Mahmud Pasha, the second chief of the General Staff and members of the General Staff office donated one fourth of their six month salaries and it was decided to name the first aircraft to be purchased *Ordu* (Army).[21] Sultan Mehmed Resad participated in the aid campaign with a thousand gold liras. He ordered that the airplane to be bought be called *Osmanlı* (Ottoman) and be brought to the ceremonies on the occasion of the anniversary of his accession to the throne.[22] Captain Fesâ and Lieutenant Kenan completed their training in the Blériot aviation school in France and received their licenses in 1912. They were employed in the aviation school and two two-seat Depérdessin airplanes and a training aircraft were handed over to them. The Turkish pilots were first to make trial flights in these airplanes. But, due to the strong storm that came up at night, the airplanes were destroyed completely. Therefore, it was not possible to make any flights in the first aircraft that were purchased. A few months later, on the anniversary of Sultan Mehmed Reşad's accession to the throne, the first Turkish airplane was flown.

According to the contract signed with the French R.E.P. factory, an airplane was ordered and bought for 30,000 francs to participate in the festivities to be held on the accession ceremony on April 27, 1912. The airplane arrived Istanbul steered by the pilot Gordon Bell, on April 26 one day before the accession ceremony. On that day it took off from Yesilköy and made a 45 minute trial flight over Istanbul. On the day of the ceremony Sultan Mehmed Resad arrived at Hürriyet-i Ebediye hill to watch the military procession. At that time, the airplane, piloted by Gordon Bell, took off from Yeşilköy at 13.20. It landed at the spot of the ceremony at 13.30, hovered over the military units and joined the procession.[23]

In line with a decision taken by Minister of War Mahmud Şevket Pasha, on 6 May 1912, Lieutenant-colonel Süreyya Bey [İlmen], head of the commission for aircraft, and Mehmed Ali Bey were sent in a delegation to conduct studies in Europe.[24] Several European companies offered attractive suggestions to this Ottoman board with the aim of selling airplanes and balloons. As the board could not reach a final decision regarding which model was more suitable, it preferred to buy one sample each of the existing models in Europe at that time. In conclusion, seventeen airplanes and a balloon were ordered for an amount of 50,000 gold liras.[25]

An aviation school was built on the site of the Tayyare Merkezi (Aviation Centre) in Ayastefanos (Yeşilköy). The contract for the construction project was awarded to contractors and started in July 1912.[26] The teaching staff and curriculum of the Tayyare Mektebi (Aviation School) were also prepared and Monsieur Breson was invited from France to be its director. A smith and a carpenter were brought from France to work at the school.[27]

The following sections were planned to be built at the Aviation Center: a directorate, departments for officers, aircraft mechanics and watchmen, a small hospital, a repair shop, warehouses, underground gasoline tanks, six hangars each in the capacity of storing four airplanes, garages for cars and stables.[28]

In the Aviation School, training would be given three terms a year, each term lasting three months. Fifteen to twenty candidates would be admitted each term. Those who graduated as pilots would be organised in squadrons and be sent to the army. Flight training would be given theoretically on land; after the candidate got familiar with flying, he would pilot the airplane by himself. A class of navigation officers (*râsıd sınıfı*) would be established within the school to assist the pilots in determining the locations for take-off and landing.[29] Later, theoretical courses on engines and aircraft were also taught. The first lieutenant, aviator Midhat Nuri [Tuncel] Bey published the first Turkish technical book on flying titled *Vesâit-i Tayeran*.[30]

The flights of Turkish pilots for the purposes of demonstration or reconnaissance were received with great enthusiasm and these pilots were awarded medals and monetary prizes. They were tolerated even when they were often involved in accidents, and caused great damage.

During the first stages of military aviation in the Ottoman Empire, an aviator of Hungarian origin by the name of Oszkàr Asbôth (1891-1959) personally applied to the Ottoman government by submitting a project. This attempt has a special place in Turkish history of aviation. Oszkàr Asbôth submitted this detailed project about the organization of an air force in the Ottoman army and a related preliminary budget proposal to the Ministry of War. In his project Oszkàr Asbôth first of all suggested founding a school to train pilots and a repair shop for airplanes and engines. Later, it would be expanded and turned into a factory for the manufacture of airplanes.

The project also included special items concerning the establishment of departments of aviation in the army and the need for related buildings, machines, materials and equipment. The project attached great importance to the models and characteristic of airplanes which would be purchased from abroad. In Oszkàr Asbôth's words, the Ottoman military authorities appreciated this project greatly and the Ministry of War invited him to Istanbul through the mediation of the director of the Istanbul branch of Ganz Motors Factory. Oszkàr Asbôth came to Istanbul with his friend, who was a lawyer, on August 13, 1913. As a result of the negotiations, the entire project was accepted and a written agreement was prepared in a period as short as one week. However, despite all efforts, this project of Oszkàr Asbôth was not implemented. The discussions went on in full secrecy, but somehow, the members of the press and the representatives of the foreign states in Istanbul learned about the project.[31]

Strong reactions to Oszkàr Asbôth's project arose in the press. A serial titled 'Memleketimizde Tayyarecilik Doğmadan Öldürülecek mi?' [Will aviation be killed before being born in our country?] was published in an esteemed Ottoman newspaper of the time called *Tasvir-i Efkâr*. These articles emphasized that Ottoman aviation faced the danger of complete disappearance; touching upon the admiration of Ottoman people for the French, they stated that the only solution was to request assistance from France. The Ottoman General Staff was even subjected to criticisms in an insulting manner because they had accepted such a project.[32] In addition to the opposing attitude of public opinion, this project came to naught when the French ambassador, exploiting the admiration for French aviators, asserted the issue of the capitulations. The ambassador prevented the Ottoman government from signing such an agreement with Oszkàr Asbôth and convinced them that France could supply the necessary airplanes for the Ottoman army.

On September 28, 1913 the French Aviation Club organized a journey in the destination of Paris-Istanbul-Cairo. Three French aviators participated in this journey. First, Pierre Daucourt with his aircraft mechanic Henri Roux set off from Paris on October 20 and arrived in Istanbul on November 20 in his Borel type airplane.[33] Daucourt made a few demonstrations in Istanbul and took off from Yeşilköy on November 20. He moved toward the southeast in the destination of Adapazarı, Eskişehir, Akşehir, Konya and Ereğli. But while he was flying over the Taurus Mountains he had to make an emergency landing in Pozantı due to a defect in the engine and the airplane suffered great damage. He went to Adana to get assistance, but while a peasant standing near the airplane was trying to light his cigarette, the airplane caught fire and burnt out completely. Thus, Daucourt returned to Paris.[34]

The arrival of French aviators aroused great enthusiasm among the people and the newspapers wrote about these exciting meets for days. During these meets, high-ranking Ottoman military and civil officials accompanied the Turkish pilots in their public demonstrations organized at the aviation headquartes in Yeşilköy. On December 2, 1913, Fethi Bey took off from Yeşilköy in his airplane called *Osmanlı*, together with Belkıs Hanım who was entitled as the first Turkish and perhaps the first Muslim woman to fly in an airplane in the Ottoman and Islamic world. During the flight, Belkıs Hanım dropped messages encouraging Ottoman women to help in buying airplanes for the army.[35]

Secondly, Marc Bonnier and the aircraft mechanic Barbier took off in a Nieuport type airplane and came to Istanbul in the beginning of December. At that time, the third pilot Jules Védrines who set out from Nancy on November 20th in his one-seat Blériot of 80 horsepower engine landed in Yeşilköy on December 5, 1913. After making a few demonstration flights in Istanbul, Bonnier and Védrines left Istanbul on December 17th and 18th, respectively. Védrines was the first to arrive Cairo (December 29, 1913). The French pilots who were welcomed with a great ceremony made demonstration flights in Cairo and carried passengers in return for money.[36]

The Ottomans greatly admired the successful French pilots. During the Balkan War, Ottoman pilots did not achieve any success; they did not fly except for reconnaissance activities. It was decided to organize a journey between Istanbul-Cairo in order to erase these bitter memories, to demonstrate the power of the Ottoman state and restore its broken prestige and direct the admiration away from French aviators to Turkish pilots.

The Minister of War Enver Pasha ordered that two teams join the journey. The Blériot airplane called *Muavenet-i Milliye* and the Depérdessine *Prens Celaleddin* which were in the possession of the army participated in the journey. Fethi Bey and the navigation officer Sadık Bey would fly in the Blériot; Nuri Bey and Captain İsmail Hakkı Bey would fly in the Depérdessine. Although it was a typical adventure trip, ground services were well organized, gasoline and spare parts were sent to the centers that were determined beforehand. Moreover, mechanics would follow the airplanes by land and repair any breakdowns.[37]

The program of the journey would be organized en route from Istanbul-Beirut-Cairo to Alexandria. The airplanes would travel for a total of twenty-five hours covering a distance of 2,500 kilometers [38]. The trip started in Yesilköy on February 8, 1914, in rainy conditions. First, the *Prens Celaleddin* airplane took off at 09.10, and two minutes later the *Muavenet-i Milliye* airplane took off, piloted by Fethi Bey. During the air show in Beirut on February 19 the *Muavenet-i Milliye* airplane had a breakdown. Its pilot Fethi Bey had to deal with the repair. He left Beirut on February 24 and arrived at Damascus.

Fethi Bey and the navigation officer Sadık Bey set off from Damascus on the way to Jerusalem on February 27. Unfortunately, their airplane crashed in a rocky place called Valley of Hell in the vicitinity of the Sea of Galilee (Lake Tiberias); Fethi Bey and Sadık Bey were martyred.

Foreign aviators praised the heroism of the Turkish pilots. The whole country lamented over the martyrs who were brought to Damascus and, following a crowded funeral ceremony, buried in the enclosed graveyard of Selahaddin Eyyubi's tomb near the Umayyad Mosque of Damascus.[39] On the same day Nuri Bey and his navigation officer reached Damascus.

This accident which led to great grief in Istanbul almost turned into a legend among the people. A renowned Egyptian poet by the name of Hafız İbrahim wrote a eulogy for the Ottoman pilots. It would be recited at the ceremony of welcome to be organized upon their arrival in Egypt. This poem described the Ottoman pilots as the first Muslims to fly an airplane in the Orient and considered them as the 'travelers on Burak'— the traditional name of the horse which carried the Prophet Muhammad for his ascent.[40] Upon the death of Fethi and Sadık Beys, Hafız İbrahim also wrote an elegy, expressing the sorrow of the Egyptian people.[41]

The Ministry of War decided that a third team must set off. The pilot Salim Bey and navigation officer Kemal Bey set out on their journey on March 6, from Istanbul in a Blériot type airplane called *Ertugrul*. They could fly as far as Edremid on the West Anatolian coast. Nuri Bey and his navigation officer came to Beirut from Damascus on the same day, flying along the coast, they reached Jaffa on March 9 and set out on March 11. However, the airplane crashed into the sea during takeoff. Nuri Bey was drowned and buried in Damascus together with the other martyrs. Navigation officer İsmail Hakkı Bey returned to Istanbul.[42]

A fourth journey was to be organized instead of the trip that was cancelled because the third airplane which set out from Istanbul was damaged in Edremid. Thus, the flight would continue from Jaffa where the aviators were martyred and the trip to Cairo would be completed. The Blériot type airplane called *Edremid* was bought with the contributions of the people in Edremid and was transported to Beirut on a ferry[43] The parts of the airplane were assembled there and after a few trial flights, Salim Bey and

Kemal Bey flew to Jerusalem on May 1, 1914. The airplane arrived at Cairo on May 9, 1914.

The Ottoman pilots were welcomed with joy and enthusiasm in Cairo. In his eulogy, titled *İstikbal*, the Egyptian poet Ahmed Shawki Bey, known as 'the prince of poets' in the Arab World, praised the *Edremid* airplane and the Ottoman pilots.[44] Following several air shows in Cairo, the Ottoman pilots moved to Alexandria and arrived in Istanbul by ship on May 22.[45] Salim Bey and Kemal Bey returned to Istanbul on May 24. The aim of this journey was fulfilled. Those who lived on Ottoman lands saw the airplanes of their own, thus the Ottomans were able to show their power in aviation to their people. Commemorations were held every year for the martyred pilots and a great monument was built on the site near Lake Tiberias where the airplane had crashed.

The Ottoman government felt the need to expand and reorganize the Aviation School and applied to France for assistance. Upon the advice of the French Government, a delegation, which was sent to Europe under the chairmanship of Veli Bey, director of the new organisation of aviation met with the famous pilot Marquis Louis de Goys de Mezeyraki about the subject of aviation. De Goys came to Istanbul in May 1914 and signed a two-year contract to be employed in the Ministry of War. The French pilot would be in charge of training, repair and construction activities in this period.[46] Under the administration of De Goys, a law was enacted regulating the use of Ottoman airspace and the organization of voyages.[47] This was the first law on aviation that was legislated in the Ottoman Empire. Uniforms were also designed specifically for the Ottoman aviators.

During the time of De Goys, all of the airplanes that were necessary for the reorganization of the Air Forces were bought from France. Thus, the Ottomans became more dependent on the French in the field of aviation. A delegation was sent to France which bought a total of forty-five airplanes of various models and types. Among them, thirty were airplanes and fifteen were naval aircraft. Thus, De Goys attempted to build a new air force.[48]

Ottoman aviation during the First World War

When the First World War broke out, France was aware of Ottoman sympathies for Germany and recalled the French pilots and experts and seized the airplanes and naval aircraft which the Ottoman army had ordered from France.[49] The organization of Ottoman aviation was connected to the general headquarters; squadrons were constituted and the Ottomans entered the war. Since the French had left the country, Ottoman aviators in the Aviation School tried to carry on the training of pilots and the preparation of the new candidates for flight. These activities failed, however, because of the need for pilots on different fronts during the war, thus, in a period when they were most needed, the training activities in the Aviation School were halted completely.

In early 1915, twelve airplanes arrived from Germany with pilots and technicians. First lieutenant Erich Serno, who came with this fleet, was appointed at the head of the Ottoman air forces with the rank of Captain. On February 3, 1915, he became the director of the Aviation School.

Turkey entered the First World War just at the time when the program for the improvement of aviation was being implemented. At the onset of war, there was a squadron of six airplanes at the Aviation School in Yeşil-köy.[50] When the Ottoman government declared the state of war, the air forces included four airplanes.[51] Throughout the War, 450 airplanes in total served in the Turkish army including the 150 airplanes of the German *paşa bölüks*.[52]

During the First World War Turkish armies engaged in air operations in the fronts of the Caucasus, the Straits, Iraq-Iran, Sinai-Palestine and the Hijaz. Despite several difficulties, Turkish pilots had successful flights at Çanakkale, Sinai-Palestine, Iraq, Medina, and the Caucasus.[53]

Çanakkale

During the reconnaissance flights, which continued in September and October 1914, Ottoman pilots found that the British forces were getting prepared for an attack on the Dardanelles.[54] Compared to the Eastern front,

Ottoman aviators had greater success and luck in the Çanakkale wars (3 November 1914 - 9 January 1916) during which the forces of the Ottoman army showed heroic courage. In fact, the battles that took place on the Dardanelles were the first difficult examination which the Ottoman military aviation faced in its period of naissance.

In the beginning of 1915, two Nieuport naval aircraft which were not functioning were brought back to Yeşilköy. The Blériot airplane *Ertugrul* was sent to the Dardanelles. Until March 18, *Ertugrul* gave important services as a reconnaissance plane.[55] As a result of reconnaissance flights Turkish and German airplanes found out that the fleet of the Allies was approaching the Dardanelles on March 18. Thus, they helped the units of the Turkish army, which were in a state of alarm, to take the necessary steps.[56]

As the landing of the Allies began in Çanakkale on 25 April 1915, the Turkish forces had a squadron of four planes including three airplanes and a seaplane. After July 1915, this squadron was ordered to cooperate with the Fifth Army and gave useful services of reconnaissance, observation, and support.[57]

During the air operations between 25 April - 6 July 1915, the Allied air forces were by far superior to the Turkish-German air forces in terms of the quantity of arms and personnel. But, the first squadron under the command of the Fifth Army fulfilled its duties of reconnaissance and bombardment.[58]

During the air operations that took place between 7 July - 6 August 1915, three of the five Gota airplanes from Germany arrived in Istanbul. Two of them were sent to the Dardanelles and constituted the Deniz Tayyare Müfrezesi (Seaplane Detachment). But, since these seaplanes were not armed for defense, they would carry on their duties of reconnaissance and attack at night.[59]

The Çanakkale operations set a new example of the effective role which the air forces played in wartime. The mounting of a torpedo on the British Short seaplane gave the seaplane the capability of attacking ships. This weapon was used successfully for the first time in August 1915 when it sank a Turkish steamer in the Sea of Marmara.[60]

During the Anafarta operations, which were led by Colonel Mustafa Kemal and started on 7 August, lasting four days, the first squadron made an extraordinary effort, despite great difficulties, to support the Fifth Army. The four aircraft of this squadron flew for 51 hours during the battle. Turkish forces countered the attacks by the aircraft of the Allies armed with short guns and pistols. None of the Turkish aircraft crashed during these battles.[61]

On September 27, 1915, a Turkish aircraft hit the first aircraft of the Allies and caused it to fall. In October 200 photographs of the enemy's positions were taken and important information was collected. Toward the end of November, Turkish aviators won their first victory in the air.[62] Between 8-9 January 1916, the Allies evacuated Seddülbahir. The battles on land ended, but the naval and air operations continued.[63]

The Turks established air bases in Istanbul, Uzunköprü, Çanakkale, the Gallipoli peninsula, and Izmir to counter the air network which the British established over the Aegean Sea. Thus, the Turkish aircraft countered the air operations of the enemy and controlled their bases. The Sixth Fokker squadron (Altıncı Tayyare Bölüğü) which was brought from Germany began to fight against the Allies on the Gallipoli peninsula.[64]

During the air battles that took place around Çanakkale in 1916, the pilots of the Fokker Tayyare Bölüğü were very successful. Particularly, the German Captain Bodeckke was praised with medals of honour by Turkish and German authorities. The Turkish naval air unit (Türk Deniz-Hava Birliği) fulfilled very important duties in sweeping and detecting mines.[65]

It was only in 1916 that Ottoman aviation began to improve and proved itself in warfare. During November - December 1916, the number of airplanes, pilots and navigation officers rose to ninety, eighty-one, and fifty-eight respectively. The air forces were divided into twelve squadrons. These squadrons participated in the war and had important successes on various fronts. Until the end of the First World War, the number of airplanes in the Ottoman fleet fluctuated. Since the staff of the Kuvva-yi Havâiye Müfettiş-i Umumiliği (General Inspectorate of the Air Forces) consisted of Germans, the administration of the organization of aviation remained completely under their influence.[66]

Iraq-Iran

When the War began in 1914, the Ottomans could not send any aircraft to the Iraqi front. However, one squadron was sent to Iraq in December of 1915.[67] Battles took place between the British forces which occupied Basra and moved to the north at the end of 1914, and the Ottoman forces. During these battles, the Turkish units seized two aircraft and numerous provisions related to airplanes. Turkish pilots, who took off from an airport situated near Baghdad and surrounded with palm trees, engaged in air operations.[68] Before the arrival of the second squadron in Baghdad in December 1915, Turkish pilots collected information about the British forces through reconnaissance flights. These flights continued toward the end of December.[69]

In the beginning of January 1916, Turkish land and air forces detected operations by British forces on the front through continuous reconnaissance flights. The second squadron, sent from Yeşilköy, served in the battles against the British. However, the first Paraso aircraft of the squadron broke down during the trial flight.[70] Turkish forces continued the reconnaissance flights by repairing the present airplanes. At that time the squadron consisted of three Albatross and one Fokker type aircraft.[71]

By April 1916, after moving towards Baghdad, the army of 14,000 British and 3,700 Arab troops was besieged at Kut-el-Amara. The Royal Flying Corps (RFC) force (now consisting of fourteen aircraft) attempted to restock the expeditionary force by air, but failed. The troops had to surrender on 29 April.[72]

While the Turkish forces defeated the British troops in Kut-el-Amara, the Russians had started to attack Iran. In 1915 Turkish-German forces started to act against the Russian army.[73] The Second Squadron, which served under the command of the Sixth Army around Tigris River, did not have sufficient staff and aircraft to fight with the Russian army which was advancing from Iran to Iraq. For this reason, the Sixth Army Command requested that pilots and aircraft be sent to the area. Following violent battles, the Turks took Khaniqin and the enemy retreated towards Qasr-i Shirin. During the reconnaissance flight carried out by Turkish aircraft around Qasr-i Shirin on 9 June 1916 the Turkish pilots saw that the enemy was retreating from this city.[74]

On the morning of 27 June 1916, the Turkish units started a general attack and the Russians began retreating. Although most of the Turkish aircraft were out of service, and flew over steep and broken terrain, they managed to fly for long hours and presented detailed reports about the situation of the enemy. Thus, the 13th Army Corps Command praised them and gave them monetary prizes.[75] Due to continuous air operations, however, the Turkish army could not send any aircraft to Iran in December of that year.[76]

In 1917, the 6th Army and 13th Army Corps Commands presented letters of appreciation to the 2nd and 12th squadrons for fulfilling their duties of reconnaissance, observation and attack successfully.[77]

The battles that took place in Iraq between Ottoman and British armies continued until 1917 during which the Ottoman air forces served actively. The Turkish forces began to retreat when the war ended in favor of the British.

Caucasia

Ottoman aviators were unsuccessful in the Eastern (Caucasian) front in 1914. Two airplanes were sent out on ships to Trabzon, but the Russian fleet attacked the convoy and sank the ships. Thus, the Ottomans could not send any aircraft to this front. Turkish airplanes were not active in Caucasia in 1915 because they served in the Dardanelles.

Despite great difficulties, the Ottomans were able to transport aircraft to this front in 1916. Upon demand by the 3rd Army Command, an air formation consisting of four staff and two aircraft were sent to this front.[78] It was necessary to reinforce the 7th squadron, which was on duty in this front, with new airplanes and staff. Thus, the Inspectorate of Aviation sent two Albatross-C (III) aircraft to Ulukışla. One of them broke down during landing, but the other reached Suşehri. German First Lieutenant Fünfhausen was appointed commander of the 7th squadron. The reconnaissance flights of the Turkish aircraft under the order of the 3rd Army Command were successful. They flew above the positions of the Russian army and collected important information.[79]

Sinai - Palestine

In 1914, at the beginning of the First World War, the Ottomans were not able to send any airplanes to Palestine due to a lack of aircraft.[80] In 1915 during the preparations for the operations on the Suez Canal, a squadron of four aircraft was allocated under the command of the Fourth Army. However, one of them (a Ponnier aircraft) broke down during an emergency landing and the rest of the aircraft were in bad condition.[81]

In 1916 as the preparations for the operations on the Suez Canal were going on, a small-scale operation was planned in order to disturb the enemy and prevent them from moving their forces away from the Canal. During this operation two war planes of the 300th Paşa Tayyare Bölüğü would support the offensive at hand. Although Turkish forces were few in number, owing to the high qualifications of the aircraft and pilots they were very successful in fulfilling their duties.[82]

In 1917, the Fourth Army Command was worried that the British side which reinforced their air forces could establish superiority in the air. They communicated this in a report submitted to the headquarters of the chief command. Thus, the old Parasol aircraft were replaced with the modern Albatross D II fighter planes, and the new type Rampler C aircraft were equipped with machine guns.[83]

During the first battle at Gaza the 300th Paşa Bölüğü carried out continuous reconnaissance flights and collected information about the enemy. Although a Turkish flight team destroyed the water pipes that carried water from the Nile, which was used by the British army, these were soon repaired.[84]

During the second battle at Gaza the British forces, which had reinforced their positions very well, started an offensive on 18-19 April. British artillery and aircraft bombed Turkish positions. Despite all efforts, the Turks could not complete the construction of the remaining parts of the roads and railways.[85]

During the operations that took place in the preparatory period of the third battle at Gaza, Turkish and German air units engaged in continuous activity over and behind the British front. Both sides carried out reconnaissance flights, and air battles continued between May and August 1917.[86]

While the battles continued in Palestine, the British occupied Baghdad on 11 March 1917. The Yıldırım Army Group was constituted in order to regain Baghdad and the German General Von Falkenhayn was appointed its commander. During July-August 1917, four squadrons were allocated to the Yıldırım Army Group.[87] The Turkish army remained on the defense on the Baghdad front and started an offensive on the Palestine front by reinforcing the army with the Yildirim Army Group.[88] Towards the end of October, for the first time on this front, the British side used S.E. 5 reconnaissance planes in addition to Bristol fighter planes.[89]

During the third battle at Gaza, on the morning of 31 October 1917, the British started an offensive on the entire front, capturing Beersheba on 31 October and Gaza on the night of 6-7 November.[90] The Turkish-German squadrons which arrived and began to settle in Palestine during the last half of October were trying to complete their preparations. At that time, the British air attack started. The Turkish squadrons had Albatross C IV and A.E.G. aircraft which were much better compared to British aircraft.[91]

During the operations one fourth of the Turkish aircraft were lost. Heavy battles raged between the two sides. The Yıldırım Army Group commanded the Eighth Army to retreat to the line extending from Jaffa to Lod and the Seventh Army in the direction of Jerusalem. On 9 December British forces entered Jerusalem. The Yıldırım Army Group stopped the front operations of the British forces on the line extending from the north of Jaffa to southern Nablus and the hills east of Jericho. The Turkish defense continued and incurred heavy losses on the British.[92]

On 1 May 1918, the Fourth Army Air Troops Command was formed and the third, fourth and fourteenth squadrons were detached to it. Towards the end of May, these squadrons made reconnaissance flights when weather was favorable. The photographs taken from the air showed that there was much activity in the positions of the enemy. The Turkish aircraft continued their activities between June and August.[93]

The report of the Fourth Army Air Troops Command, dated July, stated that the third and fourth squadrons did not have any staff and aircraft left, and only the 14th squadron served at the front. Its duty was to carry out reconnaissance, observation and bombardment around the east of Jordan river and both sides of the Damascus - Ma'an railway.[94]

Although the British aircraft were superior in number, the British could not gain complete air superiority. Between 16-25 July 1918, fifteen British aircraft attacked the station and the airport, held by the Turks, in Amman, dropping eighty bombs. Turkish and German staff were presented with medals and letters of appreciation as a reward for their success in the air operations.[95]

From the last week of August onwards, in the face of the crushing British superiority, Turkish aircraft could not carry out any important reconnaissance flights. During the night of 18-19 September, British aircraft started to bomb the station and airport at Affula. Following the pitched battle of Nablus that took place on 19 September, Turkish units started their retreat.[96] During the retreat, several Turkish aircraft broke down or were destroyed due to the lack of spare parts and particularly due to the deplorable state of the airports as a result of the air raids carried out by British forces.[97]

On 19 September 1918, British forces started a major offensive against the Turkish army and the British air forces, which were by far superior, caused great damage to the Turkish war-effort. A large Turkish column, which was discovered by reconnaissance while moving through Wadi el Fa'a was defeated by air attack. As a result of these heavy losses the armistice was enforced on 31 October.[98]

Yemen

While the Ottomans did not have the means to send any aircraft to Yemen, the British sent a small number of aircraft of the Royal Naval Air Service to Aden in 1916. They engaged in reconnaissance flights, leaflet-dropping sorties and occasional bombing attacks against Turkish positions in this area.

A permanent Flight was established in Aden in 1917. The Flight continued its operations against the Turks until the Armistice.[99]

Hijaz

Because of its railway, the Hijaz was a very important strategic area for the Ottoman state. At the beginning of the First World War, Sherif Hüseyin tried to stir up the local population into a rebellion against the Ottoman government. Ottoman control was bound to the good functioning of the railway and therefore the Ottomans sent an air squadron to protect it. Ottoman aircraft could take off from the plains located around Tabuk, Eilat, Medina and Mecca. Thus it would be possible to control the highways and the Hijaz railway from the air.[100]

Pilot First Lieutenant Orhan was sent to Medina where he contacted the commanders, gave information about the qualities of the aircraft which would be sent to this area, the locations of the airports and the weather conditions. The results of his investigations along with the report of the Fourth Army Commander were communicated to the Inspectorate of the Air Forces.[101] In summary, the report underlined the necessity to send more speedy aircraft to this area, capable of flying long distances, and suitable to fly in hot weather.

A detachment of three aircraft from the Third Squadron was sent to Medina. However, since the inexperienced pilots and mechanics of this squadron only had a training of a few months in Germany they often caused accidents.[102]

Upon the suggestion of the Fourth Army Command, four aircraft and some spare materials were sent from the 300th Paşa Tayyare Bölüğü. Moreover, the unsuccessful commander of the Third Squadron was replaced by First Lieutenant Pilot Fazıl. On September 1916, he arrived in Medina and joined the squadron. In a short time, the aircraft which were not capable of flying were prepared for flight. They carried on successful duties of reconnaissance and bombardment between October and December 1916.[103]

XI

212

In the defense of the Hijaz against the rebellious Arab forces, the protection of the Hijaz railway was important. The Third Squadron was in charge of the safety of this railway. Although the commander of the squadron Captain Fazıl applied to the authorities, stating that due to the lack of fuel and spare parts it was not possible to fly in March, he did not receive the desired reply.[104]

The Third Squadron which was located in Ma'an prepared for flight and acted against the rebels. Between September 1917 and January 1918, the aircraft of this squadron carried out fifty-one flights of reconnaissance and bombing. They were not completely successful, however, because the aircraft were located in different places; leadership and administration was difficult at best.[105]

The War of Independence

Following theFirst World War, Ottoman air forces and the organization of aviation was completely dissolved. When the Allied Forces occupied Istanbul (13 November 1918) after the armistice of Mudros (30 October 1918) they seized the aircraft and equipment in Yesilköy airport. Some of these were hidden at the Maltepe quarter on the Asian side of Istanbul by Ottoman officers before the occupation. In this way, they prevented the English from seizing the aircraft.[106] When the Turkish War of Independence started in 1920, several of the officers went to Anatolia to join the national forces, but they could not take any airplanes with them.[107]

During the War of Independence, Turkish aviators gained military successes despite limited means and the limited number of aircraft. Following the Armistice of Mudros (30 October 1918), the Allies had occupied the fortifications of the Dardanelles in order to gain safe access to Istanbul. Mustafa Kemal Pasha started preparations to fight against the Allies so as to regain all the territories lost. Turkish pilots made reconnaissance flights and bombardments with some old airplanes at hand. During the preparations for Büyük Taarruz (The Great Attack) air forces were strenghtened, and on 7 March 1922, ten airplanes were sent to the front for reconnaissance flights. Meanwhile, according to the Treaty of Ankara which was signed with the French Government, Turkish forces took over fourteen reconnaissance planes — ten of them were Bregent planes — which were in the possession of the French on the Southern front. These airplanes were brought from Adana to Konya. Only four of them were prepared for flight and transported from Konya to Akşehir on May 21, 1922. Twenty Spad XIII type fighter planes were acquired from the Italians. Since there were no weapons on these airplanes, they were equipped with machine guns from the old German military aircraft. During the Great Attack, Turkish aviators made reconnaissance and patrol flights and successfully hindered the attacks of Greek airplanes, captured a few Greek airplanes and included them in the Turkish air fleet. Following the Great Attack on 30 August 1922, which ended with victory, Mustafa Kemal liberated Izmir (9 September 1922) and the Turkish army moved towards Istanbul. By mid-September the army reached Çanakkale, a town inside the region occupied by the Allies. To prepare for the inevitable confrontation, the region had to be reinforced notably by units of the British Royal Air Force (RAF). Thus, an RAF troopship on its way to Mesopotamia was moved to Çanakkale. The following forces were also sent to the Dardanelles: On 9 September, the carrier HMS Pegasus with her five Fairey IIID aircraft sailed into Çanakkale; on 22 September, the carrier HMS Argus arrived with five Nieuport Nightjar fighters of No 203 Squadron and another seventeen IIIDs. Further reinforcements joined this force in late September and early October: No 4 Squadron (Bristol F2Bs), No 25 Squadron (Sopwith Snipes), No 207 Squadron (DH9 As), No 208 Squadron (Bristol F2Bs) and No 56 Squadron (Snipes), all arriving from the United Kingdom, in addition to No 267 Squadron from Malta.[108]

This considerable force established its headquarters in Istanbul. The Aircraft Park and the main operating base were set up at Yeşilköy. There was also another base at Kilia on the Gallipoli Peninsula opposite Çanakkale. Despite the unfavourable conditions of winter, the British aircraft carried out reconnaissance, surveillance and general patrol missions as Turkish forces prepared to attack the British positions.[109]

Although the French and Italian contingents retreated from the Dardanelles, the British remained in the neutral zone. Finally, the British

accepted Mustafa Kemal's demands and the Armistice of Mudanya was
signed between the two sides (October 11, 1922). Its conditions were later
ratified in a formal peace treaty signed at Lausanne on 24 July 1923. Thus, a
possible clash between the British and Turkish forces was avoided.[110]

After the air operations ended on October 11, 1922 the Government of
the Turkish Grand National Assembly took the necessary steps in fostering
the training of Turkish aviators. Following the armistice, signed at Mudanya, Harry, an aviator of Hungarian origin was brought to Izmir. He
would work as instructor to train new pilots and to complete the training of
the old pilots. Harry hesitated to start the training in view of the neglected
state of the available aircraft and materials. After adding cabins to the jerry-
built military and reconnaissance aircraft, which were repaired anew, he
began the flight training. After the Hungarian instructor left, French instructors were employed due to the beginning of the period of rapproche-
ment with France; their activities yielded successful results.[111] Following
the proclamation of the Turkish Republic on October 29, 1923, this legacy of
the Ottoman air forces lay the foundations of the new state in this field, as it
was the case in several other fields.

Conclusion

In conclusion when one looks at the main features of the Ottoman attempts
to acquire the modern technology of aviation it is evident that the Ottomans followed the Western developments in regard to aviation continuously. They were informed of the newest developments through daily newspapers as well. Air shows in Istanbul started at the same time as they took
place in many European countries. Ottoman aviation started with the purchase of airplanes by collecting aid from the people and sending a few officers to Europe for education. This first reception did not lead to the development of a policy for the transfer of this new technology, and thus they
were not able to transfer the technology of manufacturing aircraft. In a
short while, Ottoman aviation became dependent on European countries,
particularly France. The first attempts in this field focused on the establishment of a school for training pilots and an airport, but the foundation of

factories for the manufacture of airplanes was neglected. However, the Ottomans preferred to meet their military needs urgently instead of transferring aircraft technology. In the later periods, obstruction by foreign missions and major European states led to the failure of the Ottoman attempts to develop a domestic aircraft industry. Due to the admiration for France in public opinion, Ottoman aviation completely depended on France in the first period, and on Germany after the First World War broke out. The major factors that obstructed the development of Ottoman aviation, which was still in its preliminary stage, were the outbreak of the First World War immediately after the Balkan War and the beginning of the Turkish War of Independence. The policy of aviation in the later period, after the proclamation of the Turkish Republic, did not change until the 1930s and the administrators of the Republic preferred to train pilots rather than aeronautical engineers.

Notes

1. Ekmeleddin İhsanoğlu, 'Some Remarks on Ottoman Science and its Relation with European Science and Technology up to the End of the Eighteenth Century,' *Journal of the Japan-Netherlands Institute: Proceedings of the First International Congress, Transfer of Science and Technology Between Europe and Asia Since Vasco da Gama (1498-1998) (5-7 June 1991, Amsterdam)* III (1991) 45-73; 'Ottoman Science in the Classical Period and Early Contacts with European Science and Technology,' in: E. İhsanoğlu ed., *Transfer of Modern Science and Technology to the Muslim World: Proceedings of the International Symposium on 'Modern Science and the Muslim World'* (Research Centre for Islamic History, Art and Culture, IRCICA; Istanbul 1992) 99, 1-48; 'Ottomans and European Science,' in: Patrick Petitjean et al. eds., *Science and Empires* (Dordrecht 1992) 37-48.

2. E. İhsanoğlu, 'Osmanlı İmparatorluğu'nun Teknoloji Çağını Yakalama Gayretleri,' in: E. İhsanoğlu ed., *Çağını Yakalayan Osmanlı!* (Research Centre for Islamic History, Art and Culture, IRCICA; Istanbul 1995) VII-XVI.

3. E. İhsanoğlu, 'Osmanlı İmparatorluğu'nun.'

4. E. İhsanoğlu, 'Osmanlı Havacılığına Genel Bir Bakış,' in: E. İhsanoğlu ed., *Çağını Yakalayan Osmanlı!*, 497-554.

5. Abdurrahman el-Caberti, *Tarih-i Acayib el-Asar fi el- Teracim ve'l-Ahbar* II, (2nd ed.; Beirut 1978) 29-230; Pierre Oberling, 'A History of Turkish Aviation: Part 1, Aerostation Among the Ottomans,' *Archivum Ottomanicum* IX (Wiesbaden 1984) 139-40.

216

6. Mehmed Ali Beyhan, *Câbî Ömer Efendi, Câbî Tarihi (Tarih-i Sultan Selim ve Sultan Mahmud-i Sanî) Tahlil ve Tenkidli Metin* (I.Ü. Sosyal Bilimler Enstitüsü, basılmamış doktora tezi; Istanbul 1992).

7. Oberling, *'A History of Turkish Aviation,'* 159.

8. Pierre Oberling, 'The State as Promoter of Technology Transfer: The Early Years of Ottoman Aviation,' *Journal of Turkish Studies* 8 (1984) 209-214.

9. This balloon was ordered to the German firm Parseval Luftfahrt Flugzeug Gesellschaft in order to be used in the battlefield. It made its first ascent on 6 November 1912 during the Balkan War. Oberling, *'A History of Turkish Aviation,'* 156-157.

10. The impressions of Mustafa Subhi Bey, the correspondent of the newspaper *Tanin*, about the first aviation race, held in the city of Rheims in France, with a prize of 100,000 Francs clearly indicates the interest shown in aviation in Europe and how the studies undertaken in this field influenced the Ottoman intellectuals.

11. Yavuz Kansu, Sermet Sensöz, Yılmaz Öztuna, *Havacılık Tarihinde Türkler, 1* (Ankara 1971) 114.

12. *Yeni Tasvir-i Efkâr*, no. 183, 2 Kânûn-i Evvel 1909, 3-4.

13. *Yeni Tasvir-i Efkâr*, no. 184, 3 Kânûn-i Evvel 1909, 4-5.

14. *Yeni Tasvir-i Efkâr*, no. 187, 6 Kânûn-i Evvel 1909, 6.

15. *Yeni Tasvir-i Efkâr*, no. 182, 1 Kânûn-i Evvel 1909, 4-5.

16. *Yeni Tasvir-i Efkâr*, no. 194, 13 Kânûn-i Evvel 1909, 6.

17. *Yeni Tasvir-i Efkâr*, no. 194, 13 Kânûn-i Evvel 1909, 8; *Tanin*, no. 460, 13 Kânûn-i Evvel 1909, 1.

18. *Türk Silahlı Kuvvetleri Tarihi*, III, Genel Kurmay Harp Tarihi Baskanlığı, Resmi Yay. no. 2, Ankara Genel Kurmay Basımevi, 1971, s. 488.

19. The first bombs were dropped by the Italian pilot Gavotti on 1 November 1911 at Ayn-i Zara and Tacura. Orhan Koloğlu, 'Dünya'da İlk Hava Savaşi,' *Tarih ve Toplum* 50 (February 1988) 16.

20. *Sabah*, no. 8078, 27 Rebiyülevvel 1330, 4 Mart 1329/17 March 1912, 1.

21. *Sabah*, no. 8077, 26 Rebiyülevvel 1330, 3 Mart 1329/16 March 1912, 1.

22. *Sabah*, no. 8080, 29 Rebiyülevvel 1330, 6 Mart 1329/19 March 1912, 1.

23. *Türk Silahlı Kuvvetleri Tarihi* III, 490.

24. *Türk Silahlı Kuvvetleri Tarihi* III 490.

25. *Yeni Gazete*, no. 1411, 14 July 1912.

26. *İktiham*, no. 138, 12 July 1912, 2.

27. Kansu, Sensöz, Öztuna, *Havacılık Tarihinde Türkler*, 156.

28. Kansu, Sensöz, Öztuna, *Havacılık Tarihinde Türkler*, 157.

29. Kansu, Sensöz, Öztuna, *Havacılık Tarihinde Türkler*, 157.

30. Mithat Nuri (Tuncel), *Vesâit-i Tayerân*, Istanbul Matbaa-i Osmaniye, 1330/1914, 94 p

31. George Hazai, 'Macar Havaci Oszkàr Asbôth'un Türk Havacılığına Ait bir Projesi,' *Çağını Yakalayan Osmanlı!*, 491-495.

32. *Tasvir-i Efkâr*, no. 925/126, 29 Zilhicce 1331, 16 Teşrin-i Sânî 1329, 29 November 1913, 2.

33. *Tasvir-i Efkâr*, no. 909/110, 10 Zilhicce 1331, 28 Teşrin-i Evvel 1329, 10 November 1913, 1.

34. *Tasvir-i Efkâr*, no. 924/125, 28 Zilhicce 1331, 15 Teşrin-i Evvel 1329, 28 November 1913, 1.

35. Belkis Şevket Hanım, 'Uçarken,' *Kadınlar Dünyası Dergisi*, no. 100, 30 Teşrin-i Sânî 1329. 3-4.

36. *Tanin*, no. 1774, 8 Muharrem 1331, 3 Kânûn-i Evvel 1913, 1.

37. *Tasvir-i Efkâr*, no. 971, 21 Safer 1332, 6 Kânûn-i Sânî 1329, 19 Kânûn-i Sânî 1914, 3.

38. *Tasvir-i Efkâr*, no. 987, 9 Rebiyülevvel 1332, 24 Kânûn-i Sânî 1329, 6 February 1914, 5; no. 989, 11 Rebiyülevvel 1332, 26 Kânûn-i Sânî 1329, 8 February 1914, 3.

39. *Sabah*, no. 8782, 3 Rebiyülâhir 1332, 16 Şubat 1329/1 March 1914, 3.

40. *Divan-ı Hafız İbrahim*, haz. Ahmed Emin, Ahmed al-Zeyn, İbrahim al-İbrayî, cüz II, Kahire, Matbaat Dar al-Kutub al-Mısriye, 1937, s. 76-81.

41. *Divan-ı Hafız İbrahim*, 179-80.

42. *Sabah*, no. 8801, 22 Rebiyülâhir 1332, 7 Mart 1330, 20 March 1914, 2; no. 8815, 7 Cemaziyelevvel 1332, 21 Mart 1330, 3 April 1914, 2.

43. *Sabah*, no. 8844, 6 Cemaziyelahir 1332, 19 Nisan 1330, 2 May 1914, 2.

44. Ahmed Şevki, *Al-Sevkiyat*, haz. Mehmed Hüseyin Heykel, cüz I, Mısır 1970, s. 217.

45. *Tanin*, 1954, 7 Recep 1332, 19 Mayıs 1330, 1 June 1914, 1.

46. *Sabah*, no. 8847, 9 Cemaziyelâhir 1332, 22 Nisan 1330, 5 May 1914, 2.

47. *Takvim-i Vakayi*, no. 1847, 20 Receb 1332, 1 Haziran 1330, 14 June 1914, 1.

48. *Sabah*, no. 8847, 9 Cemaziyelahir 1332, 22 Nisan 1330, 5 May 1914, 6..

49. *Türk Silahlı Kuvvetleri Tarihi*, III, 497; Kansu, Şensöz, Öztuna, *Havacılık Tarihinde Türkler*, 159.

50. Cemal Akbay, *Birinci Dünya Harbinde Türk Harbi*, I: Osmanlı İmparatorluğu'nun Siyasî ve Askerî Hazırlıkları ve Harbe Girişi (Genelkurmay Basımevi; Ankara 1970) 168.

51. *Türk Silahlı Kuvvetleri Tarihi* III, 497.

52. Holzhausen, *Çanakkale Havacıları ve Birinci Dünya Savaşında Türkiye'deki Hava Destek Kuvvetleri*, (München 1962) Turkish trans. 1962, Gnkur. Harp Tarihi Bşk. Unpublished works, cited in Kansu, Şensöz, Öztuna, *Havacilik Tarıhinde Türkler* I, 1971, 167.

53. *Birinci Dünya Harbi*, IX: *Türk Hava Harekâtı* (Genelkurmay Basımevi; Ankara 1969) 9.

54. *Türk Silahlı Kuvvetleri Tarihi*, III, 498.

55. *Birinci Dünya Harbi*, IX, 26.

56. Kansu, Sensöz, Öztuna, *Havacılık Tarihinde Türkler*, 192.

57. *Birinci Dünya Harbi*, 39.

58. *Birinci Dünya Harbi*, 49.

59. *Birinci Dünya Harbi*, 49-50.

60. Charles Messenger, *The Century of Warfare: Worldwide Conflict from 1900 to the Present Day* (London 1995) 69.
61. Kansu, Sensöz, Öztuna, *Havacılık Tarihinde Türkler*, 202.
62. Kansu, Sensöz, Öztuna, *Havacılık Tarihinde Türkler*, 204.
63. *Birinci Dünya Harbi*, 57.
64. *Birinci Dünya Harbi*, 85.
65. *Birinci Dünya Harbi*, 86.
66. *Türk Silahlı Kuvvetleri Tarihi* III, 499-500.
67. *Türk Silahlı Kuvvetleri Tarihi*, III, 498-499.
68. *Birinci Dünya Harbi*, 63.
69. *Birinci Dünya Harbi*, 63.
70. *Birinci Dünya Harbi*, 92-94.
71. *Birinci Dünya Harbi*, 102.
72. Michael Armitage, *The Royal Air Force: An Illustrated History* (revised ed.; London 1995) 29.
73. *Birinci Dünya Harbi*, 103.
74. *Birinci Dünya Harbi*, 104.
75. *Birinci Dünya Harbi*, 107, 109.
76. *Birinci Dünya Harbi*, 110.
77. *Birinci Dünya Harbi*, 184.
78. *Birinci Dünya Harbi*, 73.
79. *Birinci Dünya Harbi*, 79-80.
80. *Türk Silahlı Kuvvetleri Tarihi*, III, 498.
81. Kansu, Sensöz, Öztuna, *Havacılık Tarihinde Türkler*, 210.
82. *Birinci Dünya Harbi*, 111-112.
83. *Birinci Dünya Harbi*, 190.
84. *Birinci Dünya Harbi*, 192-193.
85. *Birinci Dünya Harbi*, 194.
86. *Birinci Dünya Harbi*, 197-199.
87. *Birinci Dünya Harbi*, 199-200.
88. *Birinci Dünya Harbi*, 201.
89. *Birinci Dünya Harbi*, 202.
90. *Birinci Dünya Harbi*, 203.
91. *Birinci Dünya Harbi*, 204.
92. *Birinci Dünya Harbi*, 205.
93. *Birinci Dünya Harbi*, 246, 247.
94. *Birinci Dünya Harbi*, 248.
95. *Birinci Dünya Harbi*, 249.
96. *Birinci Dünya Harbi*, 250.
97. *Birinci Dünya Harbi*, 251.
98. Armitage, *The Royal Air Force*, 29.

99. Armitage, *The Royal Air Force*, 43-46.
100. *Birinci Dünya Harbi*, 15.
101. *Birinci Dünya Harbi*, 120.
102. *Birinci Dünya Harbi*, 121.
103. *Birinci Dünya Harbi*, 121, 123.
104. *Birinci Dünya Harbi*, 205, 206.
105. *Birinci Dünya Harbi*, 207.
106. *Türk Silahlı Kuvvetleri Tarihi* III, 500.
107. *Türk Silahlı Kuvvetleri Tarihi*, 500.
108. Bernard Lewis, *The Emergence of Modern Turkey* (2nd ed.; London 1968) 254; Armitage, *The Royal Air Force*, 36.
109. Armitage, *The Royal Air Force*, 36-37.
110. Armitage, *The Royal Air Force*, 37; Lewis, *The Emergence of Modern Turkey*, 254.
111. Yavuz Kansu, 'Milli Mücadelede Hava Kuvvetlerimiz,' *Hayat Tarih Mecmuası*, Jan. 1974, year 10, no. 1, 60.

XII

MODERN TURKEY AND THE OTTOMAN LEGACY

The Ottoman Empire, with its vast geography and long history of more than six centuries, had a multi-lingual and religious structure, comprising various nations, groups and communities of different ethnic origins. The central place and importance of Turks among these ethnic groups become clear when we consider this vast land, which was peacefully ruled by the Ottomans and called *Pax Ottomanica* by European historians. It was composed of three main parts, namely today's Turkey, the Balkans and the Arab states. The Ottoman Empire was seen by contemporary European states and writers as 'the Turkish Empire', 'the Turkish State' and 'Turkey'. Accordingly, from an early period on the Ottoman state has been called in European languages Turquie/Turekia/Turkey, and books written by European writers on Ottoman history have described it as a Turkish state.[1] In addition to these, the earliest translations of the Qur'an have called the sacred book of Islam the 'Gospel of the Turks' or the 'Book of the Prophet of the Turks'. All of these show clearly how the Ottoman state was regarded by foreigners as a Turkish state.[2]

In fact, the Ottomans have also considered themselves to be Turks, and, as Nejat Göyünç points out, fifteenth-century Turkish texts made use of the word 'Turk' for the Ottomans. Such derisive expressions as *etrak-i bi-idrak* (Turks without a thinking mind) were used for the nomadic Turkmen tribes of Anatolia who gathered around Shah İsmâil, the founder of the Safavid Empire, and worked for his regime. We see the use of similar expressions in historical texts describing the Celali riots at the beginning of the seventeenth century.[3] Therefore, it would be wrong to claim, on the basis of such expressions, that the Ottomans did not consider themselves as a Turkish nation.

---

[1] While such European historians as Hammer and D'Ohsson have called the Ottoman state the Ottoman Empire, many Europeans have preferred to use the name/adjective 'Turk'. As an example, we may refer to the British historian Richard Knolles' *General Historie of Turkes* published in 1604, and to the Italian priest Abbe Toderini's *De la Litterature des Turcs*, which is about the Ottoman culture.

[2] The title of the first German translation of the Qur'an published in 1616 is *Alcoranus Mahumeticus, das ist: der Turcken Alcoran, Religion unde Aberglauben,...* For this translation made by Salomon Schweigger, see İsmet Binark-Halit Eren, *World Bibliography of Translations of the Meaning of the Holy Qur'an: Printed Translations 1515-1980*, ed. by Ekmeleddin İhsanoğlu, (Istanbul, 1986), p. 222. Many European writers have used the world 'Turk' in a somewhat misleading way. See, for instance, Alexander Ross's four volume *A Complete History of the Turks, from their origin in the year 755, to the year 1718...* published in 1719 in London, which contains also a translation of the Qur'an. See *ibid.*, p. 110.

[3] Nejat Göyünç, 'Bazı Osmanlı Tarihi Eserlerine ve Arşiv Belgelerine Göre Türk, Kürt ve Arap Deyimleri Hakkında', *Anadolu Dil Tarih ve Kültür Araştırmaları Dergisi*, no. 1 (Afyon, 1996), pp. 1-5 and *idem*, 'Osmanlı Devleti Hakkında', *Cogito*, no. 19 (Istanbul, 1999), pp. 86-92.

The Turkish nature of the Ottoman state was felt both within and outside its borders, without any imposition by the Ottomans. Until the beginning of the twentieth century, the Muslim Albanians and Bosnians living in the Balkans and coming from different ethnic origins regarded themselves as a kind of 'Turkish' community, even after the Ottomans pulled out of these areas. It is known that these people defined their identities by saying 'Thank God we are Turks'. This 'Ottoman-Turkish' identity was a super-identity, arising from the historical interpenetration of Islamic and Ottoman elements on the one hand, and the identification with Turkish culture and rule on the other. This concept was different from what Ziya Gökalp calls the 'Turkish' identity, which has come about as an offshoot of modern nationalism.

The fact that the Ottomans were considered, by themselves and by outsiders, as a Turkish nation, and that other Muslims of different ethnic origins perceived themselves as Turks does not, however, mean that the Ottoman state had a nature and structure similar to today's nation-states. At this point, it is important to distinguish between the Ottoman-Turkish identity which flourished within the Ottoman cultural melting pot, and the Turkish identity which came into being with the Turkish Republic, based on the modern nation-state notion. It goes without saying that especially in today's Turkey, where conceptual confusion is rampant, one has to be careful not to confuse these concepts but to deal with both the old and new notions in their own terms.

It is, both in principle and historically, wrong to reduce the name of a state to a particular nation or race. It would certainly be wrong to call the Roman Empire an Italian state, or the Byzantine Empire a Greek state, or the Abbasid state an Arab state. The same holds true for the Ottoman Empire, which is comparable to the aforementioned states insofar as the vast land it ruled and the multiplicity of ethnic groups it comprised are concerned. The fact that the founding dynasty or the members of the succeeding rulers belonged to the same family does in no way justify the reduction of a state to a single ethnic group. As a result of this, it is both anachronistic and unhistorical to look for the existence of nationalism and the nation-state system within the multi-national and multi-religious structure of the Ottoman Empire.

In the light of the preceding remarks, we may say that the Ottoman dynasty contributed not only to the establishment and expansion of the state, but also to the philosophy of political rule and the formation of various state institutions. Nevertheless, the Turkish character was the dominant factor in the Ottoman state, as Turkish traditions were the foundation stones for the rise of the idea of state in the Ottoman Empire. It is also beyond doubt that the acceptance of Turkish as the official language of the state was an important factor in the emphasis laid on the Ottoman's 'Turkish' characteristics.

Today, the Turkish Republic, which is celebrating the seven hundredth anniversary of the foundation of the Ottoman Empire, has accepted itself as an heir to this heritage. It is interesting to note that none of the states that came into being in the Balkans, the Middle East or North Africa in the wake of the collapse of the Ottoman Empire have undertaken to celebrate this anniversary or claimed to be the inheritors of this heritage. In fact, a great majority of these states look at the Ottoman legacy with disdain, and some even see it as having been a blow to their progress and national identity. The Ottoman state has become the target of nationalist sentiments that regard one nation as superior over others, and at the same time function as a source of legitimacy for the establishment of modern nation-states. In order for the European powers to gain a bigger share in the Ottoman legacy, their efforts towards the eradication of the legitimate sovereignty of the Ottoman state, which they called the "sick man" during its last century, were fostered by the above mentioned hostile attitude. One direct result of this attitude has been the rise of the 'ideological' image of the Ottomans that we encounter in those states established in the former lands of the Ottoman Empire – an attitude totally at odds with the scholarly study of history.[4] Even though it is beyond the scope of this survey, the most recent and the clearest example of this is the Bosnian and Kosovan wars of the 90s. The Serbs, who claim to find the roots of their ultra-nationalism in the triumph of the Ottoman armies over the Serbian army and in the death of Murad I in the 1389 Kosovo war, have tried to come to terms with this six-century old 'legacy' by means of ethnic, religious and cultural cleansing. This is the most pernicious example of the misinterpretation of the Ottoman legacy on the basis of political and ideological interests.

As the natural inheritors of the Ottoman heritage, it would be

---

[4] For the Ottoman image and its change in historical textbooks used in Turkish schools since 1930, see Etienne Copeaux, *Tarih Ders Kitaplarında (1931-1993) Türk Tarih Tezinden Türk-İslam Sentezine*, tr. by Ali Berktay, (Istanbul, 1998). For some other aspects of this issue, see *Tarih Eğitimi ve Tarihte 'Öteki' Sorunu*, Ali Berktay and Hamdi Can Tuncer (eds.), (Istanbul 1998), and Salih Özbaran, *Tarih, Tarihci ve Toplum*, (Istanbul 1997). For the textbooks used in Egypt as an illustration of the situation in the Arab world, see Ekmeleddin İhsanoğlu, '1912-1980 Yıllarında Mısır Okullarında Okutulan Tarih Kitaplarına Göre Osmanlı Devleti Tarihi ve Arap Dünyası ile İlişkileri', and for the Turkish summary of this article, see *Studies on Turkish-Arab Relations*, no. 1 (Istanbul, 1986), pp. 331-340; for the Arabic version, see *ibid.*, pp. 85-118.

5

interesting to investigate our own position and responsibility towards this tradition and to see whether our position differs from the positions of the Balkan and Arab states. This, however, is beyond the confines of our present study. It should be noted at the outset that the analysis of the position of 'modern Turkey' towards its Ottoman heritage is predicated upon a contradiction, because the concept of 'modern Turkey' is an outcome of the late period of Ottoman history. In other words, modern Turkish history has two major phases: the first phase is the Ottoman period, the second the Republican era. Here we shall focus mainly on what Republican Turkey has inherited from Ottoman Turkey and which intellectual currents have been instrumental in this transition from the first to the second phase.

The generation that was responsible for the formation of intellectual life at the time of the foundation of the Republic was the same group of people who were at the centre of intellectual activity in the second Meşrutiyet period. This generation had both positive and negative influences, which have led, in turn, to a number of important events in post-Meşrutiyet Turkey. First of all, the ideas of the people's democratic participation in the political administration of the country and the freedom of expression of political views enjoyed by intellectuals were crystallised during this period. Seen in this light, it would not be wrong to say that the political, social and legal foundations of the Republican era were laid in the Meşrutiyet period. It is beyond any doubt that this parliamentary experience, with an interval of 33 years, has played a very significant role in the democratisation process, notwithstanding some of its negative consequences.[5]

The idea of limitless freedom that came with the Meşrutiyet was attached to a sort of anarchism, which caused considerable disturbance in society. As was seen in the Balkan War and the wars that followed, the idea of 'Ottoman' identity was not sufficient any more to keep different ethnic groups together In a similar vein, the project of 'Islamism' failed due to the nationalist movements that had become prominent among the Muslim Albanians and Arabs for various political and military reasons. In the period of transition from the Meşrutiyet to the Republican era, only two political movements, 'Westernism' and 'Turkism' had a legitimate basis, as opposed to the other rival movements, 'Ottomanism' and 'Islamism'. Although the Westernization project was deemed necessary for the preservation of the Ottoman state, with the new nation-state it assumed a new meaning and came to signify a civilizational transformation that would be accompanied by a total reform of

---

[5] Orhan Okay, 'Batılılaşma Devri Fikir Hayatı Üzerine bir Deneme', in *Osmanlı Devleti ve Medeniyeti Tarihi*, II, E. İhsanoğlu (ed.), (Istanbul, 1998), pp. 216-217.

society. In order to achieve this goal, the 'Ottoman' and 'Islamic' thought system, and even the way of life, had to be changed, and ties with the past had to be cut off. With the establishment of the new state, the idea of 'Turkism' limited the old ideal of *Turan* to the *Misak-i Millî* borders. In addition to these currents of thought, that were mainly of an ideological, didactic and even catechistic character rather than being philosophical as a result of the freedom brought about by the Meşrutiyet (and we may add here the influence of politics and war), there were also some other intellectual movements whose impact was to be seen in the long-term.[6]

It was within this political and intellectual climate that the negative attitude of the new Republic against its Ottoman heritage began to take form. According to this new position, as stated in very strong terms by Ziya Gökalp in his celebrated book *The Principles of Turkism*, which was published in the same year as the declaration of the Republic, the concepts of 'Turkish' and 'Ottoman' are set against each other as terms with irreconcilable differences in meaning: "What is the cause of this strange situation peculiar to our country? Why are these two models, i.e., the Turkish and Ottoman models, so far away from each other?"[7]

With such a division, how could the members of the newly born state look back to the Ottoman heritage under whose rule and civilisation they had lived? How relevant could the answer of this question be to those who were actually undergoing this transformation? Gökalp, on his part, provided an answer by asking the following question: "Why is it that everything related to the Turkish example is good, and everything belonging to the Ottoman model is ugly?"[8]

Such was the judgement; everything Turkish was seen as beautiful and everything Ottoman was deemed to be ugly. As a result, the 'Ottoman' and 'Turkish' elements have been regarded as two different entities. According to Gökalp, Turks did not need to inherit the ugly heritage of the Ottomans, who were not Turks, because they were of a different nation. In order to separate the Turks from the Ottoman legacy, Gökalp offered the following argument:

"The Sunni Turks who shared the same religion with the Ottomans were not culturally subjected to Ottoman imperialism. Rather, they created

---

[6] *Ibid.*, pp. 211-219.
[7] Ziya Gökalp, *Türkçülüğün Esasları*, (Ankara 1339/1923), pp. 33.
[8] *Ibid.*

their own cultural sphere and remained entirely independent of the Ottoman civilisation."[9]

It should not be hard to estimate the new generation's view of the Ottoman legacy when these ideas that were prevalent in the intellectual life of the formative years of the Republic penetrated into the essential elements of the official view of history. In spite of such opposition to the Ottoman inheritance, the new state inherited this legacy willy-nilly, while watching the fading away of the old state. The fact that the new Turkish State defined itself as the inheritor and a continuation of the old state is usually overlooked. The Grand National Assembly (TBMM), in its decision of October 30, 1922, which officially ended the Ottoman state, described the newly established state as the heir of the Ottoman state in a legal framework and a very clear language.[10] According to a recent study, Republican Turkey inherited 93 percent of the Ottoman army and 85 percent of its administrative structure.[11]

This new state of mind set itself against the Ottoman legacy, and such a denial has created a rupture in the historical consciousness of new generations. As a result, this has deprived them of the idea of continuity and a holistic perspective, which lie today at the root of their own conflicted relationships amongst themselves.

Some of these ideas and positions which were carried into the Republican period by those who were brought up in the atmosphere of neglect of the unifying historical perspective of Ottoman culture on the one hand, and the polarising attitudes of the Meşrutiyet era on the other, continued until our time. Some other nations, which went through a number of radical changes

---

[9] *Ibid.*, p. 34.

[10] The Heyet-i Umumiye (general delegation) issued a resolution on 30 Teşrin-i evvel 1338 (October 30, 1922), no. 307, in which the dissolution of the Ottoman Empire and the establishment of the government of the Turkish National Grand Assembly were announced. The text of this resolution reads as follows: "It has been decided that the Ottoman Empire has officially been dissolved and that the government of Grand National Assembly has been established. The new Turkish government is established in the place of the Ottoman Empire as its heir within its national boundaries. According to the *Teşkilat-ı Esasiye Kanunu* (the Constitution), the power of governance has been entrusted to the people itself and consequently the sultanate in Istanbul has become obsolete and part of history. Since there is no longer a valid and legitimate authority in Istanbul, the Grand Assembly and its officers have been entrusted with the governance of Istanbul and its vicinity. It has also been agreed upon that the office of the caliphate (*makam-i hilafet*) which is the legitimate right of the Turkish Government will be salvaged from the hands of foreign powers." *Düstur*, 3. Tertib, C. III (2nd edition), (Ankara 1953), p. 99.

[11] Dankwar: A. Rustow, "The Military Legacy", *Imperial Legacy: The Ottoman Imprint on the Balkans and the Middle East*, L. Carl Brown (ed.), (New York, 1996), p. 257.

before or after us, have not attempted to sever their historical ties with the past, turning tradition into a tree with roots dangling in the air. Although in the early stages of every revolution such extremes have occurred, as the fire of revolution calmed down, the historical roots of the nation were re-established and the state and society were reunited with their historical heritage. Once the aforementioned distinction had been made between the Turkish era as good, bright and progressive (and even progressivist), and the Ottoman era as bad, dark and backward (and even regressive), the Turkish people who wanted to separate themselves from the first stage of Turkish history began to search for different roots. They tried to relate themselves to times, nations and civilisations which had nothing to do with their historical ancestors. Among these attempts, the re-interpretation of linguistic and historical facts have claimed supreme importance. For instance, a relation of kinship was established among the Turks, the Hittites and Sumerians, who have no racial or geographical proximity to the Turks, while at the same time the Turkish language, which is the most important binding factor between the Anatolian and Central Asian Turks, was diverted from its historical evolution process and transformed into an artificial language under the pretext of 'Turkish brotherhood'. Consequently, many new words were concocted as part of the new Turkish language, which is now totally unintelligible to the newly independent Turkic people of Caucasia and Central Asia. The highly advanced Ottoman Turkish which became the lingua franca of both literature and sciences was subjected to an ethnic cleansing, in addition to a pruning of various Turkish words used for centuries by the Ottoman Turks. Such words as 'özgürlük' (freedom), 'uygarlık' (civilisation) and 'bağımsızlık' (independence) replaced 'hürriyet' (freedom), 'medeniyet' (civilisation) and 'istiklal' (independence), which are understood and used today by the Central Asian and Caucasian Turks as well. This replacement was forced, and these newly invented words remained the work of a group of people, devoid of any linguistic cogency or scholarly value. If, as it is said, the vocabulary of the Ottomans is still being used today from the 'Great Wall of China to the Adriatic Sea' and is understood by both the professor in the university and the shepherd on the mountains, we have to reconsider the value and importance of the Ottoman heritage, beginning with the question of language and extending it to other domains of history. Or at least, we should see that we have no right to impose these strange inventions, created in the name of the Turkish people, on other nations of Turkic origin.

The Composite Structure of the Ottoman Legacy

The history of civilisations, as Fernand Braudel points out, is a history of their exchanges over the centuries. In this interactive process, however, every civilisation preserves its distinct characteristics, adds to borrowed elements its own cultural elements, and then passes it on to others.[12] This holds true without doubt for the Ottoman Empire, which presents a perfect example of this process as described by Braudel. The Ottomans borrowed or inherited many elements from the cultural traditions of various nations that they had ruled, and created their own synthesis of a civilisation by imprinting their identity on it. This pattern was then followed by nations under Ottoman rule, and the influence of this synthetic culture was carried to lands outside Ottoman rule; the value of this culture has been admired both in the East and the West. We think, therefore, that the most important and conspicuous aspect of Ottoman culture is this unifying and synthesising structure. This is both a synthesis and a symbiosis. Said differently, Ottoman civilisation has functioned both as a form of distillation and as a melting pot, integrating many different elements into a coherent whole on the one hand, and creating a tapestry and symbiosis of diverse nationalities and allowing them to co-exist peacefully on the other. It is this synthesis that needs to be examined closely.

The Ottoman culture, which was one of the most successful and longest living cultural syntheses in the Mediterranean area, should be analysed in comparison with other preceding civilisations. Having been united under Assyrian rule, the Near East entered the Hellenistic cultural milieu and began to flourish with Alexander the Great (336-23 BC) only after the rule of many succeeding Iranian kings (521-485 BC). The Greek language and culture was the historical foundation of this civilisation. Although some elements of this cultural identity were altered due to a number of political and administrative changes, and because of the rise of new sub-cultures at the time of the Roman and Byzantine empires, its underlying characteristics remained intact. In the wake of the advent of Islam and its rapid expansion (632-42), most of the Near East came under the rule of Islam, paving the way for the creation of a new cultural synthesis. The great civilisation of Islam was based on the blending of a number of factors, the most important of which can be stated as follows: the teachings of Islam, the Arabic language and its literature, the Hellenistic heritage of philosophy and science, some elements of Persian, Indian and, to a lesser extent, Chinese culture and civilisations. During the ascent of the Abbasids, the Arabic-Islamic version of this great civilisation extended from Andalusia and Central Asia to India and China, thus exerting a considerable influence upon the non-Islamic world.[13]

---

[12] Fernand Braudel, *A History of Civilisations,* tr. by Richard Mayne, (London: Penguin Books), 1995, pp. 3-8.
[13] *Ibid.,* pp. 41-84.

The Karakhanid (840-1212) and Tulunid (875-905) states, which were established under the umbrella of the Abbasid Empire, the Seljukid state, its branches and other small states that came into being after its collapse, all stayed within the confines of this Abbasid-Arabic-Islamic civilisational axis. In addition to this, the Persian element began to occupy a more central place in this synthesis during the Seljukid reign. It is clear that the 'Turkish' element had no discernible presence in Islamic civilisation before the advent of the Ottomans. With the rise of the Ottomans, an Ottoman version of Islamic civilisation was created, ushering in a new synthesis of Ottoman Turkish-Islamic civilisation; included in this synthesis were also the previous Abbasid-Arabic-Islamic and Arabic-Persian-Islamic elements. As with other syntheses along the Mediterranean rim, language played the central role in this newly formed civilisation. This time, however, the lingua franca was Ottoman Turkish.

The Ottoman civilisation, while appropriating the heritage of the Abbasid culture, had taken many new elements from the cultural traditions of those areas which had not been part of the Islamic world before, such as Asia Minor, the Balkans and the Caucasus, and had generated a new civilisation by blending these elements with its own Turkish identity. The turning point in this crucial process was the conquest of Istanbul by Mehmed II (the Conqueror) in 1453. One of the important factors in the creation of this new civilisational force was the centralisation of imperial rule by Mehmed II and his attempt to gather around himself all the scholars and artists who had come from different parts of the Islamic world as well as from Europe. Consequently, many facets of Ottoman cultural heritage, such as language, literature, architecture and cuisine which can be delineated as various forms of this synthesis, were created in the Palace. It is incumbent upon us to explore and analyse the different aspects of this multi-faceted legacy. Here, in order to show the formation of this historical synthesis, we shall focus on two examples which will explain how the Ottoman synthesis was shaped. This, we hope, will also clarify the position of the 'Turkish' element in this process. These examples will provide, we believe, a clue for understanding the Ottoman attitudes towards other cultural traditions.

Sûdî of Bosnia (d.1591) and Ibn Hamza of Algeria (d.1614), two Ottoman intellectuals of the sixteenth century, came from two countries that had no contact with Turkish culture before the Ottomans. The former became famous for his Turkish commentaries on Rûmî's *Mathnawi*, Sa'di's *Bustan* and *Gulistan* and Hâfız's *Diwan*, in addition to his Turkish translation of Ibn al-Hajib's *Shafiyah* and *Kafiyah*, one of the most important works in the Arabic language. This Bosnian writer contributed to Ottoman and new Islamic humanism by translating the most significant works of Arabic and Persian literature into Turkish. His books have also been republished outside Istanbul, in Egypt, showing the vivid continuity of this cultural tradition.[14]

Ibn Hamza, also known as *'Mağribi'* (North African), was born in the Maghreb, came to Istanbul, completed his education there and in fact became a professor in the *medreses* of Istanbul. After being appointed as the religious judge (kadi) of Algeria, Tripoli, Yemen and Tunisia, he came back to Istanbul to teach and in 1591, while in Mecca, he wrote in Turkish one of the most important books on Ottoman arithmetic.[15] Remarkably, this Ottoman intellectual, religious judge, professor and mathematician served the Ottoman state in its vast land while at the same time writing a book on mathematics in Turkish. Both Sûdî and Ibn Hamza are two examples of the synthesis of Ottoman civilisation that we have been expounding – a point that needs further clarification and scholarly analysis.

The Ottoman intellectuals, in their encounter with the West, tried to strike a balance between their own civilisation and that of the West in order to learn the secrets of the power and progress of this rival world. The indirect contact of such Ottoman thinkers as Kâtib Çelebi, known in the West as Khaji Khalifah, and Kethüdazade Arif, both of whom were educated within classical Ottoman culture, was succeeded and complemented by such scholars as Şanizâde Atâullah Efendi and Ishak Efendi, who knew some European languages and, accordingly, culture and sciences better. This new encounter opened up new horizons within Ottoman culture. During the modernisation process, Western culture became part and parcel of the worldview of Ottoman intellectuals, in a way similar to the synthesis and symbiosis of classical Ottoman culture. By putting the lingering debate over the 'old-new' conflict aside, many of the Ottoman statesmen, artists, poets, scientists and scholars were able to maintain a balance between these two cultural forms. From among the countless number of people that fall within this category, we shall focus here only on some. A great majority of these intellectuals knew at least one European language well, French being particularly important, in addition to their knowledge of Arabic, Persian, the Ottoman culture, and the religion of Islam. One of these figures, who had an intimate knowledge of both the East and the West, was Salih Münir Pasha (1859-1939). A descendant of Çorlulu

---

[14] For Sûdi, see Nazif M. Hoca, *Sûdî, Hayatı. Eserleri ve İki Risalesinin Metni,* (Istanbul, 1980) and Kathleen Burrill, "Sûdî", *EI,* IX, p. 762.

[15] Ekmeleddin İhsanoğlu, Ramazan Şeşen et. al., *Osmanlı Matematik Literatürü Tarihi,* I, (Istanbul, 1999), pp. 118-123.

Ali Pasha, a composer and son of the famous historian Mahmut Celaleddin Pasha, he acquired a first-hand knowledge of classical Ottoman-Islamic culture in his private education, after which he learnt about Western culture in his high school education at Galatasaray Lyceé. A remarkable statesman and diplomat, he served the prince Abdülhamid Efendi (Abdülhamid II) and worked for many years in the State Department and at the Paris embassy. Even a cursory look at his Turkish and French works will reveal a great deal of his intellectual caliber, immense knowledge and erudition in Islamic, Turkish and European history, diplomacy and political history.[16]

Another example of people with a profound knowledge of Eastern and Western cultures is the famous couple Seniha Sami Hanım (1886-1982) and Mehmed Rauf Bey (1882-1918). Descended from two long-established Ottoman families, this couple is probably one of the best examples to illustrate the main characteristics of later Ottoman intellectuals. Mehmed Rauf whose father (Ferik Atif Pasha) was a soldier and mother (Fahriye Hanım) was a poet, knew nine languages, including Arabic, Persian, French, English, Italian and Greek. He worked as a civil servant in Istanbul, and taught mythology and the history of Greek and Italian literature at Darülfünun, while at the same time writing books on these subjects. One of the most interesting figures of the late Ottoman history, Mehmed Rauf was interested in painting, music and poetry. He wrote some plays and was also the editor of *Resimli Kitap*, published after 1908.[17] In a similar way, his wife Seniha Sami Hanım had an excellent knowledge of Persian, French and English languages and literature from her private education. In her youth, she taught language to the young princes in the palace. In addition to her translations and works on history and literature, she worked at the Topkapı Palace Museum after the declaration of the Republic, making her immense culture and linguistic ability available to the public. Upon the request of the Ministry of Education, she wrote a book called *Iran Edebiyat Tarihi* (History of Persian Literature) and translated Bartold's *Moğol Tarihi* (History of the Mongols) and Samuel N. Kramer's *Sümer Edebiyat Tarihi* (History of Sumerian Literature).[18] The success of this Ottoman synthesis and its meaning for later generations can be fully appreciated only when we

---

[16] For Salih Münir Paşa, see *Salih Münir Paşa'nın Ünlü Eseri: Kur'an'a Göre Din ve Ahlak Kuralları*, Taha Toros (ed.), (Istanbul, 1986) and Yılmaz Öztuna, "Münir Paşa", *Türk Ansiklopedisi*, XXV, (Ankara 1977), p. 33.

[17] One should not confuse Mehmed Rauf, to whom we have referred here, with the novelist Mehmed Rauf (1875-1931) who was a member of the New Literature movement (Edebiyat-i Cedide). For Mehmed Rauf (1882-1918), see T.H. Menzel, 'Mehmed Rauf', *İslâm Ansiklopedisi*, VII, (Istanbul, 1972), pp. 610-612 and *Türk Dili ve Edebiyatı Ansiklopedisi*, VII, (Istanbul, 1990), p. 289.

[18] For Seniha Sami, see Taha Toros, "İlk Kadın Müzecimiz: Seniha Sami", *Skylife*, no. 132 (April, 1994), pp. 68-72.

undertake a thorough study of the contributions of later Ottoman intellectuals who had a first-hand knowledge of Eastern and Western civilisations. All we can do here, however, is to allude to this 'second synthesis' by stressing the importance of this cultural development.

The examples given above point to the synthetic character of Ottoman civilisation on the one hand, and the cultural richness of Ottoman intellectuals on the other. This rich cultural tradition was not a closed system, open only to members. It was a tradition in which, to use Braudel's phrase, there was both taking and giving. The impact of Ottoman civilisation was quite visible in the territories under its rule and in Europe, with which it had close ties. Since this is a virgin field in need of serious scholarly attention, thus making it impossible to draw a general picture here, we have to be content with providing some examples in order to illustrate our point. The examples that we shall choose from the field of arts will show how Ottoman civilisation was admired and accepted by Europeans. The interest in Ottoman-Turkish taste, art works and life-style, which was a result of economic and diplomatic relations, can be traced as far back as the fifteenth century. The sixteenth century marks the zenith of European interest in Ottoman decorative arts. With the rise of continuing interest in Turkish clothes and manners in the eighteenth century, we see an increase in the number of people adopting the Turkish way of dress and life. Among those who followed the Turkish fashion, we can mention the celebrated paintings of Lady Mary Wortley Montague and Lady Mary Gunning, among the British aristocracy, in which they wear traditional Turkish clothes.[19] In the same way, the collections of Turkish rugs in European and American museums and the paintings of interiors attest to the fact that Turkish rugs have been admired and extensively used outside the Ottoman world. Most of these Turkish rugs were produced in western Anatolian cities, such as Uşak, Bergama, Ladik, Milas, Konya and Akşehir and exported to Europe and America for mass distribution.[20]

In addition to rugs, Turkish textiles have also been widely used and collected by many Europeans. The Europeans living in the Ottoman Empire bought Turkish textiles in bulk, and all the major European cities had centres where the textiles imported from Ottoman lands were sold. Some of these textiles were used to make royal gowns and robes for priests. Also, the

---

[19] Ernest J. Grube, "Introduction", *At the Sublime Porte Ambassadors to the Ottoman Empire (1550-1800)*, London 1988, pp. 5-9. For an example of these various influences, see J. Strzgowski-H. Glück-Fuat Köprülü, *Eski Türk Sanatı ve Avrupa'ya Etkisi*, tr. by A. Cemal Köprülü, [Ankara].

[20] Esin Atıl, "Osmanlı Sanatı ve Mimarisi", *Osmanlı Devleti ve Medeniyeti Tarihi*, II, Ekmeleddin İhsanoğlu (ed.), (Istanbul, 1998), p. 470.

14

well-preserved carpets that were being passed on from one generation to another as precious souvenirs were either hung on walls or used on tables. Although these carpets were used in the Ottoman Empire and had thus become worn, they were considered to be precious and expensive items and preserved diligently as rare pieces of royal or church treasures up to our own day.[21]

Music, which is one of the most neglected areas of Ottoman culture, can be seen as one of the most advanced and articulate products of Ottoman civilisation. Having a wide impact outside the Ottoman world, classical Turkish music influenced not only the Muslim and non-Muslim populations under Ottoman rule, but also such eastern countries as Iran and India. Its effect on European music, on the other hand, was felt in both classical and military music. With the introduction of such *mehter* (Ottoman military band) instruments as the *davul* (tympani), *zil* (cymbals) and *çelik üçgen* (steel triangle), the symphony bands were supplied with an unprecedented dynamism, leading first opera and then symphony composers to write pieces with Turkish themes. Even though European states had initially used only wind instruments for military music, they began to include various forms of percussion after seeing its impact on the enemy soldiers in wars which they had lost to the Ottomans. It was after this encounter that the European states, from 1742 on, began to form bands in imitation of the Ottoman military band. As a result of this, many European composers began to compose musical works in the Turkish style. This is certainly one of the ways in which Ottoman music exerted a remarkable influence on European music, whose traces are still visible today.[22]

These aspects of Ottoman civilisation, which we have tried to describe in our preceding analysis, became much clearer in a congress held by IRCICA on April 12-15, 1999, under the title Learning and Education in the Ottoman World. This was the first international conference held for the celebration of the 700th anniversary of the establishment of the Ottoman Empire. The works presented in this congress, in which scholars from all over the world participated, showed not only the incredible richness of Ottoman cultural heritage, but also provided a unique opportunity to uncover many dimensions of Ottoman culture that were formerly neglected.[23] One of the surprise events

---

[21] Esin Atıl, ibid., p. 467 and also Veronika Gervers, *The Influence of Ottoman Turkish Textiles and Costume in Eastern Europe*, (Toronto, 1982).

[22] Cinuçen Tanrıkorur, "Osmanlı Mûsikîsi", *Osmanlı Devleti ve Medeniyeti Tarihi*, II, Ekmeleddin İhsanoğlu (ed.), (Istanbul, 1998), pp., 509-511.

[23] For the academic program, socio-cultural activities and speeches given by Prof. Ekmeleddin İhsanoğlu, the general director of IRCICA, Prof. Haris Silajdzic, prime minister of Bosnia-Herzegovina, Prince Hasan bin Talal, of the Hashimite Kingdom of Jordan, and Süleyman Demirel, president of Turkey, see *IRCICA Haber Bülteni*, special issue, no: 48 (April, 1999).

of the conference was the piano recital/demonstration by Vedat Kosal, entitled 'Classical Western Music in the Ottoman Empire'. This presentation made it clear that the Ottoman interest in classical Western music goes back to a very early date and that the official interest shown during the reign of Mahmud II gained a new dimension and impetus with the establishment of the Royal Band (*müzika-yi hümayun*). We also learnt that, beginning with Abdülmecid I, many Ottoman sultans and members of the dynasty composed, in addition to Turkish classical works, many new pieces in the Western form. Outside the palace, the first Turkish opera was staged in 1840, and four years later the first libretto was written by Hayrullah Efendi, the father of the celebrated poet '*şair-i azam*' Abdülhak Hamid.[24] In the light of our present knowledge, we can say that the works composed by the Ottomans in the nineteenth and twentieth centuries highlight the aforementioned synthetic power of Ottoman civilisation and prove the success and significance of this new synthesis, which was a result of contact with the Western culture.

As our examples taken from fashion, textiles and music show, Ottoman culture had a value of its own, respected even by other nations living outside its territories. At this point, it is extremely important to study the impact of Ottoman culture on peoples of different religious, linguistic and cultural origins under its rule, as well as on the East and the West outside its domain.

The Legacy of Language

Language is one of the most important elements of the Ottoman legacy. As we have stated before, the Ottomans have usually been accused of neglecting Turkish and giving priority to Arabic and Persian. Historical records, however, show beyond any doubt that the development of Turkish as both the official language of the state and a language of literature was the result of the insistence of Ottoman sultans. Many important works were translated into Turkish upon their request, creating thus a very rich literature in Turkish.[25] The following couplet by Aşık Pasha (d.1332) is probably the best evidence to show the status of Turkish in the early phases of Ottoman history:

Türk diline kimesne bakmaz idi
Türklere hergiz gönül akmaz idi

---

[24] For the piano recital and talk given by the pianist Vedat Kosal in the international congress held on Learning and Education in the Ottoman World on April 12-15, 1999, see 'The Program of the Piano Recital', and also Vedat Kosal, "Önce Müzik Hayatı Batılılaştı", *Hürriyet Gösteri Sanat Edebiyat Dergisi*, no. 212 (July-August, 1999), pp. 62-63.
[25] Günay Kut, "Anadolu'da Türk Edebiyatı", *Osmanlı Devleti ve Medeniyeti Tarihi*, II, Ekmeleddin İhsanoğlu (ed.), (Istanbul, 1998), p. 21.

> No one used to pay attention to Turkish
> No heart used to be given to Turks

Anatolian Turkish, which was a poor language compared to the rich literature of Arabic and Persian, became an advanced language thanks to the Ottomans, and was used as the official language of the state from Budin and Algeria to the Yemen. The richness of classical Ottoman literature can hardly be overemphasised. As some recent studies show, Ottoman Turkish was also used as a language of science, the best examples of which will be found in the fields of astronomy and mathematics. The extensive use of Turkish as a language of science will become clear when we look at the studies done under the series of Ottoman Science Literature, in which 2,438 astronomical works belonging to 582 Ottoman scholars and 963 mathematical works belonging to 491 Ottoman mathematicians were published.[26]

With the translation of European, primarily French works on science into Turkish in the nineteenth century, Ottoman Turkish became rich enough to express modern scientific ideas. Being a language of literature, art and science, Ottoman Turkish reached a supreme level of maturity at the beginning of this century. Geoffrey Lewis, one of the highest authorities on the Turkish language, has even claimed that the rich vocabulary of Turkish can be compared only to English, emphasising one more time the points we have been explaining so far.[27] In a lecture delivered at Istanbul University in 1992, Lewis described the situation of Turkish with the following words: 'What befell Turkish is a successful disaster.'[28] It is due to the hostile attitude towards Turkish that the Turkish language today has become an enigma. Recently, a similar remark was made by Talat Halman, a writer, scholar and expert on Turkish and western languages. He describes the current state of Turkish as follows:

> For seventy years, our language is in a state of shattering, which is beyond the ken of imagination. In recent centuries, no major language of the

---

[26] Ekmeleddin İhsanoğlu, Ramazan Şeşen et. al, *Osmanlı Astronomi Literatürü Tarihi*, I-II, (Istanbul 1997) and Ekmeleddin İhsanoğlu, Ramazan Şeşen et. al, *Osmanlı Matematik Literatürü Tarihi*, I-II, (Istanbul, 1999).

[27] Geoffrey Lewis, a member of the British Academy and professor emeritus at Oxford, has a number of important works on Turkish grammar and Dede Korkut, in addition to his quite successful translations from Turkish. For his views that we have just alluded to, see his "The Ottoman Legacy in Language", in *Imperial Legacy: The Ottoman Imprint on the Balkans and the Middle East*, L. Carl Brown (ed.), (New York, 1996), pp. 214-223.

[28] From the speech delivered by Prof. G. Lewis in March 19, 1992 on the occasion of the conferral of an honorary doctorate by İstanbul University.

world has undergone such a 'linguistic quake': The alphabet, the writing system, the expulsion of old grammar through foreign words, the invention of new words and terms, the frenzy of broken sentences, the structural change in expressions and style, and endless mistakes of usage in writing, in speech and in loud reading,... I wonder if Turkish has been so impoverished and acculturated in any part of the twentieth century. In schools, universities, the media, and daily conversations, our language is like a wreckage. All of us are quake-stricken, victims of the language quake.[29]

Having influenced the Balkan languages under its rule, as well as Arabic and Persian from which it benefited, Ottoman Turkish is considered today 'Arabic' by the Turkish youth. Consequently, with the exception of a few experts, the new generation in Turkey is not able to read and understand the writings of their fathers and grandfathers. Given this state of things, it is as if the libraries and museums appear to belong to other nations rather than being the cultural institutions of the Turkish youth.

The basic drawback of the Turkish experience in this domain will become clear when we look at the parallel cases of the French, Russian and Chinese societies that underwent radical revolutions before or after Turkey. The French language has never become the subject of 'ethnic cleansing', i.e., the pruning from French of Greek, Latin or Arabic words borrowed in the Middle Ages, nor has it been among the aims of the revolutionaries, who destroyed the monarchy and established the new French republic, to create a so-called pure French. Since French was never subjected to such polarisation, the French language, just like the rest of the world languages, was left to its natural course of evolution. The authority on language and related matters was always in the hands of the academia, and it has not been the subject of political and ideological preferences, as unfortunately has been the case for the Turkish language.

At this point, it would not be incorrect to say that the tragic outcome of our brief analysis of the 'heritage and the inheritors' is as follows: whereas the children of all nations, whether they have been through any revolutionary process or not, have inherited the heritage of their ancestors and as such have been able to understand and transmit it for posterity, unfortunately the 'modern Turks' happen to be the exception to this rule.

[29] Talat Halman, "Dil Depremi", *Milliyet Gazetesi,* October, 3 1999.

The Sediments of the Heritage

It would be a nostalgic and even ideological attitude to claim that everything inherited from the Ottomans is perfect and ideal. It is a plain fact that there are some elements in this heritage which effect us in a number of negative ways. We believe that this point needs the scholarly attention of historians and social scientists. Here we can only point to two important examples of the negative consequences of the Ottoman heritage. The first example is related to the tough style of intervention by the Ottoman state. It was certainly a mistake on the part of the Ottoman Empire to see itself as a sacred being, placing itself above other values, viz., as a *'devlet-i ebed müddet'* (the eternal state) and consequently able to oppress those who did not show it absolute obedience. A typical example of this is the abolition of the Janissary Corps in 1826.[30]

As a reaction to the iron hand of the state, some opposition groups and intellectuals went so far as to jeopardise the very existence of the state, and this tension between the two extremes is another aspect of this negative tradition bequeathed to us from the past. To use a fashionable expression, this can be called the other side of the coin. In this regard, it is very important to understand why and how the Young Turks weakened the internal structure of the Ottoman state by collaborating with foreign powers and how they tried to conquer the state from within at all costs. It is also important to see how this movement effected the collapse of the Ottoman Empire.[31] This pernicious legacy, which we still see among some intellectuals and which continues to shake the society by challenging the state and creating polarisation among the educated, has now become part and parcel of Turkish society at the end of the twentieth century. As we have pointed out, one discernible impact of this legacy is the superior and oppressive attitude of the state on the one hand, and the extreme opposition of some intellectuals to this aggression on the other. This has led to a great deal of disturbance in contemporary Turkish society. It seems clear that Turkish society, with its socio-political achievements and a new class of people who are willing to preserve its cultural values, need more time to overcome such bitter aspects of its inherited legacy from the Ottomans. We hope that this will take place before the celebrations of the eight hundredth anniversary of the Ottoman state, in the year 2099. We also hope that the

---

[30] For the abolition of the Janissary Corps and the criticisms raised against its uprooting in a bloody manner, see Ahmed Cevdet Paşa, *Tezâkir*, IV, Cavid Baysun (ed.), (Ankara, 1967), pp. 218-222. For a historical account of this important event, see Enver Ziya Karal, *Osmanlı Tarihi*, V, Ankara 1983, pp. 142-150 and Stanford J. Shaw, *History of the Ottoman Empire and Modern Turkey*, II, (Cambridge, 1977), pp. 19-26.

[31] For the Young Turks, see M. Şükrü Hanioğlu, *Bir Siyasal Örgüt Olarak Osmanlı İttihad ve Terakki Cemiyeti ve Jön Türklük (1989-1902)*, (Istanbul, 1985) and also, *The Young Turks in Opposition*, Oxford 1995.

problematic aspects of the Ottoman legacy will be overcome by placing an emphasis on the need for democratisation and civil governance, both of which also constitute an essential feature of the Ottoman heritage.

***

Today, Turkey is celebrating the seven hundredth anniversary of the establishment of the Ottoman state and has a different perspective on the Ottoman legacy. It is true that some Turkish intellectuals believe that this celebration is a contradiction of the Republican revolutions and that the new Turkish Republic has nothing to do with the Ottoman tradition. They have even gone as far as to write open letters to the president, protesting against the celebrations. Despite this, we believe that the general public is not in favour of such polarisation and conflict.

A clear example of this is the support given to the celebrations by various state authorities, and especially by the president, Süleyman Demirel, who made it clear in his speeches that Ottoman and Republican values are not necessarily on a collision course with each other. Even though not discussed publicly, the army has also participated in the celebrations, which has had a very positive effect on the general public. Accordingly, the voice of those opposing factions have not been taken seriously. It is worth quoting here the following remarks from the president's speech delivered to the Grand National Assembly on October 1, 1999:

> The Ottoman Empire, of which we are the founders and the inheritors and whose seven hundredth anniversary of its establishment we are celebrating this year, has had a decisive role in the formation of the Mediterranean and European cultural milieus for 624 years. The fact that we are the heirs of a world empire that has ruled over the meeting point of three seas, three continents and various cultural traditions has a direct impact on our past as well as on our future. It is therefore an imperative to understand our history properly and grasp its place in the history of humanity.[32]

We can confidently say that the position of present day Turkey towards the Ottoman legacy has a promising level of objectivity. In order to convert this perspective into a constructive attitude, we have to revive the integrity of historical consciousness in the mind of the Turkish people. In other words, the historical rupture that came about during the transition from the

---

[32] For the text of the speech given by President Süleyman Demirel in October 1, 1999 on the occasion of the annual opening of the Grand National Assembly, see the website of the presidency: www.tccb.gov.tr

Ottoman to the Republican era should be eradicated and the ties of the Turkish people with their past be re-established. Achieving this goal requires supplying the Turkish people with opportunities to learn about their past during their formal education. Stated more clearly, we have to provide our children with an education that will enable them to read and understand the literature and cultural works of their ancestors. No nation of the world today is in need of interpreters when visiting its national library, museums or other cultural institutions; only Turkey is.

It may sound nostalgic for some people to talk about history and historical heritage. They may even have the right to dive into the cloudy world of imagination and spend their time there. This group of people, however, has no right to turn their nostalgia into an ideology and impose it on other members of the society, the result of which is certainly social unrest and conflict. In the same way, no one, we believe, has a right to turn a particular period of Turkish history, whether the Ottoman or the Republican era, into a nostalgic ideal and impose it forcefully on others.

The legacy of the past should be transmitted into posterity in a natural manner. No one has a right to enforce this legacy on anyone else. By the same token, no one has the right to bar a people from their past or to conceal history's keys from them. Although revolutions and radical transformations may have recourse to some drastic measures, the waters are expected to return to their natural course after a while and the river of time continues to flow from pre-eternity to eternity.

Turkey will certainly reach new horizons when it reclaims the positive heritage of the Ottoman culture. This will help Turkey catch the spirit of being a 'great state' again in the outside world. This will also eradicate the internal conflict on the meaning of the Ottoman legacy on the one hand, and the social tension we witness today on the other. We hope that the superior and oppressive attitude of the state and the destructive and nihilistic opposition of intellectuals like the Young Turks will be among the elements of this heritage with which we will have dispensed as we enter the new millennium.

It is quite likely that Turkey in the next millennium, when celebrating the eight hundredth anniversary of the Ottoman Empire, will have a better grasp of Ottoman heritage and will make a more productive use of this tradition. Most probably, the experience of our generation and of the generation preceding us will be seen by posterity as a strange episode of our history.

INDEX

War
    College (Mekteb-i Harbiye):V 238
    industry: V 239
    technology: I 50
Warfare: X 50
Water clock: I 55
Western
    civilization: VIII 96; XII 14
    Culture: XII 12
    Europe: I 50; VI 59; X 56, 57
    ideas: VIII 94
    medicine: I 64, 66
    music: XII 16
    observations of the Ottoman medreses: I 48
    science: VIII 87; IV 160; VIII 94
    science and technology: I 71
    style educational system VI 49; VIII 95
    Technology: X 61
    Westernization: XII 6–7
Wooden quadrant III 20
Wright, Orville: XI 192
Wright, Wilbur: XI 192

Yahya Naci Efendi: IV 159
Yakup Paşa: I 64
Yalova: X 58
Yaltkaya, Mehmet Şerefeddin: VI 63 64, 65, 66, 67, 68, 69, 70

Yanya (Janina): IV 157, 158
Yazd: III 13
Yazdigird: II 8
Yemen: III 15; XI 210; XII 12, 17
Yeşilköy airport: XI 212
Young Turks: XII 19
Yusuf Nesim Efendi: II 28, 29

Al-Zahrawi, Abu'l-Kasim: III 22
Zárate, Agustín de: I 60
Zeyrek: VI 79
Zeytinburnu Iron Factory: X 59
Zīj: II 4, 6–9; III 18–20, 26
    -i: Frengī: II 8, 9
    Gurgani: III 21
    of Cassini: II 32
    of Cezmī Efendi: II 9
    al-Hîākimī: II 8
    of Ibn Yūnus: II 8
    of Kepler: II 9
    of Lalande: II 32
    of Lansberge: II 6, 9
    of Tycho Brahe: II 9
    Shahinshahi: III 20
    Ulug Beg: III 23
Ziyaeddin Efendi, Prince: XI 194
Zodiac: II 4, 7, 34
Zonaras, Johannes: III 26
Zulmetten Nura: VI 55